Bioelectrochemistry IV

Nerve Muscle Function—
Bioelectrochemistry, Mechanisms,
Bioenergetics, and Control

NATO ASI Series

Advanced Science Institutes Series

A series presenting the results of activities sponsored by the NATO Science Committee, which aims at the dissemination of advanced scientific and technological knowledge, with a view to strengthening links between scientific communities.

The series is published by an international board of publishers in conjunction with the NATO Scientific Affairs Division

A	**Life Sciences**	Plenum Publishing Corporation
B	**Physics**	New York and London
C	**Mathematical and Physical Sciences**	Kluwer Academic Publishers
D	**Behavioral and Social Sciences**	Dordrecht, Boston, and London
E	**Applied Sciences**	
F	**Computer and Systems Sciences**	Springer-Verlag
G	**Ecological Sciences**	Berlin, Heidelberg, New York, London,
H	**Cell Biology**	Paris, Tokyo, Hong Kong, and Barcelona
I	**Global Environmental Change**	

Recent Volumes in this Series

Volume 263 — Angiogenesis: Molecular Biology, Clinical Aspects
edited by Michael E. Maragoudakis, Pietro M. Gullino,
and Peter I. Lelkes

Volume 264 — Magnetic Resonance Scanning and Epilepsy
edited by S.D. Shorvon, D.R. Fish, F. Andermann, G.M. Bydder,
and H. Stefan

Volume 265 — Basic Mechanisms of Physiologic and Aberrant Lymphoproliferation
in the Skin
edited by W. Clark Lambert, Benvenuto Giannotti,
and Willem A. van Vloten

Volume 266 — Esterases, Lipases, and Phospholipases: From Structure to
Clinical Significance
edited by M.I. Mackness and M. Clerc

Volume 267 — Bioelectrochemistry IV: Nerve Muscle Function—
Bioelectrochemistry, Mechanisms, Bioenergetics, and Control
edited by Bruno Andrea Melandri, Giulio Milazzo, and Martin Blank

Volume 268 — Advances in Molecular Plant Nematology
edited by F. Lamberti, C. De Giorgi, and David McK. Bird

Volume 269 — Ascomycete Systematics: Problems and Perspectives in
the Nineties
edited by David L. Hawksworth

Series A: Life Sciences

Bioelectrochemistry IV

Nerve Muscle Function—
Bioelectrochemistry, Mechanisms,
Bioenergetics, and Control

Edited by

Bruno Andrea Melandri

University of Bologna
Bologna, Italy

Giulio Milazzo

Late of Instituto Superiore di Sanita
Rome, Italy

and

Martin Blank

Columbia University
New York, New York

Plenum Press
New York and London
Published in cooperation with NATO Scientific Affairs Division

Proceedings of a NATO Advanced Study Institute / Twentieth Course of the International
School of Biophysics on
Bioelectrochemistry IV: Nerve Muscle Function—Bioelectrochemistry, Mechanisms,
Bioenergetics, and Control,
held October 20–November 1, 1991,
in Erice, Italy

NATO-PCO-DATA BASE

The electronic index to the NATO ASI Series provides full bibliographical references (with
keywords and/or abstracts) to more than 30,000 contributions from international scientists
published in all sections of the NATO ASI Series. Access to the NATO-PCO-DATA BASE is
possible in two ways:

—via online FILE 128 (NATO-PCO-DATA BASE) hosted by ESRIN, Via Galileo Galilei,
I-00044 Frascati, Italy

—via CD-ROM "NATO Science and Technology Disk" with user-friendly retrieval software in
English, French, and German (©WTV GmbH and DATAWARE Technologies, Inc. 1989). The
CD-ROM also contains the AGARD Aerospace Database.

The CD-ROM can be ordered through any member of the Board of Publishers or through
NATO-PCO, Overijse, Belgium.

Library of Congress Cataloging in Publication Data

Bioelectrochemistry IV: nerve muscle function: bioelectrochemistry, mechanisms,
bioenergetics, and control / edited by Bruno Andrea Melandri, Giulio Milazzo, and
Martin Blank.
    p. cm.—(NATO ASI series. Series A, Life sciences; v. 267).
  "Proceedings of a NATO Advanced Study Institute, Twentieth Course of the International School of Biophysics on Bioelectrochemistry IV: nerve muscle function—bioelectrochemistry, mechanisms, bioenergetics, and control held October 20–November
1, 1991, in Erice, Italy"—T p. verso.
  "Published in cooperation with NATO Scientific Affairs Division."
  Includes bibliographical references and index.
  ISBN 0-306-44813-0
  1. Bioelectrochemistry—Congresses. 2. Electrophysiology—Congresses. 3. Neuromuscular transmission—Congresses. I. Melandri, Bruno Andrea. II. Milazzo, Giulio. III.
Blank, Martin, [DATE.] IV. North Atlantic Treaty Organization. Scientific Affairs Division. V. International School of Biophysics (20th: 1991: Erice, Italy). VI. Title: Bioelectrochemistry 4. VII. Title: Bioelectrochemistry four. VIII. Series.
  [DNLM: 1. Nervous System—physiology—congresses. 2. Neuromuscular Junction—
physiology—congresses. 3. Muscles—innervation—congresses. 4. Muscles—physiology—congresses. 5. Biochemistry—congresses. 6. Electrochemistry—congresses. 7.
Biophysics—congresses. WL 102 B6136 1994]
QP517.B53B5425 1994
612.8'1—dc20
DNLM/DLC   94-30805
for Library of Congress   CIP

ISBN 0-306-44813-0

©1994 Plenum Press, New York
A Division of Plenum Publishing Corporation
233 Spring Street, New York, N.Y. 10013

Printed in the United States of America

PREFACE

by G. MILAZZO and M. BLANK

This book contains the lectures of the fourth advanced course *Bioelectrochemistry IV: Nerve-Muscle Function: Bioelectrochemistry, Mechanisms, Energetics and Control*, which took place at the Majorana Center in Erice, Italy, October 20th to November 1, 1991. The scope of the course was international in terms of both sponsorship and participation. Sponsors included the Bioelectrochemical Society, NATO, International Union of Pure and Applied Biophysics (IUPAB), the World Federation of Scientists and the Italian National Research Council. One-third of the sixty participants were from Italy, but the majority came from eighteen other nations.

Since the course was part of the International School of Biophysics, the biophysical point of view was emphasized in integrating the biology with the electrochemistry. Lecturers were asked to use a quantitative approach with accepted standards and proper units, since this is absolutely essential for developing an effective common language for communication across disciplines. Participants were also urged not to forget that biological systems could also be considered as physical systems. Ion channels are proteins and their properties as polyelectrolytes contribute to the specific biological properties. The existence of families of channels, with very similar structures but different selectivities, suggests that the specificities arise from slight variations of a general basic design. These perspectives on nerve-muscle function helped to make the school course a unique treatment of the subject.

The fourth course on Bioelectrochemistry differed somewhat from the preceding ones because of greater emphasis on physiological phenomena. The course was organized in terms of the biology. The major topics were: calcium ion fluxes and homeostasis, action potentials, excitation-contraction (E-C) coupling, the electrophysiology of E-C coupling in cardiac muscle and the modulation of pacemaker activity, the contractile elements of muscle, the neuro-muscular junction and neuro-transmitter substances, and the bioenergetics of muscle contraction. The texts of the lectures are compiled in this book. What could not be presented here were the afternoon sessions devoted to demonstrations and discussions of experimental techniques, and the *tutorial hours* of free discussions between students and speakers to clarify points in the lectures. Most of the lectures are accompanied by an abundant bibliography of references to the original literature.

We did introduce a few changes in order to reproduce, as far as possible, the content of the course, and to make this book more useful, and we also took the liberty of changing some symbols and acronyms from those originally used by speakers, in line with the following rules:

— use of the same definitions, the same units and the same symbols for the same quantities appearing in different chapters by different authors;

– use of the SI system;

– elimination of name of quantities having the same linguistic root as the name of their unit (metrological general rule);

– use of distinguishing symbols for quantities defined in one point and difference of the numerical values of these quantities in two spatially different points. For example, it is not correct, but rather confusing, to call simply (*electric*) *potential* what really is an (*electric*) *potential difference* between two points;

– use of the same acronyms for the same compounds.

We hope that this publication will contribute to a better and deeper understanding of the topics taught in this course.

Prof. Milazzo's death left the editing of this book still unfinished. I have tried to complete the work following the intentions of the original editor, as a memento of a long standing friendship.

Bruno Andrea Melandri

OBITUARY

Giulio Milazzo (1912-1993)

Giulio Milazzo, the father of Bioelectrochemistry, died of natural causes on January 6, 1993 at his home in Rome. He often pointed out that the roots of the subject go back two hundred years to Galvani and Volta, and that he had only resurrected the science. But the Bioelectrochemistry Milazzo started was certainly different from the Natural Philosophy of the past, and in many ways quite different from parallel modern developments.

At a time when science is becoming more narrowly focused and scientists more specialized, Giulio Milazzo catalyzed the formation of an interdisciplinary grouping that was broad in scope and inclusive in its organization. He envisaged Bioelectrochemistry including all aspects of the overlap of biology and electrochemistry and, more importantly, including scientists from all over the world.

– All scientific approaches were welcome – theoretical as well as experimental, high tech instrumental investigations of technologically advanced countries along with simple potentiometric measurements from East bloc or third world countries. The only criterion was that the research be of high quality and relevant.

– All scientists were welcome. Milazzo believed that science is international and that one must use all available means to foster cooperation across national barriers. The only criterion he insisted on was that a scientist be of good character.

Facilitating contacts between scientists was a central motif in his scientific life. At meetings one could hear him switching easily between English, French, German and, of course Italian, so that he could converse with scientists in their own languages. He was fluent in all, and because of his heavy accent I always found it easier to follow his French than that of native speakers. Milazzo had a focal role in establishing national groups in the Bioelectrochemical Society. This permitted the participation of scientists from East bloc or third world countries who could not pay dues because of currency restrictions. Students from all over the world were given scholarships in the four Erice courses in Bioelectrochemistry that he organized. Even his insistence on the use of approved SI symbols and units, in the Journal and in all books he edited, was because of his desire to develop a precise scientific language that could be understood by all scientists, and used as an effective means of communication.

All of this remarkable activity started shortly before retirement from his Professorship at the Instituto Superiore di Sanità in Rome. In 1971, he organized the First International Symposium on Bioelectrochemistry in Rome, the first of many regular meetings. This catalyzed the organization of the Journal, Bioelectrochemistry and Bioenergetics, of which he was the founding editor, and the founding of the Bioelectrochemical Society, where he was the first President. In addition to his work on the Journal, he edited many books including three containing the lectures of Bioelectrochemistry courses in Erice (this being the fourth), and five in the series Topics in Bioelectrochemistry and Bioenergetics. At the time of his death he was completing the editoral work on a six volume Treatise on Bioelectrochemistry that is also due out shortly.

We tend to think of Milazzo in the context of Bioelectrochemistry, but this was the focus of his last twenty years only. From 1932, when he received his PhD in chemistry from the University of Rome, he had pursued a conventional and successful career as a chemist. He had been active in inorganic and physical chemistry, and had published papers on mineral colloids, UV spectroscopy and electrochemistry. His textbook on Electrochemistry had been translated into several languages. Bioelectrochemisty became his major interest late in his career, but it was the focus of his major achievements. At his eightieth birthday, his friends and colleagues presented him with a Festchrift in recognition of his leadership and his achievements. We are all pleased that he was able to see our appreciation of his accomplishments.

Giulio Milazzo was from the "old school", an inner-directed person complete with high standards, self control and good taste. In all that he did, his aim was excellence and he was always the gentleman, or as he put it "correct". When on occasion we could invoke his internal boss as an ally in discussions against a position he had adopted, he would yield graciously. These qualities endeared him to those of us who worked closely with him, and continue to serve as a standard that we can aspire to. We shall miss him, but we shall always have his rich legacy.

Martin Blank

CONTENTS

SYMBOLS AND ACRONYMS

For the sake of consistency and to ensure immediate understanding, the symbols of the most frequently occurring quantities and the acronyms of the organic chemicals are collected here. Consulting this list, attention must be given to the following points:

1. Only the most common symbols are included. Some, only seldom used, are not included to avoid confusion. Their meaning is given in the text.
2. Since the number of all quantities symbolized in chemistry, physics, biology, *etc.*, and officially accepted by the corresponding International Unions, even using different characters (roman, boldface, italic, *etc.*) is remarkably larger than the number of available letters, it occurs that the use of the same symbols for different quantities becomes sometimes unavoidable, and was accepted by the International Unions (for example the symbol G for the free enthalpy and for the electric conductance, or the symbol A for the area and for the optical absorbance). But the quantity to be correctly considered unambiguously results from the text.

Latin alphabet

[...]	concentration of the species (mol / dm^3)	Bgt (Btx)	bungarotoxin
A	absorbance, area, preexponential factor	BLM	bilayer membrane, black lipid membrane
A	ampere	BR	bacteriorhodopsin
a	activity	C	capacitance
a.c.	alternating current	C	coulomb
AcCh	acetyl choline	°C	Celsius degree
AcChR	acetyl choline receptor	c	concentration
ADP	adenosine diphosphate		centi
ANS	anilino–naphthalene sulphonate	cyt	cytochrome
a.p.	action potential	F	farad
ATP	adenosine triphosphate	F	Faraday's constant
		$f(...)$	function of...

f	femto	*m.w.*	molecular weight
FCCP	carbonylcyanide – *p* – trifluoro- – methoxy – phenyl hydrazone	*N,n*	number of...
		N	newton
FPLC	fast protein liquid chromatography	NEM	N – methyl maleinimide
		n.m.r.	nuclear magnetic resonance
FTIR	Fourier transformed infrared spectroscopy	*P*	permeability
		p	pico
G	free enthalpy (Gibbs free energy); conductance	,p.a.g.e.	polyacryl amide gel electrophoresis
GTP	general insertion protein	PEP	processing enhancing protein
H	enthalpy	PERS	protein electric response signal
h.p.l.c.	high performance liquid chromatography	p.i.x.e.	proton induced X ray emission
		pm	purple membrane
HR	halorhodopsin	*Q*	electric charge
Hz	hertz	*R*	electric resistance, gas constant
I	current intensity; light intensity	*r*	radius
i.c.p.a.e.s	inductively coupled plasma atomic emission spectroscopy	RNA	ribonucleic acid
		RR	resonance Raman
I.R.	infrared	S	siemens
J	joule	s	second
j	current density	*sat*	(subscript) saturated
K$_m$	Michaelis Menten constant	SDS	sodium dodecyl sulfate
K	Kelvin degree	SKL	serine – lysine – leucine
k	Boltzmann constant, rate constant	SR	sarcoplasmic reticulum; sensory rhodopsin
lg	logarithm, decadic	SRP	signal recognition particle
ln	logarithm, natural	SV	slow vacuolar
LDAO	dodecyldimethyl amine	*T*	Kelvin temperature
M	concentration (mol / dm^3, molar)	*t*	time
		TEMED	tetramethylethylene diamine
m	meter, milli	TMPD	tetramethyl – *p* – phenylene diamine
m.c.d.	magnetic circular dichroism		
MES	morpholino – ethane sulfonic acid	TPMP$^+$	triphenylmethyl phosphonium
		TPP$^+$	tetraphenyl phosphonium
min	minute	*U*	electric potential difference
MIT	monoiodotyrosine	UV	ultraviolet
MOPS	N – morpholino – propane sulfonic acid	V	volt
		v	velocity, reaction rate
MPP	mitochondrial processing peptidase	VDAC	voltage dependent anion channel

W	watt
w	weight
z	ionic charge

ζ	zeta potential
θ	angle
λ	wavelength
μ	dipole moment
$\bar{\mu}$	electrochemical potential
ϱ	resistivity
Σ	sum
τ	time (as special quantity)
φ	internal electric potential

Greek alphabet

α	polarizability
Δ	difference
ε	dielectric constant

CODES FOR AMINO ACIDS

Amino acid	Three – letter abbreviation	One – letter symbol
Alanine	Ala	A
Arginine	Arg	R
Asparagine	Asn	N
Aspartic acid	Asp	D
Cysteine	Cys	C
Glutamine	Gln	Q
Glutamic acid	Glu	E
Glycine	Gly	G
Histidine	His	H
Isoleucine	Ile	I
Leucine	Leu	L
Lysine	Lys	K
Methionine	Met	M
Phenylalanine	Phe	F
Proline	Pro	P
Serine	Ser	S
Threonine	Thr	T
Tryptophan	Trp	W
Tyrosine	Tyr	Y
Valine	Val	V

THE PERSPECTIVE OF BIOELECTROCHEMISTRY

MARTIN BLANK

President, of the Bioelectrochemical Society, 1988 - 1992

*Dept. of Physiology, College of Physicians
and Surgeons, Columbia University
New York NY 10032 U.S.A.*

Bioelectrochemistry IV
Edited by B.A. Melandri *et al.*, Plenum Press, New York, 1994

The subject matter of this school *Nerve-Muscle Function* is usually considered in the domain of physiology, but it can be taught in other contexts depending upon the desired emphasis. For example, the context can be anatomy if the emphasis is on structure, biochemistry if the emphasis is on the molecular components and their reactions, biophysics if the focus is on the kinetics of the membrane potential changes and the energetics of the component processes, *etc.* Because of the interdisciplinary nature of the subject, it is important to present the problems from many points of view and to integrate them. Bioelectrochemistry aims at this type of integration catalyzed by the addition of a strong input from a non-biological discipline, electrochemistry. Electrochemistry emphasizes the physical chemistry of the system and helps us to see what happens when one treats the problems as if the detailed structures were unknown and only the electrical potentials, charge densities, *etc.* determine the properties. This approach is really a first approximation, but it can often provide considerable insight.

Let me give an example of the bioelectrochemical approach from my own research on the properties of the hemoglobin molecule. Many of the physical properties of the molecule can be understood in terms of the energetics of the interface of the protein and the aqueous solution, *i. e.*, the area of contact and the surface charge density. Because the free energy of such a system can be approximated easily, differences between the free energies of two states enable estimation of the equilibrium constant for the transformation between the two states. This approach has led to a calculation of the tetramer-dimer disaggregation constant, the variation of the oxygen binding constants with pH (BOHR effect), an explanation of the HILL coefficient for the oxygen binding reaction in terms of a GIBBS surface excess, and even the variation of viscosity of hemoglobin solutions with concentration.

The example of hemoglobin is particularly relevant to the subject of this school, because the bioelectrochemical approach has been used to account for channel opening (*i. e.*, disaggregation) and closing (*i. e.*, aggregation).

The gating currents that open potential-difference-gated channels change the charge densities and lead to changes in the protein disaggregation equilibrium. In addition, the known thermal changes that accompany protein aggregation-disaggregation equilibria account for the observed cooling on channel opening with relatively small changes in area (~ 3 %).

Another aspect of the bioelectrochemical approach applied to the subject of this school is an initial focus on the physical properties of ion channels rather than the specific properties. For example, it is well known that there are super-families of channels. The calcium channel is structurally similar to the sodium channel and the potassium channel, and there may be common properties based on these similarities. Often is the electrical properties that are critical, *e. g.* the need for an aminoacid of a particular charge at a specific position.

The need to examine the general physical properties of a type of molecular organization, and in particular the electrical properties, is the special perspective of bioelectrochemistry. This should be the starting point of an analysis, and in some cases it may prove to be sufficient with regard to understanding mechanism.

CALCIUM HOMEOSTASIS IN SKELETAL MUSCLE

EDUARDO RÍOS

Department of Physiology, Rush University
School of Medicine. 1750 W. Harrison St. Chicago, Il 60612, U.S.A.

Contents

Bioelectrochemistry IV
Edited by B.A. Melandri *et al.*, Plenum Press, New York, 1994

1.   Introduction

Everybody has probably seen already too many lists of the multiple roles of changes in $[Ca^{2+}]_{int}$ in the control of cell metabolism and in cell function. In fact, these lists have become so encompassing that I have now seen bold generalizations claiming that $[Ca^{2+}]$ increase is *the* common final pathway to effect metabolic change. It is not sure that this is the case, but the purpose here is to point out that Ca^{2+} is in fact a very dominant intracellular messenger. Just as an example, our laboratory has come across two roles of Ca^{2+} in E-C (excitation contraction) coupling that had not been described before (see following Chapter). Moreover, Ca^{2+} has what one might call structural roles, functions that it plays merely by being present, hence not requiring changes in $[Ca^{2+}]_{int}$ or *Ca-transients*. These roles have to do with the stability of proteins and membranes, and with steady block of ion channels. In summary, Ca^{2+} has many known static and dynamic roles, and probably many more that are still ignored.

This chapter will deal with the fundamental mechanims that set the level of $[Ca^{2+}]_{int}$ and that restore it to resting values after its functional changes. Let us start from an even more fundamental (and elementary) position, and give an overview of some chemical and physical properties of Ca^{2+} that are at the basis of all the phenomena that will be discussed later. Roughly the first part of this chapter will be devoted to this overview. The second part will be devoted to homeostasis proper, tackling the subject in a very partial way, speaking mostly about the homeostasis mechanisms in muscle, while the generalization will be treated in the following chapter .

2.   Overview of properties of Ca^{2+}

A good guide to the interesting aspects of what would otherwise be an unwieldy field is the proposition of DAN URRY: why is Ca^{2+} useful?, *i.e.* what makes Ca^{2+} so uniquely suited among many possible ions for so many regulatory functions? The answers to these questions are mostly due to URRY himself, plus ADRIAN PARSEGIAN and H.R. KRETSINGER. They were given in the mid-seventies, (KRETSINGER [1], URRY [2]) and still work quite well, although some ideas need to be changed.

2.1. Ca^{2+} ions in solution

In general, metal ions in solution are highly solvated, surrounded by water molecules with the oxygen pointing inward. It is true that ions with the smallest crystal radius have the greatest solvated radius. Figure 1 represents the Ca^{2+} ion, four molecules of the first hydration layer and a fifth water molecule. Two other water molecules, off the plane of the page, would also be found close to the ion. Now, it is possible to imagine most compounds of metal ions with organic molecules as analogs of

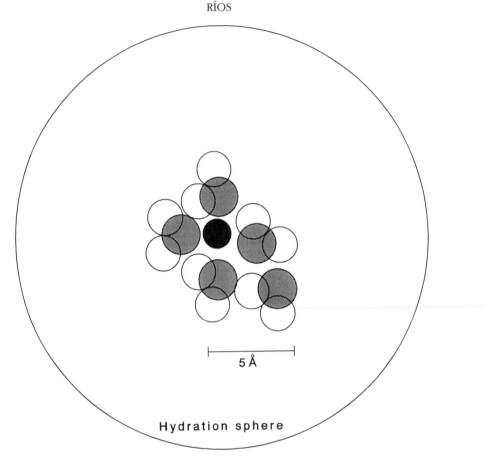

Fig. 1. Ca^{2+} and its hydration sphere. Four water molecules of the first hydration layer are represented. Normally two other molecules would be in the first shell. In the scheme the water molecules are represented more separated than they really are (there would be more interpenetration of oxygens and hydrogen-bonding hydrogens). The hydration sphere, including all water molecules significantly oriented by the ion, extends over a ≈ 10 Å radius and includes ≈ 90 molecules.

the solvated ion in which water oxygens are replaced by other atoms, usually oxygens as well, sometimes nitrogens. This bonding is of variable nature, it can be mostly ionic or it can be mostly covalent, but in all cases the groups surrounding the metal ion give electrons to the ion, hence the nomenclature is usually that of donor-acceptor, or coordination bonds, in which the metal is invariably the acceptor and the ligand the donor. Thus one important characteristic is the coordination number, or number of atoms directly combined to the ion. For Ca^{2+}, this number is usually 6 like in the hydration drawing, but sometimes it can be 7 or 8. In the most important cases of bonding to proteins, the proteins work as multidentate ligands, with one molecule providing many coordinating atoms (usually all 6 to 8 needed).

However, neither the type of bonding nor the coordination numbers are very different for most divalent cations; then, we repeat, why Ca^{2+}? The answers that have been given to this question are basically three: that Ca^{2+} is middle-of-the-road as an ion, both in charge – it is 2, not 1 or 3 – and in size (among the divalents it is intermediate in ionic radius, bigger than beryllium and magnesium ions, but not larger than barium or radium ions). Another part of the answer, due to KRETSINGER [3] is that nature or evolution found a good way to bind Ca^{2+}, reversibly and with high affinity and selectivity. These three aspects will be discussed below.

2.2. Interaction energies of ions in bilayers

To describe the interaction energies[1] of an ion, four contributions are introduced by HONIG et al. [4]:

$$E_{Tot} = E_{Born} + E_{image} + E_{dipole} + E_{neutral} \tag{1}$$

The first three terms are electrical, the fourth is the sum of all other contributions to the free[2] energy, including VAN DER WAALS, hydrophobic, specific chemical bonding, etc. The BORN and image energies are sometimes collectively termed self energies, and have related origins.

2.3. Ionic self energies

It is important that Ca^{2+} is divalent. For being divalent it goes through a whole class of ion channels with considerably more difficulty than monovalents. This is so because, with a greater charge and a smaller radius, its short-range electrical interactions are much more intense. A more quantitative formulation is the following: similar charges repel, thus a *self energy* can be defined, as the energy needed to put the charge of an ion within the volume of the ion, starting from an ideal state of infinite dispersion and zero energy. This also termed BORN energy, is proportional to the square power of the charges, thus for divalents it is fourfold greater than for monovalents. Now, more interesting things occur when the medium is polarizable, the ion induces polarization, which has the consequence of screening the concentrated ionic charge and reducing the self energy, which in a medium of dielectric constant ε is given by equation (2):

$$\text{self energy} = \frac{ze^2}{2\,\varepsilon r} \tag{2}$$

with r the ionic radius, z the valence and e the electron charge. thus, in water at 25 °C the energy is reduced by 1/81 but in lipid only by 1/2.

---

[1] *Editorial.* The terms image energy and *image force* are introduced here for quantities not identical with those defined with these terms in classical electrostatics. But, since the author's description is quite clear, it seems appropriate to leave them unchanged.

[2] *Editorial.* The HELMHOLTZ and GIBBS free energies are again somewhat different quantities, but since author is quoting this from HONIG's et al. review, the terms used by HONIG are left here unchanged.

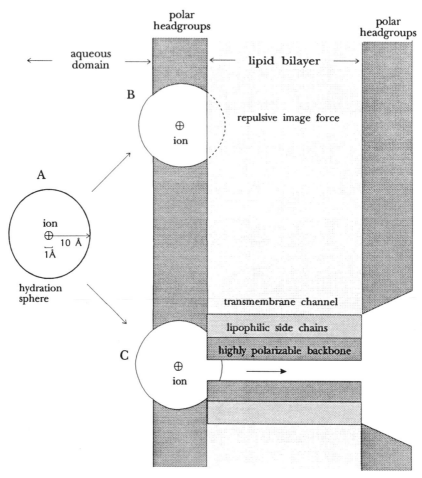

FIG. 2. Origin of the image force. From URRY [2].

Now, as an ion surrounded by water approaches a lipid bilayer, part of the water atmosphere will be displaced by lipid. Since the lipid will have a smaller polarizability (a smaller ε) its energy-reducing effect will be less, and this will amount to an energy barrier, that makes a smooth transition from the low value in the aqueous medium to the full BORN value (or close to it) in the middle of the membrane lipid. The contribution of the edges is termed here *image energy* (Fig. 2). Additionally, the phospholipids have a dipole moment, associated mainly with the carbonyls that link the fatty acid chains to glycerol. Since the electronegative oxygen points toward the aqueous interface, the result is an increased energy for cations inside the bilayer lipid.

---

[3] E and F are the denominations of two α-helical segments in parvalbumin.

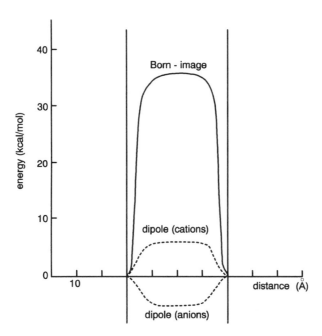

FIG. 3. Two components of the energy of an ion. The energies are represented as a function of distance across a 40 ångstrom bilayer. The BORN-image term is calculated for a monovalent, 1 Å radius cation. Thi dipole energies (dashed lines) are electric potential terms, proportional to charge in the particle. The total energy would include the terms shown and a (neutral) term describing non-electric interactions. Modified from HONIG *et al.* [4].

Figure 3 represents as a continuous line the combined BORN-image energies for a small monovalent cation and the dipole energy. The self energies are of about 35 kcal/mole and the dipole energy about 5 kcal/mole. The dipole energy is simply proportional to the charge of the ion, that is, it can be described as an electrical potential (dipole potential), thus it is positive (repulsive) for cations and negative for anions, explaining for instance the substantially higher permeability of some large anions like ClO_4^-. The BORN-image energies, however, are proportional to the square of the charge, thus the 35 kCl/mole contribution will be quadrupled for divalents and increase nine-fold for trivalents, making the problem of bilayer crossing respectively four and nine-fold greater.

In the case of a transmembrane channel, polarizable groups at the mouth of the channel generate an atmosphere that can replace water without great increase in energy. Without any details of channel structure, it should be clear that the large energy barriers make trivalent ion passage almost impossible, and that of divalents generally more difficult. Thus, for large fluxes involved in fast action potentials cells rely on monovalent ion channels, membrane Ca^{2+} channels being much less capable of large currents. Ca^{2+} channels are in this view more suitable for smaller but highly modulated movements.

These reasons were given by URRY in '78 [2], and do not look too good 13 years later, as gap junction channels, feet or SR Ca^{2+} release channels and bacterial porins seem to violate the above rule, providing large Ca^{2+} conductances. The reasoning of URRY however remains totally valid for the highly selective channels of the plasma-lemma. (The exceptions above are understandable as the size of those channels is so large that the considerations of lipid-ion energies become irrelevant and the whole bilayer can be neglected altogether when discussing these channels).

2.4. Selective calcium binding

The very same reasons that make it more difficult for Ca^{2+} to go through the selective membrane channels give Ca^{2+} the ability for very selective interactions inside the aqueous domain of the cell. These interactions have to be highly selective, have to result in conformational changes (so they will have a messenger value) and have to be reversible. Again URRY showed that a simple polypeptide chain would wrap around Ca^{2+} better than around monovalents, coordinating especially the carbonyl oxygens. The preference can be interpreted in terms again of the higher electrostatic field near the Ca^{2+} ion. Interestingly, this simple polypeptide wrap also favors Ca^{2+} over Mg^{2+}, even though Mg^{2+} is a smaller ion (with higher local electric field). The explanation here is that the polypeptide chain would have to bend too much to put all the coordinating atoms at the right di-

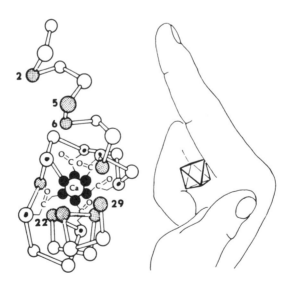

FIG. 4. The EF hand or calmodulin fold. From KRETSINGER et al. [3].

stance from this smaller ion. Thus, the final reason for preferring Ca^{2+} may be one of serendipity: apparently Ca^{2+} just happens to have the ideal size for a polypeptide to wrap around it and satisfy its coordination appetites.

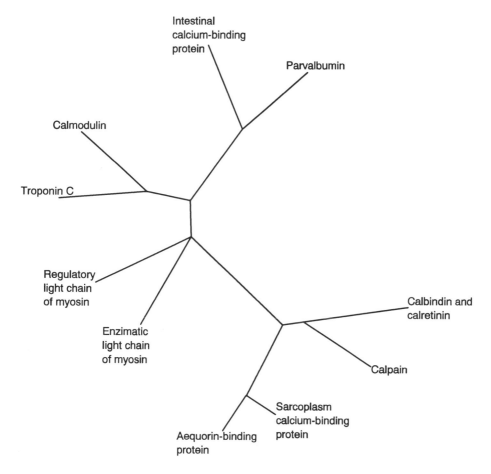

Intestinal
calcium-binding
protein

Parvalbumin

Calmodulin

Troponin C

Regulatory
light chain
of myosin

Enzimatic
light chain
of myosin

Calbindin and
calretinin

Calpain

Sarcoplasm
calcium-binding
protein

Aequorin-binding
protein

FIG. 5. Ten protein families share the EF hand motif. From KRETSINGER *et al*. [3].

A similar but more elaborate explanation was arrived at by R.H. KRETSINGER, who showed in 1975 the existence of a striking structural motif in the intracellular Ca^{2+}-binding proteins. The E^3 *hand* or *calmodulin fold* (Fig. 4) is a strand of 29 peptide residues consituting two α-helical segments (thumb and forefinger) joined by a loop that may wrap around Ca^{2+}, establishing six coordination bonds. This structure, which occurs in pairs usually joined by their hydrophobic side chains, is now found in 10 protein families (KRETSINGER *et al*. [3]) and in many other intracellular proteins not belonging to families. These families are displayed in Fig. 5 in a phenogram, indicating closeness in structure and probably in evolution, and included here to show the tantalizing possibilities of this structural analysis for the understanding of the evolutive process.

More modernly the EF hand concept has shown limitations: many extracytosolic proteins that bind Ca^{2+} do not have EF hands. However, the concept continues to guide the interpretation of structures as they become known. It is not known whether the Ca^{2+} binding sites in channels and pump proteins have the EF an motif; the DHP-sensitive Ca^{2+} channel of skeletal muscle does not contain EF hands but contains regions with some resemblance to EF hands (TANABE *et al.* [5]).

2.5.   Diffusion of calcium

Once inside the cytosol, the delivery of Ca^{2+} to target proteins occurs by diffusion. Diffusion obeys FICK's law: $\vec{M} = -D\,(\partial c/\partial x,\ \partial c/\partial y,\ \partial c/\partial z)$, where $\vec{M}$ is the flux density, D is the diffusion coefficient, and the derived equation:

$$\frac{\partial c}{\partial t} = D\left(\frac{\partial^2 c}{\partial x^2} + \frac{\partial^2 c}{\partial y^2} + \frac{\partial^2 c}{\partial z^2}\right)$$

which in one dimension reduces to:

$$\frac{\partial c}{\partial t} = D\,\frac{\partial^2 c}{\partial x^2}$$

These are analogous to the equations of heat conduction, and generations of biophysicists have taken advantage of this isomorphism by applying directly the solutions of heat problems provided by CARSLAW and JAEGER [6]. One of the useful problems for the understanding of Ca^{2+} movement is that of an *instantaneous plane source*, meaning that a quantity is released at a certain point in space ($x = 0$) and time ($t = 0$). A time t later the distribution is gaussian, and the mean distance (6) is expressed by equation (3)

$$\sigma = \sqrt{(\bar{x}^2)} = \sqrt{(2Dt)} \tag{3}$$

In other words, the average distance is proportional to the square root of time. Another manner in which this appears to the experimentalist is the converse: for example, it is possible to work with cut segments of muscle cells and obtain diffusion of buffers, ion indicators or drugs from the cut end. After cutting the segment, by mistake, one mm long instead of 0.5 mm, instead of waiting 1 hour for the diffusion of a Ca^{2+} sensitive dye one has to wait four hours.

A molecule with the size of glucose will have $D \simeq 4 \times 10^{-6}$ cm^2/s. From the previous equation, the time it takes, on average, for particles released at one point to reach another point at distance l will be $\simeq l^2 / 2D$. Thus, for glucose to diffuse 1 µm $t \simeq 10^{-8}$ cm$^2/(8 \times 10^{-6}$ cm^2/s$^{-1}) \simeq 10^{-3}$ s. Hence the usual association 1 µm $\leftrightarrow$ 1 ms which of course is a bad way of expressing the dependence as the function is non-linear.

Practical applications of this dependency with direct bearing on the contents of this course are, for example:

1. How long would the activator Ca^{2+} take to reach deep portions of a muscle fiber (radius 50 μm) if it diffused from the periphery? this problem was formulated by A.V. HILL in 1948, and his answer – that the time would be too long to account for normal activation – can be considered the birth of the field of EC coupling.

2. How long would it take for a transmitter at the triadic junction to span the transverse tubule-sarcoplasmic reticulum (T-SR) gap (of roughly 100 Å)? Even an approximate answer should make obvious to the student that the diffusion time is no obstacle for a transmitter-mediated mechanism to operate in EC coupling (cf. following Chapter).

2.6.   Diffusion of Ca^{2+} ions in muscle

A final consideration on diffusion is that Ca^{2+}, whose diffusion coefficient in water at 20 °C is 7,5 x 10^{-6} cm²/s, is at least 50-fold less diffusible in skeletal muscle cytoplasm (Fig. 6 KUSHMERICK and PODOLSKY [7]). By contrast, the diffusivities of K^+, Na^+, ATP^{3-}, SO_4^{2-} and uncharged solutes sorbitol and sucrose, are all ≈ 2-fold less than in water.

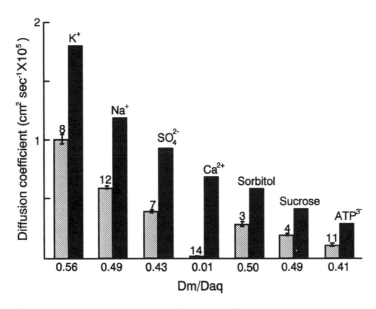

FIG. 6. Diffusibility of ions in muscle cytoplasm. Diffusion coefficients in muscle cytoplasm (hatched bars) and in water (filled bars). The figure below each set of bars is their ratio. From KUSHMERICK and PODOLSKY, [7].

Normally when an ion diffusing in a medium is partly bound to a stationary molecule, its diffusion becomes slower. This is simply because the free ion is the only diffusible species. The consequence, expressed mathematically, is that FICK's laws still apply, but with a different (*effective, apparent*) diffusion coefficient (D') given by equation (4)

$$D' = \frac{D}{1 + K} \tag{4}$$

where K is the binding constant [bound solute/free solute with large excess of binding sites]. Thus, it is straightforward to interpret the above results as implying a high ratio of bound Ca^{2+}/free Ca^{2+} in muscle. KUSHMERICK and PODOLSKY argued that the similarity of reduction in diffusibility for all the solutes tested other than Ca^{2+}, goes against a binding mechanism for the effect on those solutes, and in favor of a more physical, steric mechanism, which they envision as due to the high concentration of macromolecules (filaments, proteins). These components would simply make the path of diffusion longer, and thus delay equally the diffusion of all small molecules.

In any case, if the simple equation (3) is applied to estimate the time required for Ca^{2+} to diffuse from the release sites to the target sites in troponin, a rough estimate would be $\simeq$ 100 ms, and that is clearly too much. In other chapters, ways out of this contradiction are shown.

3. Calcium homeostasis in muscle

The following considerations will start with general notions, applicable to all cells, and go on to discuss the situation in muscle, especially skeletal muscle. There the goal will be to understand the mechanisms that set resting $[Ca^{2+}]_{int}$ and, since they are multiple, their relative contributions.

A simple quantitative expression to discuss Ca^{2+} homoestasis in muscle and other cells is the following equation (5):

$$\frac{d\,[Ca^{2+}]}{dt} = \text{input flux} - \text{output flux} \tag{5}$$

(where fluxes are intended per unit volume)

This *conservation equation* quite obviously states that the rate of change of $[Ca^{2+}]$ is the result of two unidirectional fluxes. Anything that binds, sequesters and otherwise removes Ca^{2+} from free solution will contribute to the output. flux. Imput flux will be through channels that connect the cytoplasm to organelles or to the outside, plus all the reverse fluxes in the binding reactions, plus the possible reverse fluxes in the transporters. The

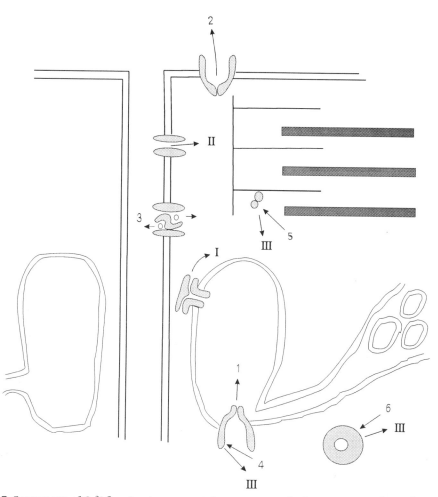

Fig. 7. Components of Ca²⁺ flux. Input components in roman numerals. Output or removal in arabic numerals. See text for key.

units of the terms in the conservation equation are in all cases concentration/time
(everything in our work is referred to the unit of cytoplasmic water volume).

   The components of input and output flux are represented schematically in
Fig.7. The main sources of output flux are numbered 1-6 and include:

1. transport by the SR Ca²⁺ pump,
2. transport by a plasmalemmal and T tubular Ca²⁺ pump,
3. transport by the Na^-/Ca^{2+} exchanger in plasmalemma and T tubules (this may become input in some situations).
4. binding to the SR pump sites, which should be distinguished from 1,
5. binding to troponin C, the functional target of Ca²⁺,
6. binding to soluble proteins, typically parvalbumin and in small quantities calmodulin and other regulatory proteins.

The input fluxes, in roman numerals, include:

I   the SR Ca^{2+} release channel,

II the Ca^{2+} channels of the plasmalemma, these include the DHP-sensitive or L channels, the T channels in some cardiac muscles and the B channels discussed below.

III the backward fluxes from fixed sites (troponin, *etc.*) and diffusible sites (parvalbumin).

The magnitudes of these fluxes are very different during functional EC coupling, when large imbalances occur, and at rest, when a steady state or balance between input and output is maintained (left side of equation (4) is zero).

First will be described, the situation at rest when a steady concentration of 10^{-7} *M* or less is maintained in skeletal muscle, and perhaps somewhat greater concentrations in cardiac and some types of smooth muscle (in which the term *rest* may have an altogether different meaning).

The fluxes of Ca^{2+} at rest are known to be very small, a point made for instance in a recent study of HIDALGO *et al.* [8]. At the resting potential difference the *U*-sensitive Ca^{2+} channels are largely closed (MA *et al.* [9]). Perhaps the only Ca^{2+} channels that contribute some flux at rest are Ca^{2+}-selective leak channels, recently described as B channels in cardiac muscle (ROSENBERG *et al.* [10], and as non-inactivating *U*-insensitive channels in skeletal muscle (*e.g.* MEJIA-ALVAREZ *et al.* [11]), channels that may be elevated in DUCHENNE muscular dystrophy, explaining the increased resting Ca^{2+} in dystrophic myocytes (FONG *et al.* [12]), see Fig. 8.

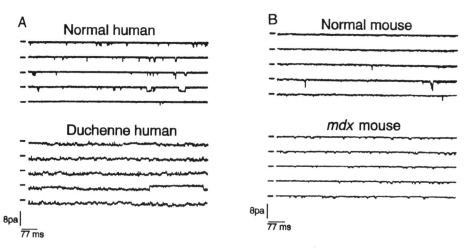

FIG. 8. Calcium leak channels in myotubes. From FONG *et al.* [12].

In the resting situation there is probably a small leak from the SR. This leak might arise through reversal of the pump (if it is working close to thermodynamic equilibrium, *vide infra*) or, as suggested recently by WANG *et al.* [13], because the pump might under some circumstances behave as open channels, or even through deviant release and other channels. Both leaks, from the SR and from the extracellular medium, are balanced by three processes: pumping into the SR, active pumping at the plasmalemma and T tubules and $Na^+- Ca^{2+}$ exchange. Each one of these will be considered separately and briefly, mostly for quantification of fluxes.

3.1. The Ca^{2+} pump of the SR

Its mechanism and role has been reviewed often. INESI's review [14] is especially useful. The pump constitutes 50 % of the membrane dry weight, it is a polypeptide of 115 000 Da arranged in dimers. There are 21 000 pump molecules/μm^2. From here one can estimate the concentration of pump per kg of myoplasmic water, using that there are some 3000 cm^2 SR membrane per cm^3 of fiber of which only 7 % is devoid of pump (the junctional SR). These calculations (BAYLOR *et al.* [15]) yield about 0.1 m*M*ol of pump per kg of muscle or ≈ 0.2 m*M*ol/dm^3 of cytoplasmic water. That is, the effective concentration of pump molecules in the cytoplasmic water is about 0.2 m*M*. Since there are two Ca^{2+} binding sites per pump, the concentration of binding sites is about 0.4 m*M*. By comparison, the total SR Ca^{2+} content is somewhere between 0.5 and 4 m*M* (expressed per kg of fiber water).

The pump molecules are capable of a turnover rate of about 10 s^{-1}, implying a maximum pumping rate of 4 mMole/s per dm^3 of cytoplasmic water; in other terms, they can refill completely the SR contents four times in one second. In the resting

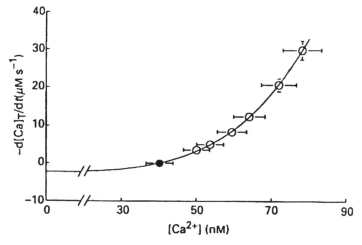

FIG. 9. Free $[Ca^{2+}]$ dependence of removal flux. The abscissa is the $[Ca^{2+}]_{int}$ level measured with Fura-2 in a cut skeletal muscle fiber of the frog. The ordinate values are removal fluxes, that is, rates of decay of total $[Ca^{2+}]$ in the myoplasm. From KLEIN *et al.* [16].

situation, however, and at a free [Ca²⁺] under 100 nM, the fluxes are much smaller. There are estimates of *total calcium leak* into the cytoplasm, which should be equal to total active pumping. In the frog at 10 °C the estimate is 2 μMol/s per dm³ of cytoplasmic water (s. Fig. 9), two thousand-fold lower than the maximum possible SR pump rate (Klein *et al.* [16]). As discussed later, part of this very low flux could be produced by other transport mechanisms. This *dynamic range* of pumping of nearly four orders of magnitude is achieved through a 4th order dependence of rate on cytoplasmic [Ca²⁺]. Since there are two Ca²⁺ binding sites per polypeptide, one interpretation of the 4th order dependence is that the functional unit is the dimer and that the four Ca²⁺ binding sites interact cooperatively, only allowing transport when they all contain bound Ca²⁺.

The pump couples transport of 2 Ca²⁺ to hydrolysis of one ATP and this hydrolysis provides 65 J/mole of free energy at the concentrations existing in muscle. To obtain the free energy required for the transfer of two moles of Ca²⁺ across the membrane at room temperature (25 °C = 298 K) equation (6) expressing this quantity must be used:

$$\Delta G = n\,RT \ln \frac{[Ca^{2+}]_{ext}}{[Ca^{2+}]_{int}} \tag{6}$$

Introducing all numerical values needed, including the conversion coefficient of natural to decadic logarithms, equation (6) becomes

$$\Delta G = 2 \times 8.3 \times 298 \times 2.303 \log \frac{[Ca^{2+}]_{ext}}{[Ca^{2+}]_{int}}$$

implying that the transfer of two moles of Ca²⁺ requires about 12 KJ per decade of the concentration ratio.

Thus, the theoretical limit is a ratio of 10⁵ in the concentration on both sides of the SR. If the resting [Ca²⁺] in the myoplasm is 100 nM, the pump will essentially stop working when the concentration inside the SR reaches ≈ 10 mM. Even though the free [Ca²⁺] inside the SR is not precisely known, it is clear that the resting is close to the theoretical limit for SR pumping.

As interesting as the ability of the pump to maintain the low myoplasmic concentrations is its very nonlinear response to concentration increases; the 4th order dependence with [Ca²⁺]$_{int}$ results in a more rapid homeostatic response (or more effective negative feedback) than would occur under a dependence of lower order. Thus the stability against increases in [Ca²⁺]$_{int}$ is insured by the high capacity and high order dependence on [Ca²⁺]$_{int}$ of the SR pump. On the other hand, what insures stability towards lowering [Ca²⁺]$_{int}$ is by no means clear.

3.2. Ca²⁺ pump of the sarcolemma

In skeletal muscle the study of this pump is made extremely difficult by the possibility of contamination with other pumps, especially from the SR. Two recent studies (HIDALGO et al. and ERVASTI et al. [17]) in mammalian muscle agree that the pump, which unlike the SR pump is sensitive to calmodulin, has a maximal rate of 10 nMol/ mg tubular protein x min). HIDALGO et al. have compared this with the rate of Na^{2+}– Ca^{2+} exchange in frog, and found it to be 10-fold smaller. One can compare these figures with the existing estimates of maximum rate of the SR pump by expressing the pumping rate in terms of cytoplasmic water. Considering that the maximum binding capacity (B_{max}) of the dihydropyridine nitrendipine in the tubular preparation of HIDALGO et al. is 125 pmol/mg tubular protein and that the binding of a dihydropyridine, PN 200-110, to depolarized muscle is 0.5 pmol/mg of muscle protein (SCHWARTZ, MCCLESKEY and ALMERS [18]), the ratio of tubular to total protein can be derived. This can be used to express the pumping in terms of fiber water, using the figure 153 mg muscle protein per gram of muscle wet weight and the figure 0.58 dm³ of fiber water/kg of muscle. The final result is 2 x 10⁻⁷ mol/s per dm³ of fiber water.

The above comparisons are suspect for involving different species. Calculations carried out at our request by Dr. CECILIA HIDALGO based on more direct comparisons of the activities of both pumps in frogs give 2 x 10⁻⁶ mol/s as the maximum rate for the T membrane pump. In any case, these estimates are three to four orders of magnitude less than the maximum rate of the SR pumps; even the highest of the two estimates is not greater than the estimate of total leak at rest at 10 °C (KLEIN et al. [16]). the bottom line is that the T membrane pump can do quantitatively very little and one cannot help but question whether they have any functional role. One possibility (C. HIDALGO, personal communication) is that they become important if and when the SR [Ca²⁺] becomes too high. Another is that the tubular pump might have a lower order dependence on [Ca²⁺]$_{int}$, which would make it relatively more important at very low [Ca²⁺]$_{int}$.

3.3. Na⁺-Ca²⁺ exchange

The first strong indication of the existence of a Na⁺/Ca²⁺ exchanger was given by the studies of LÜTTGAU and NIEDERGERKE [19] of antagonic effects of extracellular Na⁺ and Ca²⁺ on cardiac contractility and the demonstration of such mechanism came 10 years later with studies of REUTER and SEITZ [20] on cardiac muscle and BAKER and BLAUSTEIN [21] on nerve. In cardiac muscle the transporter couples the movement of three Na⁺ in one direction with that of one Ca²⁺ in the other direction. It is a completely reversible process, the direction of which is solely determined by energetics, which in turn depends only on the concentrations and the membrane potential difference. Since the transport has a net electrical equivalent of one elementary charge moving in the direction of the Na⁺ transport per cycle, it is easy to calculate an equilibrium *U*-value, by equating the electric energy difference and the free energy difference associated with the transfer of one Ca^{2+} and three Na^+

$$\mathcal{F}(\varphi_{int} - \varphi_{ext}) = RT \ln \left(\frac{[Ca^{2+}]_{ext}\,[Na^+]^3_{int}}{[Ca^{2+}]_{int}\,[Na^+]^3_{ext}} \right)$$

which is equal to -30 mV at the concentrations in frog skeletal muscle. Thus, in the resting muscle, at $(\varphi_{int} - \varphi_{ext}) \approx -80$ mV, the exchanger would work in the Ca^{2+} efflux mode. That it does work has been demonstrated recently by CASTILLO *et al.* (22), measuring the dependence of *U* on $[Na^+]_{ext}$ and the Na^+– Ca^{2+} exchange inhibitor dichlorobenzamyl on single frog fibers, and by HILDAGO *et al.* (8) on T tubular vesicles.

HILDAGO *et al.* concluded that the maximum transport rate was $\approx$ 10-fold greater than that of the Na^+ pump in the same membranes, (or between 2 x 10^{-6} and 2 x 10^{-5} mol/s per dm³ of fiber water by our previous calculations). The apparent affinity of the transporter for intracellular Ca^{2+} is 3 μM (HILDAGO *et al.* [8]) implying that the rate at 100 nM $[Ca^{2+}]_{int}$ would be 10^{-7} to 10^{-6} mol/s per dm³ of fiber water. This number is greater than the probable rate of the T tubular Ca^{2+} pump at rest. It is less than the estimate of resting Ca^{2+} leak into the myoplasm (2 μmol/s per dm³ of fiber water) but close to it.

In summary, the two active transport systems of the T tubular membrane, the ATPase and the exchanger, having a much lower maximum capability than the SR pump, possibly play a role in setting the resting $[Ca^{2+}]_{int}$. Their contribution to extrusion of Ca^{2+} should become important at very low $[Ca^{2+}]_{int}$ where the SR pump rate will be very low given its 4th order dependence, and in a situation of high Ca^{2+} contents in the SR, as the SR pump is working close to its theoretical limit of gradient. On the other hand, during an action potential the exchanger will contribute its own inward flux of Ca^{2+}.

No active processes are known that tend to increase resting $[Ca^{2+}]_{int}$. Resting $[Ca^{2+}]_{int}$ appears to be the result of three active extrusion processes trying to keep up with leaks from the SR and the exterior of the cell. These layers of insurance against increase in $[Ca^{2+}]_{int}$ demonstrate the importance of keeping a low $[Ca^{2+}]_{int}$. Maintained increases in $[Ca^{2+}]_{int}$ not only hamper short term functionality but are believed to cause long term damage (TURNER *et al.* [23]).

Acknowledgements

I am grateful to Dr. MILOSLAV KARHANEK for help in the preparation of figures and especially of the computer demonstrations that accompanied the lectures. Many of the insights on homeostasis in this chapter are due to several, animated, fax-mediated discussions with Profs. CECILIA HIDALGO (Centro de Estudios, Santiago, Cile) and HECTOR RASGADO-FLORES (University of the Health Sciences, North Chicago, Illinois), who also made manuscripts in press available. Gracias amigos. We were supported by grants from the National Institutes of Health (USA), the Muscular Dystrophy Association of America and the American Heart association and its Chicago Affiliate.

References

[1] R.H. KRETSINGER, in *Calcium Transport in Contraction and Secretion.*, E. CARAFOLI, F. CLEMENTI, W. DRABIKOWSKY and A. MAGRETH, (Editors), North Holland, Amsterdam (1975) pp. (1975),469-478.

[2] D.W. URRY, *N.Y. Ann. Acad. Sci.*, **307**, 3, (1978).

[3] R.H. KRETSINGER, N.D. MONCRIEF, M. GOODMAN and J. CZELUSNIAK in *The Calcium Channel, Structure, Function and Implications*, M. MORAD. W. NAYLER, S.KAZADA and M. SCHRAMM (Editors) Springer-Verlag, (1988), pp. 16–35.

[4] B.H. HONIG, W.L.HUBBELL and R.F. FLEWELLING, *Ann. Rev. Biophys. Biophys. Chem.*, **15**, 263 (1986).

[5] T. TANABE H. TAIKESHIMA A. MIKAMI, V. FLOCKERZI, N. TAKAHASHI, N. TANGAWA, M. KOJIMA, H. MATSUO, T. TIROSE and S. NUMA, *Nature (London)*, **328**, 313 (1987).

[6] H.S. CARSLAW and J.C. JAEGER, *Conduction of Heat in Solids*, Oxford University Press, Oxford, (1947).

[7] M.J. KUSHMERICK and R.J. PODOLSKY, *Science*, **166**, 1297 (1969).

[8] C. HILDAGO, F. CIFUENTES and P. DONOSO, *Ann. NY. Acad. Sci.*, **639**, 483-497 (1991).

[9] J. MA, C. MUDIÑA-WEILENMANN, M.M. HOSEY and E. RIOS, *Biophys. J.*, **60**, 890 (1991).

[10] R.L. ROSENBERG, P. HESS and R.W. TSIEN, *J. Gen. Physiol.*, **92**, 27 (1988).

[11] R. MEJIA-ALVAREZ, M. FILL and E. STEFANI, *J. Gen. Physiol.*, **97**, 393 (1991).

[12] P. FONG, P.R. TURNER, A.F. DENETCLAW and R.A. STEINHARDT, *Science* **250**, 673 (1990).

[13] J.S. WANG, J. TANG and R.S. EISENBERG, *J. Membr. Biol.*, **130**, 163-181 (1992).

[14] G. INESI, *Ann. Rev. Physiol.*, **47**, 573 (1985).

[15] S.M. BAYLOR, W.K. CHANDLER and M.W. MARSHALL, *J. Physiol.*, **344**, 625 (1983).

[16] M.G. KLEIN, L. KOVACS, B.J. SIMON and M.F. SCNEIDER, *J. Physiol.*, **441**, 639 (1991).

[17] C. HIDALGO, GONZALEZ-GARCIA, *B.B.A.*, **854**, 279 (1986), J.M. ERVASTI,J.R. MICKELSON and C.F. LOUIS, *Arch. Biochem. Biophys.*, **269**, 497 (1989).

[18] L.A. SCHWARTZ, E.W. MC CLESKEY and W. ALMERS, *Nature (London)*, **314**, 747 (1985).

[19] H.C. Lüttgau and R. Niedergerke, *J. Physiol.*, **143**, 486 (1958).

[20] H. Reuter and H. Seitz, *J. Physiol.*, **195**, 451 (1968).

[21] P.F. Baker and M. Blaustein, *B.B.A.*, **150**, 167 (1968).

[22] E. Castillo, H. Gonzàlez-Serratos, H. Rasgado-Flores and M. Rozycka, *Ann. NY. Acad. Sci.*, **639**, 554-557, (1991).

[23] P.R. Turner, T. Westwood, CM Regen, RA Steinhardt, *Nature (London)*, **335**,735 (1988).

CALCIUM ION FLUXES ACROSS PLASMA MEMBRANES

GUY VASSORT

U-241 INSERM, Bât. 443 - Centre Universitaire
F91405 ORSAY cedex

Contents

Bioelectrochemistry IV
Edited by B.A. Melandri *et al.*, Plenum Press, New York, 1994

1.   Introduction

Ca^{2+} entry is needed for muscle to contract. This was originally reported by RINGER [1] in 1883 and later reinforced by showing that Ca^{2+} injection by means of a micropipette brings about a local contracture. The existence of a permeability for Ca^{2+} in excitable cells was recognized nearly 40 years ago in crab muscle fibres just after the proposal of the ionic theory by HODGKIN and HUXLEY [2]. Moreover, (in 1956) CORABOEUF and OTSUKA [3] claimed that a guinea pig ventricle heart cell can generate action potentials in the complete absense of external Na$^+$. In 1964, NIEDERGERKE and ORKAND [4] reported that the amplitude of the action potential *overshoot* varied as a function of [Ca^{2+}]$_{out}$ as if the membrane behaves like a Ca^{2+}-electrode. Direct evidence of Ca^{2+} entry was given by the demonstration under *potential difference* clamping that a current could be carried by Ca^{2+} and further increased by adrenaline [5]. Ca^{2+} entry is now easily observed with the use of ion-sensitive fluorescence indicators (TSIEN [6]).

The cytoplasmic [Ca^{2+}] of most mammalian cells is maintained at a value that is at least ten thousand times lower than that of the surrounding milieu ([Ca^{2+}]$_{in}$ ≈ 50-100 nM). This, together with the fact that the membrane potential is highly negative, -60 to -80 mV, implies that sophisticated homeostatic mechanisms for maintaining a low [Ca^{2+}]$_{in}$ should be expected. Pure phospholipid bilayers are not strictly impermeable, but Ca^{2+} may also enter cells through channels activated by potential difference or different ligands. One can thus expect that one or several mechanisms are involved to extrude Ca^{2+}.

The objective of this short review is to give an overview of the major currently described mechanisms related to sarcolemmal Ca^{2+} flux, such as U-operated channels, ligand-operated channels, Na$^+$-Ca^{2+} exchange and Ca^{2+}-ATPase. The literature has been kept to a minimum by referring to other recent reviews and some specific publications.

The observations described thereafter were obtained by using several complementary techniques. They include many electrophysiological approaches: cellular membrane potential difference measurement thanks to an internal microelectrode and more precisely membrane ionic current measurement under potential-difference-clamp conditions. A major recent advance was the development of the patch–clamp technique (HAMILL *et al.* [7]) which, according to the different configurations, allows whole-cell current recording or single–channel recording (cell-attached, inside–out or outside–out). A great deal of information can be obtained with tracers (^{22}Na, ^{45}Ca) on tissues and now, more widely used, on membrane visicles. Several generations of intracellular Ca^{2+} sensor have been dealt with: these are Ca^{2+}-sensitive photoproteins such as aequorin, absorbance dyes such as arsenazo III or antipyrylazo III and the exploding field of fluorescent dyes (quin-2, fura-2 and its derivatives, and Indo-1).

2.   Ca²⁺ Conductance

It is not our aim to give here all the information relative to Ca^{2+} conductance (more than 3000 publications). A recent extensive review is by PELZER *et al* [8], typically on an U-operated channel (UOC). A comprehensive, biophysical approach of ionic channel behaviour has been given by HILLE [9].

2.1   Kinetics

The ubiquitous role of Ca^{2+} in excitation-contraction or excitation-secretion coupling and of Ca^{2+} influx in cell depolarization justified the extensive studies devoted to analysing one of the major membrane pathways for Ca^{2+}: the Ca^{2+} channel. Before the development of the patch-clamp technique, this was difficult because of the absence of easily handled tissues such as the squid axon for the Na^+ and K^+ conductances, the absence of specific inhibitors such as tetrodotoxin for the Na^+ channels, and the presence of several other poorly defined conductances. Indeed, many studies were performed on multicellular cardiac preparations for which today up to 17 ionic conductances have been described.

Three major electrophysiological approaches were devoleped to study the Ca^{2+} channels in excitable cells:

i) U-clamp experiments such as those performed in muscle fibres, or since 1964 on multicellular cardiac preparations. These studies distinguished at least two inward conductances (Na^+ and Ca^{2+}) based on ionic selectivity and kinetics. Slow responses, obtained on partially depolarized cardiac preparations (medium enriched with 15-25 mM K^+) were devoted more to pharmacological modulations of this conductance.

ii) Maior steps were achieved with the use of the patch-clamp technique (HAMILL *et al.* [7]), which allows both ionic currents on a whole cell or on a limited area of membrane to be measured and the ionic composition of the solutions on the external and internal face of the membrane to be controlled.

iii) Electrophysiological and pharmacological tools allow us to distinguish at least three types of Ca^{2+} channels: T, N and L. One or several channels can be expressed in a given cell type. A more general description of the L–type Ca^{2+} channel in cardiac tissues is given in this overall review. A detailed characterization of the three types of channels is given for neurons.

2.1.1 Activation

When muscle cells are depolarized by an U-clamp pulse from resting potential to 0 mV for several hundred ms, a macroscopic Ca^{2+} current quickly turns on (activation), reaches a peak, and then decays with a slower time course (inactivation). A

subsequent depolarization elicits the same response only if there is sufficient time at
more negative potential to allow for reactivation (Fig. 1).

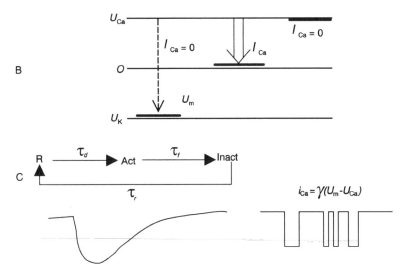

		Na$^+$	K$^+$	Ca$^+$	Mg^{2+}
A	C_{out} (mM)	140	5.6	1.8	0.8
	C_{in} (mM)	10	140	0.0005	>1
	U_x (mV)	+69	-84	106	-5.0

FIG. 1. Characteristics of the calcium current.

A) Distribution of the four major cations involved in most excitable cells. Their chemical equilibrium potential
is given by the NERNST equation:

$$U_x = \frac{RT}{zF} \ln \frac{[X]_{out}}{[X]_{in}}$$

where R, T, z and F have their usual values and $[X]_{out}$ and $[X]_{in}$ are respectively the external and internal
concentrations of a given ion.

B) Diagram emphasizing that I_{Ca} depends on both the electric potential difference and the membrane con-
ductance: $I_{Ca} = G_{Ca} (U_m - U_{Ca})$. At the single Ca^{2+} channel level $i_{Ca} = \gamma (U_m - U_{Ca})$, where γ is the elementary
channel conductance.

C) Idealized Ca current recorded under whole-cell patch-clamp conditions on applying a 200 ms depolarizing
step to 0 mV from the cell resting potential (-80 mV), with the scheme of the three main states of a channel: re-
sted, activated and inactivated. The time constants of passage from one to the other state are τ_d for activation,
τ_f for inactivation and τ_r for recovery of inactivation. These time constants are generally governed by the potential difference for most ionic channels. τ_f for the Ca channel is also controlled by Ca^{2+} ions accumulating at
the intracellular mouth of the channel, so-called Ca^{2+}-dependent Ca^{2+} inactivation.

On the right are idealized single Ca^{2+} channel openings.

When the Ca^{2+} channels are opened, Ca^{2+} ions move according to their transmembrane potential difference $U_m - U_{Ca} \approx -200$ mV, where U_m, is the membrane potential, and U_{Ca}, the NERNST equilibrium U-values for Ca^{2+} have the values -80 mV and +120 mV respectively. The Ca^{2+} current I_{Ca} is generally described on the basis of the HODGKIN and HUXLEY model: $I_{Ca} = G_{Ca} (U_m - U_{Ca})$ where the calcium conductance $G_{Ca} = \overline{G}_{Ca} \, d \, f$; *i.e.* it is the product of a maximal conductance and of the activation and inactivation variables. This original model assumed that membrane conductance resulted from the stochastic openings and closings of a *gate* with probability depending upon potential difference and time. The patch-clamp technique demonstrates that Ca^{2+} channels, recorded for the first time in neonate rat cells, have only two states, open and closed. Thus the macroscopic Ca^{2+} current $I_{Ca} = NP \, i$ where N is the number of Ca^{2+} channels in a single cell, P the opening probability and i the elementary current through a single pore. The intensity of i is very weak in the order of pA and its recording is generally achieved with the use of high concentrations of Ba^{2+}, a more permeant cation. It varies linearly with the membrane potential indicating a conductance of about 20 pS. The potential difference also controls the probability P. There is a distinct threshold U-value below which there are no channel opening; above it, open state probability increases sigmoidally to reach a maximal value (below 1) at about 50 mV above threshold. The U-dependence of elementary current intensity and of the open-state probability accounts for the fact that the U-dependence of the macroscopic current is bell-shaped. Typically a Ca^{2+} I–U relationship shows a threshold potential at -40 mV, a peak current intensity at 0 mV and a reversal U-value at +50 to +70 mV. Maximal current density (generally expressed in relation to membrane capacitance because the last quantity can be measured directly and can be considered approximately constant per unit of surface of a membrane) is in the range of 5-10 $\mu A \; \mu F^{-1}$. Assuming membrane capacitance to be $1 \mu F /cm^2$, these values allow an estimation of the density of functional Ca^{2+}-channels in the order of magnitude of 0.25-1.25 per μm^2 of sarcolemma, or $0.2-1 \times 10^4$ channels per cardiac cell.

On single cardiac cells using tight-seal pipettes, the turn-on of the whole-cell Ca^{2+} current has a distinctly sigmoidal time course. This indicates that activation of the Ca^{2+} channel involves more than a simple transition from closed to open state. Current intensity traces elicited by activating depolarizations applied to a membrane patch containing a single Ca^{2+} channel may exhibit short opening or may not (blanks). Short openings are usually followed by short closings. Groups of short openings and short closings (bursts) are separated by markedly longer closings. In cardiac cells during bursting activity at a given U-value, the distribution of the open times is mono-exponential and that of the closed times requires two exponential. This evidence suggests that activation kinetics can be schematized by three sequential states with four adjustable rate constants as follows:

$$C_1 \overset{k_1}{\underset{k_{-1}}{\leftrightarrow}} C_2 \overset{k_2}{\underset{k_{-2}}{\leftrightarrow}} 0.$$

2.1.2 Inactivation

During a prolonged depolarization, the rapid activation of I_{Ca} is followed by a slower decay (inactivation). Its time course is described by one or, more often, two exponentials, the first of these having a time constant (3-7 ms) that is five to ten times shorter then the second. Their U–dependence is more or less U-shaped. Since the discovery that Ca²⁺ channel inactivation in *Paramecium* is dependent on Ca²⁺ entry rather than potential difference, this aspect has been studied on a wide variety of other cells. There is now a large body of evidence suggesting that, in most excitable cells, both Ca²⁺–dependent and U–dependent mechanisms control Ca²⁺–channel inactivation. Thus, the rate of inactivation during a depolarization:

i) increases when I_{Ca} intensity is increased by raising external [Ca²⁺]$_{out}$;

ii) decreases when cations other than Ca²⁺ carry the Ca²⁺–channel current;

iii) decreases when Ca²⁺ entry is buffered by intracellular EGTA or BAPTA.

Further support is obtained from the U–dependence of the Ca²⁺–channel. Prepulses to U–values more positive than -50 mV reduce the intensity of I_{Ca} during a subsequent test pulse. The dependence of I_{Ca} on prepulse potential difference, termed steady-state inactivation f$_\infty$, has a sigmoidal shape with values equal to 1 at -50 mV and 0 at +10 mV. The relation is often described by the equation:

$$I_{Ca}/I_{Ca,\ max} = 1/(1 + \exp\ [(U\text{-}U_h)/k])$$

where U_h is the prepulse potential difference that reduces I_{Ca} to one half of maximal I_{Ca} and k is the slope of the relation. U_h ranged from -20 to -30 mV and k from -4 to -11 mV.

Similar results are obtained at the single Ca²⁺ -channel level even when channel behaviour is investigated after reincorporation into lipid bilayers. There seems to be a connection between channel entry and ending of clusters of bursts, long-lived closed state (*i.e.* more blanks) and the decay of I_{Ca} during depolarization. However, investigation of the macroscopic Ca²⁺ current further shows that steady-state inactivation is not complete when the prepulses reach U–values above +20 mV. Thus, the U–dependence relationship appears U-shaped. The U–dependent mechanism at positive potential difference dominates more when Ca²⁺ are substituted by other cations.

Inactivation of the Ca²⁺ current in skeletal muscle is much slower (order of s) than in most other cell types. It was suggested that this long process was due to time-dependent depletion of Ca²⁺$_{out}$ in the T-tubules. This is reinforced by the following two observations: embryonic rat skeletal muscle myotubes with sparse T-tubules show a Ca²⁺-current which hardly inactivates, and skeletal Ca²⁺ channels in lipid bilayers do not inactivate. However, there is also evidence that the more classical U–dependent mode of inactivation is present in skeletal muscle.

2.1.3 Reactivation

Ca⁺ channels, inactivated by a sustained depolarization, can be restored to their original available state by repolarizing the membrane. Reactivation, also termed recovery from inactivation or repriming, occurs faster with more negative U-values; at room temperature, it is usually several hundred ms at -50 mV and only 100 ms at -80 mV; these values decrease markedly (by a factor of 3) with increasing temperature from 25 °C to 35 °C. The reactivation process of the Ca^{2+} current is not simply the reciprocal of inactivation. It is much slower than inactivation, and variable time courses (mono- or bi-exponential, sigmoidal, oscillatory) are reported, suggesting that fluctuations in the internal $[Ca^{2+}]_{in}$ may be an important factor.

2.2. Selectivity, permeation, block

When U-gated Ca^{2+} channels open in response to membrane depolarization, Ca^{2+} ions move down the steep electric potential gradient. Under experimental conditions the open-channel flux is in the order of 10^6 ions s⁻¹. Ca^{2+}–channels manage to be highly selective and exclude K^+, Mg^{2+} and even Na^+ ions which vastly outnumber Ca^{2+} ions. Selectivity in an ion channel requires the separation of ion types that differ in charge, size, hydration and coordination chemistry. Selectivity can occur by specific binding of the desired molecule followed by elution of the bound molecule or by rejection of the undesired ions.

Measurement of reversal potentials differences (U_{rev}) is one approach to demonstrate the ability of Ca^{2+} channels to select divalent ions over monovalent ions. The permeability ratios expressed relative to Cs^+ permeability in cardiac cells are: 4200, 2800, 7700, 9.9, 3.6 and 0 for Ca^{2+}, Sr^{2+}, Ba^+, Li^+ Na^+ and Mg^{2+} respectively.

The permeability sequence derived from unitary conductances indicates that the unitary Na^+ conductance (85 pS) is much larger than the Ba^{2+} conductance (25 pS). Indeed for Ca^{2+}, Ba^{2+}, Li^+ and Na^+, the permeability measured with reversal U-value is inversely related to the flux rate, in disagreement with the independence principle. This property could be expected for a *sticky* pore because ions that bind most tightly will have the lowest mobility (conductance).

Large organic cations can be used to probe the pore size. The relative permeability (measured with U_{rev}) of ammonium and methylated derivatives decreases with increasing ion size. The largest permeant cation, tetramethylammonium, has a diameter of 6Å, three times the diameter of a Ca^{2+} ion or about the same size as a Ca^{2+} ion with an attached water molecule.

Support for ionic binding as the mechanism of selectivity comes from experiments showing saturation of current at high ion concentration and competition between ions of different affinities. An intrapore Ca^{2+} binding site would provide a possible means of selection by affinity. If the site were located within the pore and acted as a station for Ca^{2+} ions moving from outside, occupation of this site by Ca^{2+} would help the channel reject foreign cations. Moreover the more strongly an ion is bound, the more slowly does it leave the binding site. However, studies of the interaction between Ca^{2+} and Ca^{2+}–channels give widely different estimates of affinities, indicating that at least two Ca^{2+}–binding sites are associated with the Ca^{2+}–channel. Further evidence against a simple one-site channel comes from studies of current carried by ion mixtures in which the sum of two cations is held constant. A significant minimun in Ca^{2+}–channel current amplitude is seen with mixtures compared to the maximal current obtained with one or the other cation (anomalous mole fraction effect). In fact, the model (HESS and TSIEN [10]) postulates that:

i) ions move single-file through the channel;

ii) the pore contains two binding sites with high affinity for Ca^{2+} ions and low affinity for monovalents;

iii) cations at the outer and inner sites repel each other.

This two-site model predicts that monovalent currents through single channels should be blocked in discrete events as Ca^{2+} enters and exits the pore. Indeed the open time decreases linearly with Ca^{2+} concentration. Multiple (double) occupancy of the Ca^{2+} channel suggests that there should be interactions among ions in the pore. Thus, the long-lasting openings of Ba^{2+} current at high Ba^{2+} concentrations become bursts of brief openings when low concentrations of another permeant as well as impermeant cation are included in the pipette solution.

2.3   Regulation

2.3.1 Neurotransmitters and hormones

The first evidence for modulation of Ca^{2+}–channels by neurotransmitters and hormones was obtained in the heart. *Adrenaline* and *noradrenaline* increase I_{Ca} following their binding to the ß-adrenergic receptors. The metabolic cascade is now clearly established. It involved stimulation of the adenylyl cyclase via a stimulatory guanine nucleotide binding protein (Gs), production of cyclic adenosine monophosphate, cAMP, which activates a cAMP-dependent protein kinase (PKA), and phosphorylation of the Ca^{2+}–channel (Fig. 2). Membrane permeant cAMP derivatives and cAMP injection reproduce the effect while protein kinase inhibitors prevent, and phophatases antagonize the increase in current. Cholera toxin, which ADP–ribosylates the Gs-protein, causes sustained activation of the cyclase and thus increases cAMP and I_{Ca}. Whole-cell current shows that neither the reversal U-value for Ca^{2+}, nor the U-dependent kinetics (inactivation, availability) are significantly changed. Rather the increase in I_{Ca} is consequent to both an increase in P, by prolongation of open times

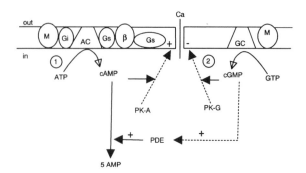

FIG. 2. Possible pathways involved in regulating the Ca-channel conductance following ß-adrenergic and muscarinic stimulations. Stimulation of the ß-adrenergic receptor might increase Ca-channel activity by two independent pathways which nevertheless both involve a stimulatory guanine nucleotide regulatory protein (G_s). Either G_s, indeed the α-GTP subunit, directly activates the Ca-channel or it activates the adenylyl cyclase (AC) to produce cyclic AMP and leads to phosphorylation by the cyclic AMP-dependente protein kinase (PKA). On the other hand, muscarinic stimulation might interfere with Ca-channel activity directly by the inhibitory protein (G_i) (not shown), by activating a cyclic GMP-dependent protein kinase (PKG), through activation a guanylyl cyclase (GC), by inhibiting cyclic AMP production by the adenylyl cyclase or by reducing the cyclic AMP level following the activation of a cyclic GMP-dependent phosphodiesterase (PDE).

and abbreviation of closed times, and an increase in Nf the number of functional channel, *i.e.* an increase in the duration of cluster of bursts (less blanks). In any case, single-channel current conductance is unchanged.

Muscarinic agonists *(acetylcholine)* that strongly increase the K^+ -conductance have no effect or a weak inhibitory direct effect on the Ca^{2+}. However, they reduce the ß-adrenergic increase in Ca^{2+} current in a number of ways. First, acetylcholine inhibits adenylyl cyclase activity through a G_i protein, thus reducing cAMP production. Adenosine acting on P_1-purinoreceptor appears to activate G_i in a similar way. Pertussis toxin ADP-ribosylates G_i and suppresses G_i activation by receptor agonists. Second, muscarinic receptor occupation increases intracellular cyclic guanosine monophosphate (cGMP) by activating a guanyl cyclase. cGMP itself stimulates cAMP hydrolysis by phosphodiesterase and on the other hand, is supposed to phosphorylate the Ca^{2+} channel at a site that induces reduction of its activity in the heart (but increases it in snail neurones). The atrial natriuretic factor (ANF) also elevates cGMP and has an effect similar to acetylcholine on the ß-adrenergic-stimulated Ca^{2+} current.

Muscarinic agonists as well as α_1-*adrenergic* agonists, *angiotensin* II etc... are known to activate the turnover of phospatidylinositol *via* stimulation of phospholipase C. This results in the generation of inositol trisphosphate ($InsP_3$) and of diacylglycerol (DAG); the latter increases the activity of Ca^{2+}-activated phospholipid-sensitive protein kinase C (PKC). PKC is known to phosphorylate a number of endogenous proteins, but there are no indications that PKC directly phosphorylates Ca^{2+} channel proteins. The PKC-activating action of phorbol esters induces stimulatory as well as inhibitory effects on I_{Ca} in neurones and heart cells. A wash-out of intracellular factors might occur in a cell under whole cell patch clamp so that no clear effect is observed. In single-channel study, the addition of phorbol esters (TPA) transiently increases the number of openings per trace.

2.3.2 Direct regulation of Ca²⁺–channel by G-protein

Neurotransmitters and hormones alter the receptor-associated G-protein that determine the levels of second messengers cAMP, c GMP, DAG.. However, G-proteins (or one of the three monomers α, ß, γ) may also regulate ionic channel activity without the involvement of cytosolic protein kinases (TRAUTWEIN AND HESCHELER [10 bis]). This was initially proposed after the demonstration that muscarinic receptor-regulated K^+ channels in cardiac myocytes are under the direct control of a PTX-sensitive G protein G_k (similar to G_i). The direct action of a G protein on Ca^{2+} conductance is proved by intracellularly applying GTPγ S, a poorly hydrolysable GTP, which then produces a pronounced stimulation. Moreover, purified G_s activated by GTPγ S reproduces the stimulation, as does the activated subunit α_s, while G_i does not. G_s can also increase the P of Ca^{2+} channels reincorporated in a lipid bilayer, but only when added on the cytoplasmic side. In the latter study, the addition of GTPγ S also increases P, and this leads us to propose that the preparation contains some G_s that must be endogenous to the *DHP receptor* isolated from skeletal T-tubule membrane.

2.3.3 Ca²⁺-antagonist and Ca²⁺–agonists

Several organic Ca^{2+}–antagonists were synthesized, because of possible therapeutic potentiality. None is highly specific, unlike tetrodotoxin on the Na^+ channel. At least three sites are revealed by binding studies. Ca^{2+}–antagonist effects are sensitive to the membrane potential, so the inhibition increases with the amplitude and the frequency of membrane depolarization. The U–dependent and use-dependent blocks are more marked with verapamil, while the DHPs induce mostly a *steady block* observed with the first depolarization. Some DHPs increase the Ca^{2+} current. Their effect is larger at lower membrane depolarization. Moreover, the same compound can reduce the Ca^{2+}–current when the holding U–value is made less negative. Extensive single-channel studies have shown that the Ca^{2+}–agonists reduce the number of blanks (mode 0) like ß-adrenergic stimulation, and induce long sustained openings (mode 2).

2.3.4 Calcium, magnesium and hydrogen ions

Whole-cell patch-clamp experiments are performed with EGTA in the pipette solution. Omission of EGTA or use of high $[Ca^{2+}]_{in}$ solution leads to a rapid run-down of the Ca^{2+} channel current. The depression of the current is also seen with Sr^{2+}. Kinetics and steady-state inactivation are unaffected. The mechanism of Ca^{2+}_{in} –induced inhibition of Ca^{2+} channel current is not understood; it might involve a Ca^{2+}–dependent phosphatase. Besides, in some studies a facilitation of I_{Ca} by $[Ca^{2+}]_{in}$ is observed as though there were a bell-shaped relation between $[Ca^{2+}]_{in}$ and the channel conductance. This is also evident in studies of post-rest, positive staircase increase in atrial I_{Ca}.

Varying the intracellular Mg^{2+} concentration, $[Mg^{2+}]_{in}$ from 0.3 to 3 mM does not affect control I_{Ca}. However, it reduces by 50% ß-adrenergic-stimulated I_{Ca} and hardly decreases DHP-increased I_{Ca}. At a higher concentration (9.4 mM), Mg^{2+}_{in} almost completely blocks I_{Ca} due to enhanced steady-state inactivation.

The Ca^{2+} channel conductance is inhibited by acidic pH with half-inhibition at pH 6.7. Internal pH is the modulator: studies require the inhibition of the Na^+–H^+ exchange mechanism. Open-state probability declines at pH < 8 with only a small effect on elementary current amplitude.

2.3.5 Membrane effects

Many compounds alter the membrane phospholipid bilayer, and this affects Ca^{2+} entry, as well as other transmembrane ionic movements. thus, in cardiac tissues, antiarrhythmic agents that are supposed to act by inhibiting the Na^+ channel also reduce both I_{Ca} and the Na^+–Ca^{2+} exchange current. This is also found with general anaesthetics including neutral ones such as alcohols. The potency of the latter increases with the chain lenght, as does the octanol/water partition coefficient, to the extent that dedecanol is as potent as verapamil (MONGO and VASSORT [11]).

The phospholipids in the sarcolemma are asymmetrically distributed over the two monolayers with negatively charged phospholipids mostly in the inner one. Exposure of cardiac myocytes to the negatively charged amphiphile docecylsuphate (DDS, 10 μM) increases (185 %) contraction, while the positively charged dodecyltrimethylammonium (DDTMA, 10 μM) decreases it (58 %). Most of these inotropic effects are related to an increase or a decrease in I_{Ca}. These variations in I_{Ca} intensity result:

1) from a change in the external surface charge; and

2) from a direct effect on the channel protein.

Moreover, plasma membranes are abundantly glycosylated. Sialic acid (N-acetyl neuraminic acid and derivatives) is a ubiquitous anionic sugar found at the extremity of glycoconjugates. Neuraminidase which specifically hydrolyses sialic acid (57 % removal) increases Ca^{2+} current (and augments cellular $[Ca^{2+}]_{in}$).

2.4   Structure

The devolopment of patch-clamp studies on single Ca^{2+} channels was paralleled by the availability of radioactively labelled probes of high affinity and specificity for Ca^{2+}–channel binding sites. Identification and isolation of the protein components of the Ca^{2+}–channel were thus possible. The complete amino-acid sequence as predicted by cloning and sequence analysis of DNA complementary to its messenger

RNA was reported initially in skeletal muscle (TANABE *et al.*, [12]); it shows structural and sequence similarities to the *U*-dependent Na⁺–channel. The Ca^{2+}–channel is an oligomeric complex (α_1, α_2, ß, γ, δ). The major subunit α_1 of 170 kDa has four repetitive units; each includes six membrane spanning α-helices, one of which, S_4, is a hydrophylic fragment (Fig. 3).

The Ca^{2+}–antagonist receptor in cardiac muscle, α_1, is a significantly larger (185 kDa, 2171 amino acids) peptide and structurally and immunologically different from the skeletal one with 66 % homology of the aminoacid sequence. Structural similari-

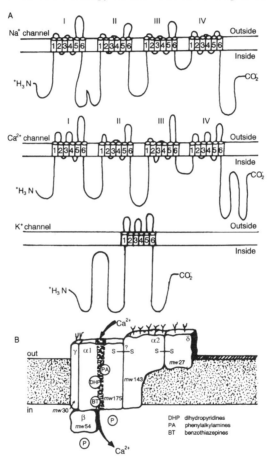

FIG. 3. Structure of the calcium channel.

A) Schematic drawing of the primary structure of the α_1 subunit of the dihydropyridine-(DHP) receptor complex as deduced from its complementary DNA sequence. The polypeptide has four distinct internal repeats that exhibit sequence similarities. Based on the hydropathy pattern each unit consists of five hydrophobic and one hydrophylic (S_4) segments. S_4 is thought to form a positively charged spiral, possibly the voltage sensor. It is actually believed that the K-channel consists in fact of four repeats. This indicates that the *U*-operated channels belong to one superfamily.

B) Schematic representation of the calcium channel including the five polypeptide subunits. Further emphasized are the different molecular weights, the three Ca-antagonist binding sites, and the two potential phosphorylation sites. (redrawn from TAKAHASHI *et.al.*, [20 bis]).

ties suggest that the skeletal and cardiac dihydropyridine (DHP) receptors have the same transmembrane topology with four potential N-glycosylation sites located on the extracellular side and six potential cAMP-dependent phosphorylation sites on the cytoplasmic side. Messenger RNA derived from the cardiac DHP receptor cDNA was sufficient to direct the formation of a functional DHP-sensitive Ca^{2+}–channel in *Xenopus* oocytes. Moreover, co-injection of the skeletal muscle α_2 subunit-specific mRNA induces substantially larger Ca^{2+}–channel currents without affecting their sensitivity to DHP or altering the peak I–U relationship. Construction of complementary DNAs, encoding chimaeric Ca^{2+}–channels, in which one or more of the four repeats of the skeletal muscle DHP receptor are replaced by the corresponding repeat derived from the cardiac DHP receptor, allows one to show that repeat 1 determines the activation rate of slow skeletal muscle or fast cardiac activation (TANABE *et al.* [13]).

The density of dihydropyridine-sensitive Ca^{2+}–channels in the heart varies with the purification level from 100 to 400 fmol/mg protein (or 8 to 30 binding sites per μm^2 sarcolemma), but remains always less than in the T-tubules of skeletal muscle. It could not exactly reflect the density of active channels if, as in skeletal muscle, less than 5 % of the sites are associated with functional channels. In cardiac tissues, this density is controlled by physiological mechanisms. The number of binding sites on myocytes in culture increases after two hours' stimulation by thyroid hormones, and after a few days' stimulation of ß–adrenergic receptors. During myogenesis the density increases and the relative distribution varies along the sarcolemma and T-tubules, but is constant during cardiac hypertrophy.

3.   Ligand - operated channels

The idea that Ca^{2+} can move into the cell through channels whose opening mechanisms are not dependent upon membrane potential is recent. Moreover, at least two major types of channels have been proposed, which depend on direct or indirect ligand activation. Besides their apparent U–insensitivity, these channels are also insensitive to organic channel blockers. No information other than these electrophysiolo gical investigations are available, particulary about their molecular structure. A more complete discussion can be found in PIETROBON *et al.* [13 bis].

3.1   Receptor - operated channels (ROC)

The existence of ROC was suggested on the basis of smooth muscle responses to acetylcholine which depend on Ca^{2+} entry while membrane potential does not change. They are somewhat similar to the pentameric nicotinic receptor at the neuromuscular junction except that ROCs carry Ca^{2+} ions together with Na^+ and other cations.

One well-characterized Ca^{2+} ROC is the glutamate-activated channel, more precisely the NMDA receptor seen in the central nervous system. This channel is permeable to Na^+ and Ca^{2+} with a reversal U-value around 0 mV. It is powerfully inhibited by Mg^{2+} although depolarization can relieve the inhibition. Channel conductance is 50 pS. Other ROCs are the channels activated by extracellular ATP or ADP. Detailed electrophysiological characterizations of ATP-activated channel have been carried out initially in smooth muscle cells (BENHAM AND TSIEN [14]) and more recently in neurons and frog heart (BEAN and FRIEL [15]).

In smooth muscle, exogenous ATP in the micromolar range opens a channel with a conductance of 5 pS and a selectivity ratio of 3/1 for Ca^{2+}/Na^+. Mg^{2+} ions have no effect on this channel, Ba^{2+} is as permeant as Ca^{2+} and three to ten times less permeant than Na^+. Similar characteristics are reported in sensory neurons and frog heart cells, but with low Ca^{2+} permeability.

Detailed studies of the ADP-induced cation channels in platelets were done with Mn^{2+} ion to quench quin-2 and fura-2 (SAGE et al. [16]). Fura-2 emission is independent of Ca^{2+} for excitation at 360 nm but fluorescent is still quenched by Mn^{2+}. This allows the selective study of Mn^{2+} entry without interference in the signal caused by Ca^{2+} release from internal stores. ADP induces Mn^{2+} entry without measurable delay (<20 ms), followed by a secondary rise in $[Ca^{2+}]_{in}$. More recently, cell-attached patch recordings have demonstrated an external ADP-activated channel that might carry the rapid phase of Ca^{2+} influx in platelets. To this type of controlled Ca^{2+} entry should be added the receptor-activation of the Ca^{2+} conductance through a G-protein as described earlier.

3.2 Secondary messenger - operated channels (SMOC)

The first evidence for these channels was probably obtained in the earliest whole-cell patch-clamp study in a heart cell. It was suggested that an increase in intracellular $[Ca^{2+}]_{in}$ induces the opening of a non-selective cationic conductance (COLQUHOUN et al. [17]).

Thus, Ca^{2+}, presumably released from intracellular store, might serve as a second messenger that regulates the plasma membrane cation channels. Such channels have been described in a variety of excitable and non-excitable cells. They show a poor selectivity for various monovalent cations and exclude anions (BENHAM et al. [18]). The intracellular $[Ca^{2+}]_{in}$ concentration necessary to activate them is generally considered to be outside the physiological range, or their Ca^{2+} permeability is so low that they may not be relevant in maintaining Ca^{2+} homeostasis. However, in the case of neutrophils whose Na^+ and Ca^{2+} permabilities are approximately similar, or more particularly in smooth muscle, with a P_{Ca}/P_{Na} ratio = 21, these channels may be important in the refilling of intracellular stores.

In many if not all cells, receptor activation often triggers hydrolysis of phosphatidylinositol to produce inositoltrisphosphate ($InsP_3$) and diacylglycerol (DAG). $InsP_3$

induces Ca^{2+} release from intracellular stores and thus increases $[Ca^{2+}]_{in}$. On its own, or in combination with $Ins(1,3,4,5)P_4$, $InsP_3$ was considered to be the major candidate for activation of the Ca-SMOCs, since $InsP_3$ has been reported to increase the opening probability of cation channels in excised patches from lymphocytes. However, several recent studies raise doubts about these observations.

Another more elaborated model for explaining Ca^{2+} influx linked to $InsP_3$–sensitive stores has been put forward by PUTNEY [19]: the capacitive model of Ca^{2+} entry. It implies that $InsP_3$ activates a Ca^{2+} channel on an intracellular organelle leading to an initial, transient phase of Ca^{2+}. Secondary to the depletion of this organelle, and by an unknown mechanism, a plasma membrane Ca^{2+} channel is opened, allowing Ca^{2+} entry during the second, sustained phase of the Ca^{2+} signal, as seen with many agonists acting on excitable and non-excitable cells.

In a recent study on rat cardiac cells, we proposed that a rather complex four-step cascade is involved to account for the opening of the non-selective cationic conductance by extracellular ATP (SCAMPS and VASSORT; PUCÉAT et al. [20]). It is suggested that the local increase in $[Ca^{2+}]_{in}$ results from the displacement of Ca^{2+} by H^+ on membrane binding sites. The intracellular acidosis is consequent to the sudden activation of the Cl^-–HCO_3^- exchanger following binding of MgATP to a new P_3-type of purinoreceptors. The link between the purinoreceptor and the exchanger is unknown. Although the agonist and the effects (depolarization) are similar in smooth and heart muscle and in neurons, the pathway in the heart (Fig. 4) seems to be much more complex than the initially suggested ATP-operated channel in smooth muscle.

4.   Na⁺ - Ca²⁺ exchange

The existence of a Na^+-Ca^{2+} exchange mechanism was first inferred from the

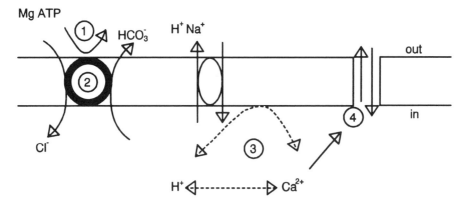

FIG. 4. Schematic diagram of a four-step cascade involved in the increase in the non-selective cation conductance by Mg ATP, as proposed on cardiac cells.

effects of varying Na^+ concentrations on Ca^{2+} efflux in guinea pig atria (REUTER AND
SEITZ [21]). Not only may Na^+-Ca^{2+} exchange be essential in mediating several Ca^{2+}-de-
pendent responses, but it may also play a role in maintaining the low cellular Ca^{2+} level in several tissues. Cells possess multiple Ca^{2+}-transport mechanisms, so that an
unequivocal assignment of a Na^+ or Ca^{2+} flux to Na^+-Ca^{2+} exchange is often difficult.
The development of techniques for the measurement of Na^+-Ca^{2+} exchange in isolated plasma vesicles has greatly facilitated investigations, together with the more recent
application of the patch-clamp making use of the electrogenicity of the exchange.
Most studies refer to cardiac sarcolemma and, to a lesser extent, to brain plasma
membrane (PHILIPSON, EISNER and LEDERER [22]).

Vesicular Na^+-Ca^{2+} exchange activity can be attributed to the plasma membrane
because its maximal activity correlates with optimal levels of 5'-nucleotidase and Na-K
ATPase. Also, the Na^+-Ca^{2+} exchange mechanism and the ATP-dependent Na^+ pump
are simultaneously present on the same membrane vesicles; Na^+ that has been acti-
vely transported into vesicles by the Na^+-K^+ pump can be exchanged with extravesi-
cular Ca^{2+}.

4.1   General properties

Cardiac sarcolemmal vesicles demonstrate initial rates for Na_i-dependent Ca
uptake up to 20-30 nmol Ca^{2+} mg (prot.)$^{-1}$ s^{-1}, i.e. comparable to the Ca^{2+} transport activity of the cardiac sarcoplasmic reticulum. Ca^{2+} loaded within vesicles derived from
plasma membranes can be easily released by increasing extracellular $[Na^+]_{out}$ inducing
a reverse mode of exchange, the Na^+_{out}-dependent Ca^{2+} efflux (Fig. 5).

The apparent K_m (Ca^{2+}) for the initial rate of Na^+_{out}-dependent Ca^{2+} uptake is
around 15-40 μM, but one group gives values at 1-2 μM. Due to competition with Ca^{2+}
for binding sites, divalent and trivalent cations inhibit the Na^+-Ca^{2+} exchange. The effectiveness of the divalent cations was related to their ionic crystal radius as compared
to that of Ca^{2+} ions: $Cd^{2+} > Sr^{2+} > Ba^{2+} \approx Mn^{2+} > Mg^{2+}$. Trivalent cations are much more
potent ($La^{3+} > Nd^{3+} > Tm^{3+} > Y^{3+}$; at 20 μM Ca^{2+}, 1 μM La^{3+} causes 50 % inhibition). In
the absence of Ca^{2+}, both Ba^{2+} and Sr^{2+} can participate in the Na^+-divalent cation exchange. Extracellular Na^+ inhibits by competing with Ca^{2+} at the Ca^{2+} site. Na^+-Ca^{2+}
exchange is very pH-sensitive. Na^+_{in}-dependent Ca^{2+} uptake is a sigmoidal function of
pH. Ca^{2+} influx is markedly inhibited at pH 6 and maximal at pH 9. This suggests the
possible importance of a histidine residue.

There is a question about how many Na^+ ions exchange for each Ca^{2+}. The $K_{1/2}$
(Na^+) is estimated between 7-32 μM. There is a sigmoidal dependence of Ca^{2+} efflux
on external $[Na^+]_{in}$ with a HILL coefficient from 2.3 to 3.2, (sometimes much more). If
the energy in the Na^+ electric potential gradient is to be used to pump out Ca^{2+} from
the cell, an important factor is how many Na^+ ions are to be coupled (coupling ratio).

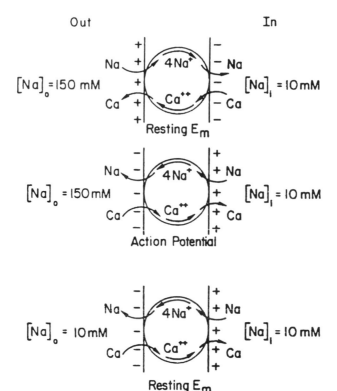

FIG. 5. Schematic diagram representing the two modes of the Na-Ca exchanger. In control conditions, forward mode, Ca ions are extruded from the cytosol at the expense of Na entry. The reserve mode can be induced either during an action potential or by reducing the extracellular Na concentration. Notice that assuming three Na ions are exchanged for each Ca ion, a net movement of one positive charge occurs; this induces an inward current (and depolarization) in the forward mode or an outward current (and byperpolarization) in the reserve mode which leads to Ca entry (in opposition to the Ca current through the Ca conductance).

The initial consideration of an electroneutral exchange does not hold since it implies that $U_{Na} = U_{Ca}$ and thus a $[Ca^{2+}]_{in}$ of about 20 μM. A coupling ratio of 3 or more Na^+ per Ca^{2+} has two effects. It greatly lowers the level of Ca^{2+} that is possible for the exchange to reach, and it makes the exchange sensitive to membrane potential. If the exchange were at equilibrium, then

$$U_m = 3U_{Na^+} - 2U_{Ca^{2+}}, \text{ and } [Ca^{2+}]_{in} = [Ca^{2+}]_{out} \frac{[Na^+]^3_{in}}{[Na^+]^3_{out}} \exp \frac{(n-2)\,\mathcal{F}\,U_m}{RT}$$

At resting U-values (U_m = –80 mV) and U_{Na} = +70 mV (with $[Na^+]_{in}$ = 10 mM), thus U_{Ca} = 145 mV, which suggests $[Ca^{2+}]_m \approx 27$ nM much below its actually measured value (70 to 100 nM). However, while thermodynamics determines the direction of the exchange it says nothing about its rate of transport. This is critically determined by

Na^+_{in} in so far as the inward movement of Ca^{2+} is concerned, as indirectly indicated by studies of contraction and relaxation in cardiac cells. Vesicular Na^+-Ca^{2+} exchange both generates a current and can be altered by imposed membrane potential. The intravesicular accumulation of the lipophile cation tetraphenylphosphonium indicates that an inside negative U-value is generated when Na^+-dependent Ca^{2+} uptake is activated. This is consistent with more positive charges leaving the vesicle (carried by Na^+) than entering (carried by Ca^{2+}) with each transport cycle. Such U-values changes in vesicles are overcome by using valinomycin to produce a K^+–diffusion U-value. Unequivocal demonstration of the electrogenicity of the Na^+-Ca^{2+} exchange has been obtained in single cardiac cells with the whole-cell patch-clamp by changing suddenly the external Na^+ or Ca^{2+} concentrations. The I-U relation of external Ca^{2+} and external Na^+-induced current showed as almost exponential U-dependence as given by the equation: $I = a \exp (rU\mathcal{F}/RT)$, where a is a scaling factor that determines the intensity of the current and r is a partition parameter of the EYRING rate theory, that rapresents the position of the energy barrier in the electrical field and indicates the steepness of the U-dependence of the current. The value of a was 1-2 $\mu A/\mu F$ and r about 0.35* for the Ca^{2+}-induced outward current. The current has a $Q_{10}*$ of 3.6-4 which is larger than that of a channel current. Only Sr^{2+} could replace Ca^{2+} in these experiments; the current was blocked by di-and trivalent cations. It requires Ca^{2+} on the internal face of the membrane. The reversal U-value was seen at a value very close to the thermodynamically expected values for 3 Na^+/1 Ca^{2+} stoichiometry, consistent with the HILL coefficient of 2.9 and 0.9 for Na^+ and Ca^{2+} ions respectively.

Under these experimental conditions, the $K_{1/2}$ values for $[Ca^{2+}]$ (1.2 mM) and $[Na^+]$ (90 mM) are similar to those obtained with tracer studies in squid axon, crab nerve and barnacle muscle, which are much larger than in cardiac vesicles. This discrepancy might originate from the difference in ionic environment. In the vesicles, one side of the membrane is exposed only to Ca^{2+} and the other only to Na^+, and it is suggested that Na^+ and Ca^{2+} compete for the binding sites.

4.2 Structure and regulation

The use of a polyclonal antibody against a partially purifield preparation of the exchanger has allowed identification of a complementary DNA (cDNA) clone encoding the Na^+-Ca^{2+} exchanger protein (NICOLL et al. [23]). The native exchanger protein has a maximal apparent size of about 170 kDa with six potential glycosylation sites. The hydropathy plot indicates 12 membrane spanning segment and a small region of the sequence similar to that of the Na^+, K^+ -ATP pump. A long cytoplasmic loop contains a calmodulin binding domain. Recent experiments on cardiomyocytes show that calmodulin antagonists (W-7, trifluoperazine) almost abolish the Na^+-Ca^{2+} exchange current. This domain may also account for the reported effects of ATP, although phosphorylation is not involved. An internal Ca^{2+}_{in} concentration of 0.6 mM is required to half-activate $[Na+]_{out}$ -dependent Ca^{2+} influx; but in the absence of ATP, Ca^{2+} has much less effect. A synthetic peptide with the deduced aminoacid sequence of the calmodulin binding site was synthesized: the concentration inducing half-enhibition, $K_i \approx 1$ μM. It inhibits the $[Na^+]_{out}$ -dependent Ca^{2+} flux in a non -competitive manner with respect to both Na^+ and Ca^{2+}, as well as the Na^+-Ca^{2+} exchange current.

By incubating with phosphatidylcholine liposomes, the cholesterol content of cardiac sarcolemma vesicles can be significantly enhanced. Such cholesterol-enrichment increases Na^+-Ca^{2+} exchange by 48 %, with an increase in affinity for Ca^{2+} ions. Cholesterol depletion induces opposite effects. Also, phosphatidylinositol hydrolysis by phospholipase C results in a stimulation of V_{max} of the exchanger. On the other hand, phospholipase A_2, which increases sarcolemmal lysophosphatidylcholine and lysophosphatidylethanolamine, causes a 50 % reduction in Na^+-Ca^{2+} exchange velocity.

Addition of low levels of anionic detergent dodecylsulphate (DDS) and anionic phospholipids or treatment with phospholipase D have been shown to stimulate the exchanger. Also, mild proteolytic treatments increase its activity.

The properties of the Na^+-Ca^{2+} exchangers are very similar in different cell types. However the exchanger of the rod outer segments has a striking dependence on K^+ (K_m activation: $1mM$). Both exchangers are modulated by the same regulatory influences, except membrane potential which has a much smaller influence on ROS.

5.   Calcium pump

Such a transporter extrudes Ca^{2+} against its electric potential gradient and hydrolyses ATP on the cytoplasmic face. This energy-dependent mechanism requires Mg^{2+} ions as a cofactor. The apparent Km for Ca^{2+} is between 0.2 and 0.6 μM at the internal face of the membrane and 1 000 to 10 000 μM at the external face. The system is inhibited by vanadate, but not by ouabain. The pump enzyme has been purified from red cell membranes by affinity chromatography on columns containing bound calmodulin, a major regulator of the pump. It appears to possess a single polipeptide chain of 130-150 kDa with about 1 200 amino acids. Amino-acid similarity and hydropathy profile comparison indicate a common transmembrane organization for Na^+, K^+-ATPase, sarcoplasmic reticulum Ca^{2+}-ATPase, and the isoform of sarcolemmal Ca^{2+} pump from brain or erythrocytes (VERMA et al.; SHULL and GREEB [24]). After reconstitution in phospholipid vesicles it retains calcium pumping and ATPase activity. Maximal pumping rates of the Ca-ATPase and of Na^+, K^+-ATPase in red blood cells are comparable (20 mmol (litre cells)$^{-1}$ hr^{-1}). The maximum velocity corresponds to a turnover rate of 30 s^{-1} at 37 °C, one-fifth the value of the Na^+, K^+-ATPase, implying that there are five times as many calcium pumps i.e. about 1 000 Ca^{2+} pumps per red cell. In cardiac cells, but not in smooth muscle cells, maximal Ca^{2+}-ATPase activity is much less (1/30) than Na^+-Ca^{2+} exchange (V_{max} 15 nmol Ca^{2+} (mg prot)$^{-1}$ s^{-1}. In both tissues, the affinity for Ca^{2+} on the cytoplasmic face is higher (Km 0.3 mM) than the Na^+-Ca^{2+} exchanger.

The Ca^{2+} pump differs from the Na^+-K^+ pump in that there is:

i) no requirement for a second ion to be transported; and

ii) a precise regulation of its activity by the level of internal Ca^{2+} ions, as mediated by the specific protein calmodulin. The V_{max} of Ca-ATPase rises and the affinity for Ca^{2+} and ATP at the low affinity binding sites increases 30 to 100 times when calmodulin is bound. This effect is not related to phosphorylation, but to the binding of calmodulin, and is not seen in reconstituted vesicles, particularly with phosphatidylserine, which maximally activates the pump. Moreover, trypsin treatment enhances pump activity. On the other hand, phospholipase C treatment and lysophosphatidylcholine reduce Ca^{2+}–ATPase activity (Na^+-K^+ ATPase is unaffected).

6.   Calcium Leak

Radioactive isotope studies of Ca^{2+} influx have shown that there is a significant influx of Ca^{2+} in every quiescent muscle even in the absence of direct stimulation of *U*-gated or receptor-gated channels. Moreover, the intracellular Ca^{2+} concentration in an excitable cell is not at the level that it would be if the Na^+-Ca^{2+} exchange were at equilibrium. In smooth muscles, there is a report of *leak* channels in fewer than 10 % of patches with a 16 pS conductance in 110 mM Ba^{2+}. Also, the elevated free Ca^{2+} concentration found in dystrophic muscle has been attributed to an increased activity of Ca^{2+} leak channels.

In many cell types (macrophages, mouse neuroblastoma cells, transformed fibroblasts), the extracellular application of ATP^{4-} causes a large increase in whole-cell membrane conductance. This results from the opening of a large non-selective channel that admits ions and solutes of molecular mass up to 0.9 kDa (GOMPERTS [25]). Permeabilization by ATP occurs very rapidly (<40 ms) and is rapidly reversible. No other nucleotide mimics this effect. It appears that gap junction channels exhibit very similar properties. On the basis that a variant of mouse macrophage which does not express connexin-43 is ATP-resistant, it has been proposed that in control cells connexin-43 forms *"half-gap junctions"* in response to extracellular ATP (BEYER and STEINBERG [26]).

In conclusion, it is assumed that Ca^{2+} leak is limited but it could be markedly increased by some agonists.

References

[1]   S. RINGER, *J. Physiol. (London)*, **4**, 29 (1983).

[2]   A.L. HODGKIN and A.F. HUXLEY, *J. Physiol. (London)*, **117**, 500 (1952).

[3]   E. CORABOEF and M. OTSUKA, *C.R. Acad. Sci.*, **243**, 441 (1956).

[4]   R. NIEDERGERKE and N.K. ORKAND, *J. Physiol. (London)*, **184**, 291 (1964).

[5]   N. REUTER, *Pflügers Arch.*, **287**, 357 (1966); O. ROUGIER, G. VASSORT, D. GARNIER, Y.M. GARGOUIL and E. CORABOEUF, *Pflügers Arch.* **308**, 91 (1969); G. VASSORT, O. ROUGIER, D. GARNIER, M.P. SAUVIAT, E. CORABOEUF and Y.M. GARGOUIL, *Pflügers Arch.*, **309**, 70 (1969).

[6]   R.Y. TSIEN, *Ti.N.S.*, **11**, 419 (1988).

[7]   U.P. HAMILL, A. MARTY, E. NEHER, B. SAKMANN and F.J. SIGWORTH, *Pflügers Arch.* **391**, 85 (1981).

[8]   D. PELZER, S. PELZER and T.F. McDONALD, *Rev. Physcol. Biochem. Pharmacol.*, **114**, 107 (1990).

[9]   B. HILLE, *Ionic Channels of Excitable Membranes,* Sinauer Associaty Inc., Sunderland MA (1984).

[10]   P. HESS and R.W. TSIEN, *Nature (London)*, **309**, 453 (1984).

[10bis] W. TRAUTWEIN and J. HESCHELER, *Ann. Rev. Physiol.*, **52**, 257 (1990).

[11]   K.G. MONGO and G. VASSORT, *J. Mol. Cell. Cardial.*, **22**, 939 (1990).

[12]   T. TANABE, H. TAKESHIMA, A. MIKAMI, V. KLOCKERCI, and M. TAKAHASHI, *Nature (London)*, **328**, 313 (1987).

[13]   T. TANABE, B.A. ADAMS, S. NUMA and K.G. BEAM, *Nature (London)*, **352**, 800 (1991).

[13bis] D. PIETROBON, F.Di VIRGILIO and T. POZZAN, *Eur. J. Biochem.*, **193**, 599 (1990).

[14]   C.D. BENHAM and R.W. TSIEN, *Nature (London)*, **328** 275 (1987).

[15]   B.P. BEAN and D.D. FRIEL, *in Ion Channels,* T. NARAMASHI, Plenum Publishing Corp. New York (1990).

[16]   S.O. SAGE, J.E. MERRITT, T.J. HALLAM and T.J. RINK, *Biochem. J.*, **258**, 923 (1989).

[17]   D. COLQUOUN, E. NEHER, H. REUTER and C.F. STEVENS, *Nature (London)*, **294**, 752 (1981).

[18]   C.D. BENHAM, J.E. MERRITT and T. J. RINK, in *Ion Transport,* D. KEELING and C. BENHAM (Editors), Academic Press, New York (1989) pp. 197-219.

[19]   J.W. PUTNEY, *Cell Calcium,* **11**, 611 (1990).

[20]   F. SCAMPS and G. VASSORT, *Pflügers Arch.* **417**, 309 (1990); M. PUCÉAT, O. CLÉMENT and G. VASSORT, *J. Physiol., (London)* **444**, 241-256 (1991).

[20bis] M. TAKAHASHI, M.J. SEAGAN, J.F. JONES, B.F. REBER and W.A. CATTERALL, *Proc. Natl. Acad. Sci. USA,* **84**, 5478 (1987).

[21]   H. REUTER and N. SEITZ, *J. Physiol. (London)*, **195**, 451 (1968).

[22]   K.D. PHILIPSON, Ann. Rev. Physiol., **47**, 561 (1985); D.A. EISNER and W.J. LEDERER,

in *Sodium - Calcium Exchange,* T.J.A. ALLEN, D. NOBLE and H. REUTER (Editors),...
Oxford (1989), pp. 178-207.

[23] D.A. NICOLL, S. LONGONI and K.D. PHILIPSON, *Science,* **250,** 562 (1990).

[24] A.K. VERMA, A. G. FILOTEO, D.R. STANFORD, E.D. WIEBEN and J.T. PENNISTON, *J. Biol. Chem.,* **263,** 14152 (1988); G.E. SHULL and J. GREEB, *J. Biol. Chem.,* **263** (8646 (1988).

[25] B.D. GOMPERTS, *Nature (London),* **306,** 64 (1983).

[26] E.C. BEYER AND T.H. STEINBERG, *J. Biol. Chem.,* **266** 297 (1991).

CALCIUM FLUXES AND DISTRIBUTION IN NEURONS

RICCARDO FESCE and DANIELE ZACCHETTI

"B. Ceccarelli" Centre and C.N.R. Centre of Cytopharmacology,
Department of Medical Pharmacology, University of Milan
DIBIT H. S. Raffaele, Milan - Italy

Contents

Bioelectrochemistry IV
Edited by B.A. Melandri *et al.*, Plenum Press, New York, 1994

1.   Introduction

As already discussed, in the first days of the course, calcium homeostasis is a central problem for the regulation of cell functions. This is true for neurons as well, but further specific aspects become important in this case. The neuron is an excitable cell, whose main specific functions are to receive and elaborate signals, conduct the signal that results from such elaboration to a different location, and transmit the signal to another nervous or effector cell. Although calcium is not involved in signal conduction over long distances, it is of crucial importance in receiving, elaborating and transmitting signals. As this course is aimed at electrochemists as well as biologists and neuroscience specialists, it seemed appropriate to give the general readers with little knowledge of neurobiology some ideas about cell excitability and signal transmission among neurons.

2.   Cell excitability

Cell excitability arises from profound electrochemical unbalance for several ionic species across the plasma membrane at rest; these electrochemical potential differences are used for the generation and amplification of nerve impulses. The ionic unbalance across the plasma membrane is generated by the combination of active mechanisms (mainly Na^+ and Ca^{2+} extrusion, either by transport or by exchange with other ionic species) and passive characteristics of the membrane (selective permeabilities): this leads to concentration differences, across the membrane, for the least permeable ion species, the related charge displacement being compensated by countermovement of other more permeable species up to their *electrochemical equilibrium*. Notice that this process necessarily involves the generation of an electrical potential difference across the plasma membrane (membrane potential, in the range -50 to -100 mV, internal negative).

The basic mechanism of cell excitability consists in the presence of voltage–sensitive Na^+ and Ca^{2+} selective membrane channels. Such channels can produce regenerative phenomena, as their opening in response to membrane depolarization produces inward currents that further depolarize the membrane and activate further channel openings. This positive feedback mechanism, combined with the relatively rapid inactivation of the channels and the delayed activation of potassium conductances, that bring the membrane potential back to its resting value, can lead to neuronal firing (generation of an *action potential,* transient inversion of membrane potential) and to the conduction of a nerve impulse over the whole membrane of the neuron without attenuation. This allows the signals generated by the neuron to be conducted with no attenuation to the nerve terminals (in some cases located at a relevant distance), where they can be transmitted to other neurons or to effector cells.

Most neurons have a cell body (*soma*), a specialized process of variable length (*axon*) which branches into nerve terminals that contact other neurons or effector cells, and a richly branched structure of *dendrites* where most nerve terminals from other neurons make synaptic contact, at specialized postsynaptic sites called dendritic spines. The transmission of signals among neurons generally involves chemical synapses, where chemical transmitters (biogenic amines, aminoacids or peptides) are released by a nerve terminal into the synaptic cleft and bind to specific receptors on the membrane of the postsynaptic neuron. Notice that generation and conduction of an action potential is not necessary for many interneurons in the central nervous system, because their terminals are sufficiently close to the soma to be reached electrotonically by electrical changes in the membrane of cell body and dendrites (for a general discussion of these themes, see Ref. 1 and 2).

3.   Calcium and excitability

Cell excitability can be seen in electrical terms: it is sufficient for each input signal be translated into a displacement of membrane potential and the spatial/temporal summation of signals will lead to a time-varying potential difference at the origin of the axon; this can be considered as the output signal elaborated by the neuron. In neurons with long axons, at this site Na^+ channels are sufficiently dense to generate action potentials, and the frequency of action potentials will essentially monitor the time-varying output signal. Notice that this *frequency modulation* transduction of the soma membrane potential makes it possible to reproduce the time-varying output signal at the remote nerve terminals with no attenuation and a high signal/noise ratio. We shall see that this is not the only cellular process exploiting frequency modulation.

The signals that come to the neuron from other neurons are translated to potential difference changes and affect its excitability by two main mechanisms: receptor-coupled ion channels and receptor-activated metabolic changes capable of altering the electrical or structural properties of the synapse.

Many receptors for neurotransmitters are directly coupled to ion channels that open in response to binding of the ligand. These may elicit excitatory or inhibitory responses, depending on the ionic selectivity of the channels opened by the ligand: sodium and calcium are much more concentrated outside the cytoplasm and have markedly positive equilibrium potentials, so the opening of channels permeable to these species will tend to depolarize the membrane and *excite* the neuron; conversely, potassium and chloride have equilibrium potentials close to the resting membrane potential, and channels permeable to such species will tend to stabilize membrane potential and reduce the depolarization produced by excitatory stimuli.

The binding of neurotransmitters and hormones to receptors can activate, directly or indirectly, enzymatic or metabolic activities within the cell, and these may determine changes in the properties of receptor-coupled or voltage–sensitive ion channels, in the structural organization of the postsynaptic dendritic spine, or in feedback secretion of modulatory transmitter substances on to the nerve terminal. All these changes will affect postsynaptic excitability or, more generally, synaptic efficiency. It may be observed that whereas direct activation of a receptor/channel by a neurotransmitter is implied in phasic transmission of rapid signals, receptors linked to metabolic changes will generally mediate effects on a longer time-scale, from tens of milliseconds, for indirect actions on membrane channels, to hours or days when structural alterations of the synaptic organization are implied.

Calcium ions are of paramount importance for cell excitability under both respects. In fact calcium ions carry charges and can mediate direct effects on postsynaptic excitability; in the meantime, calcium ions are among the most widely active regulators of metabolic and enzymatic activities in the cell, and are therefore implied in regulation of electrical, functional and structural aspects of synaptic function.

3.1.   Electrical aspects of calcium fluxes

Membrane channels permeable to calcium can be activated by the binding of specific ligands (receptor-operated channels), by changes in membrane potential (voltage–operated channels), and possibly by intracellular signalling molecules (second-messenger operated channels). Only a few calcium-permeable channels directly activated by ligand binding have been described so far, and for some of them the direct link between receptor activation and channel opening has been questioned (see, for example, VASSORT, this volume). The nicotinic acetylcholine receptor channel (mediating synaptic transmission at the neuromuscular junction) and the channel associated with a subtype of glutamate receptor (the so-called N-methyl-D-aspartate, NMDA, receptor) are the most important and best characterized calcium-permeable receptor-operated channels. Both are unselective and permeable to monovalent cations as well; thus, calcium is not the main charge carrier for these channels. However, calcium flux through the NMDA-receptor channel may be mostly important for synaptic plasticity (see below).

Several families of voltage–operated calcium channels have been described, and the number of identified and characterized subspecies of these channels is increasing every day. The different families have been defined mainly according to gating properties, cellular location and pharmacology: (i) channels with low threshold, activated and rapidly inactivated by depolarizations above -60 mV, fully recovering from inactivation only by hyperpolarization to -100 mV (so-called T channels, from *Transient*); (ii) channels with higher threshold (>-10 mV), slowly inactivating, recovering from inactivation below -10 mV and fully reactivated at -60 mV, with high conductance, blocked by di-hydro-pyridine (DHP) drugs (so-called L channels, from *Long-lasting, Large*) and (iii) channels not very different from L-type channels, but typically present on nerve cells, insensitive to DHP drugs and blocked by ω-conotoxin (so-called N channels, from *Neuronal*) (3,4) (SCHWARTZ *et al*; TSIEN *et al.*). A further type (P) has been recently described in PURKINJE neurons. These channels, like sodium channels in all excitable cells, can produce regenerative responses (action potentials), as the influx of calcium through an voltage–operated channel further depolarizes the membrane and tends to open more channels. Examples of calcium and sodium action potentials in thalamic neurons are depicted in Figs. 1 and 2 [5]: it is evident that so-

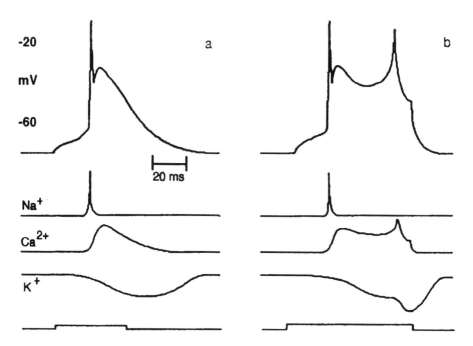

FIG. 1. Schematic diagrams illustrating electrical responses of thalamic neurones. Upper traces illustrate membrane potential, intermediate traces show a hypothetical reconstruction of the underlying ionic currents, and bottom traces illustrate the depolarizing pulse. In **a)** the depolarizing current activates a slow action potential (sustained by low-threshold calcium conductances) and a superimposed fast action potential (sustained by sodium current). The cell repolarizes as delayed voltage-dependent and calcium-dependent potassium conductances activate, while calcium conductance inactivates. Panel **b)** illustrates the behaviour of the same neuron in the presence of a blocker of potassium channels: repolarization is delayed so that a second (high threshold) slow action potential fires before the cell repolarizes. (based on data from Ref. 5).

dium currents are generally much more intense than calcium currents, and the action potentials are accordingly much faster. The coexistence of regenerative responses with different thresholds and time scales in the same cell allows for higher versatility in processing and response.

3.2. Metabolic aspects of calcium influx

Intracellular calcium is the regulator of a number of biochemical processes, either directly or by binding to calmodulin with subsequent activation of Ca^{2+}/calmodulin-dependent protein kinases, which can phosphorylate enzymes, membrane channels and structural proteins. Among the processes that are directly regulated by calcium, particularly relevant for neurons are calcium-activated ion channels, mainly K^+ and Cl^- permeable. Activation of potassium conductances by calcium inflowing through voltage–operated channels is a widespread cellular mechanism for self-limitation of calcium influx, as an increased potassium conductance tends to bring back membrane potential to its resting level and therefore to close voltage–dependent calcium channels. In many nerve cells calcium-activated potassium conductances are also responsible for marked hyperpolarizations that follow the inflow of calcium during calcium-dependent action potentials. Such hyperpolarizations can reactivate a relevant proportion of low threshold calcium channels (mostly inactivated at resting potentials of about -60 mV); when calcium has been extruded from the cytoplasm and calcium-activated K^+ channels close, the membrane potential returns to its resting value and low threshold calcium channels may open, leading to membrane depolarization, further opening of low and high threshold calcium channels (and possibly sodium channels), and a new action potential. The interplay voltage–operated calcium channels, delayed voltage–operated and calcium-operated K^+ conductances responsible for repolarization and hyperpolarization, and low threshold calcium conductances, gives rise to oscillatory behaviour in many neurons, which may spontaneously fire rhythmically (as illustrated for thalamic neurons in Fig.2) or respond preferentially to impulses at a specific frequency (resonance [6] or tuning, as described for hair cell in the cochlear and vestibular systems [7]).

The preceding discussion highlights the tight intertwining between cell excitability and metabolism. Influxes of calcium through the plasma membrane constitute one important aspect of this relationship. It may be appropriate to point out here that many other examples of reciprocal influence of membrane biophysical properties and intracellular metabolic activities have been reported. Many receptors, on nervous as well as non-nervous cells, when activated by neurotransmitters or hormones can influence cell metabolism by activating GTP-binding proteins (G-proteins) and directly or indirectly modulating intracellular levels of Ca^{2+}, inositol tris-phosphate (IP_3) or cyclic AMP; in most cases the activation of protein kinases ensues, and specific metabo-

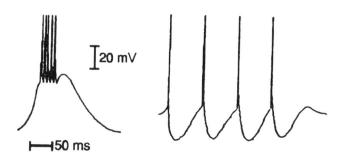

Fig. 2. Schematic diagram illustrating possible patterns of oscillation in thalamic neurones. A slow action potential sustained by calcium currents (either spontaneous or elicited after a hyperpolarization, anodic break, may elicit repetitive firing of fast action potentials (left). Alternatively, after the action potential, potassium conductances may hyperpolarize the neuron enough to de-inactivate low-threshold calcium channels; these, on return towards the resting potential, trigger a new action potential (right). (based on data from Ref. 5).

lic changes occur in the target cell [8]. These receptors have often been called *metabotropic* as opposed to the *ionotropic,* channel-associated, receptors. However, ion channels activated by G-proteins or modulated by phosphorylation have been described, so that *metabotropic* receptors may actually display relevant effects on the electrical properties of neurons, and their excitability.

4.   Regulation of intracellular calcium levels in neurons

Specialized structures are dedicated in neuronal cells, like in muscle fibres, to rapidly raising intracellular calcium concentration or to bringing it back to resting levels. These machineries must work very rapidly, like in muscle cells, but in neurons the picture is complicated by the more elaborate morphology. Figure 3 diagrammatically shows the structure of a PURKINJE neuron as a typical nerve cell, with its *soma, axon, dendrites* and dendritic *spines.* Each of these compartments may be differently equipped to regulate calcium concentration. The main structures dedicated to changing more or less rapidly calcium concentration are various types of calcium channels, operated by electric field or intra/extracellular ligands, calcium pumps and exchangers, and intracellular compartments equipped with channels, transporters and buffering proteins to store large quantities of calcium. Various types of calcium stores have been characterized, regulated either through IP_3-sensitive or through calcium- and ryanodine-sensitive receptor-channels (the latter process is called Calcium-Induced Calcium Release, CICR), or possibly through both receptors [9–11]. These structures can be located strategically to constitute specific subsets in the calcium regulation

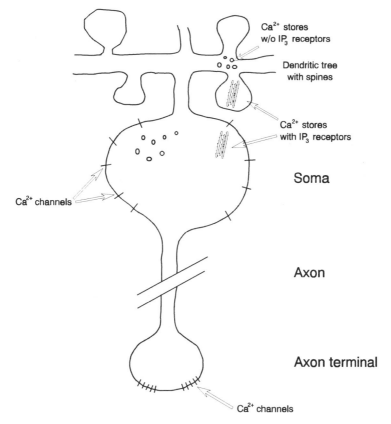

FIG. 3. The elements of calcium regulation in specific regions of a neuron. The possibility of a differential control of intracellular calcium in dendritic spines, dentrites, soma, axon and terminal are diagrammatically exemplified by differential localization of calcium channels and IP$_3$–sensitive and insensitive stores.

apparatus, related to morphology and function. As described by Rios (this volume), the diffusion in muscle cells is slower for calcium ions than for other substances or ions, by at least an order of magnitude. This is presumably the case in other cell types as well, due to the ubiquitous presence of calcium binding proteins (CBP) and to the complexity of endomembrane systems that can define differentially regulated subdomains in the cytoplasm. For neurons, in particular, the complex morphology contributes to creating multiple functionally-distinct compartments differing in terms of organelle distribution, surface to volume ratios, concentration of channels, pumps and ion-exchangers. The slow diffusion of calcium, combined with its morphological complexity, allows for strictly localized transients in cytoplasmic free calcium ion concentration *(calcium transients)*.

The most extreme case of such compartmentalization is the motor neuron, whose soma (in the spinal cord) and axonal terminals (in skeletal muscles) are located at huge distances from one-another. Thus, although the action potential allows for rapid travelling of information from the soma to the synaptic terminals, modifications of ionic concentrations in one of these compartments cannot affect the other. Indeed, even between structures as close to another as adjacent dendritic spines, differential controls of calcium regulation and different cytoplasmic concentrations of free calcium may occur. For example is illustrated in Fig. 3, in dendritic spines of cerebellar Purkinje neurons, the presence of a calcium store sensitive to IP_3 is indicated by immuno-gold labelling with antibodies against the IP_3 receptor; a different calcium storing structure, outside the spines, is instead decorated by antibodies against a low-affinity, high-capacity calcium binding protein usually concentrated in subcellular stores [12]. Such a differential compartmentalization of calcium stores may have important functional correlates, as IP_3 produced in response to activation of post-synaptic receptors, like some subtypes of glutamate receptors, might generate localized calcium transients in the spines. An efficient calcium sequestering structure in the shaft of the dendrite might then buffer calcium concentration, so preventing the transient increase in calcium concentration from affecting nearby spines [13]. On the other hand, were the latter store sensitive to ryanodine, local amplification and spatial spread of the calcium transient might ensue through CICR.

4.1. Calcium regulation at postsynaptic sites

The presence of intracellular store organelles that can release calcium when activated by IP_3 and/or by elevations of Ca^{2+} concentration itself (CICR) indicates that calcium-dependent processes can be richly modulated, in response to the activation of channel-associated as well as *metabotropic* receptors. In particular, steady elevations or rapid oscillations can be produced in free cytoplasmic calcium concentration, thus preferentially activating calcium-dependent processes with different affinity and time dependence (see below).

Among the calcium-dependent processes most intensively studied in recent years are those responsible for long-term alterations in synaptic efficiency. Cation channels associated to glutamate NMDA receptors have been shown to be implied in many of these phenomena [14,15]. The NMDA-receptor channel is present on central neurons and the pore is blocked by extracellular magnesium ions at the resting membrane potential (-60 mV); the activation of the receptor by glutamate produces calcium fluxes only when the postsynaptic membrane is depolarized from its resting membrane potential, thereby relieving Mg^{2+} block; this generally occurs following the activation of other subtypes of glutamate receptors (coupled to sodium-selective channels) located either in the same synaptic cleft or in nearby synapses. Fluxes of calcium through the NMDA receptor are therefore of minor importance in terms of phasic alterations of the electrical properties of the postsynaptic membrane; however, they appear to be of paramount importance in the generation of long term modulatory alterations of synaptic structure and efficiency, particularly long-term potentiation (LTP). LTP consists of an increase in synaptic efficiency that ensues following repetitive activation of some central synapses according to specific temporal patterns of stimulation, and can last for several hours or even longer. This phenomenon is currently considered as one of the molecular mechanisms most probably involved in the learning and memory processes [14]. A central role has also been suggested for NMDA receptors in selective synaptic depression during the development of neural networks [15] and in Ca^{2+}-related neuronal death.

4.2.   Calcium and neurotransmitter secretion

The nerve cell elaborates an electrical signal in response to the combination of all its synaptic inputs; the signal, conducted to the nerve terminals, must be back-translated to a chemical signal to be transmitted to another neuron or an effector cell. The exact mechanisms by which electrical changes in the plasma membrane of the nerve terminal are coupled to secretion of neurotransmitters have not been fully clarified. Phasic secretion for transmission of rapid signals across the synapse involves the release of multimolecular packets (quanta) of neurotransmitter, by means of exocytosis from synaptic vesicles. Influx of calcium through voltage–operated, N–type channels (ω-conotoxin sensitive) is certainly required for phasic secretion elicited by action potentials. Intracellular calcium levels are involved in modulation of the secretory response as well, but several further factors and mechanisms, including phosphorylation of synaptic-vesicle associated proteins, have been suggested to regulate secretory activity in the short, medium and long-term [16-18] (VALTORTA et al., this volume).

The importance of voltage–operated calcium channels in the presynaptic membrane, for neurotransmitter secretion, has been established by the classical works of FATT and KATZ [19], KATZ and MILEDI [20], and LLINAS and NICHOLSON [21]. Calcium is rapidly extruded from the cytoplasm of the nerve terminal, but during repetitive activation the intracellular concentration of free calcium can rise from the resting levels, and facilitate quantal release elicited by depolarization of the presynaptic membrane [22] (PARNAS and SEGEL). Calcium-dependent metabolic changes may also ensue and be responsible for longer-lasting facilitatory phenomena (see VALTORTA et al., this volume). As intense secretion of neurotransmitters can partly deplete the store of rapidly available quanta, and therefore depress subsequent release, synaptic efficiency can either increase or decrease, during repetitive stimulation, depending on the rate of release and the relative contributions of synaptic facilitation and depression. A reduction of the density and/or functionality of presynaptic calcium channels, due to binding of auto-antibodies, is responsible for a neurological syndrome known as the LAMBERT-EATON myasthenic syndrome, characterized by impaired neurotransmission and subjective muscle weakness; in this form of myasthenia, muscle weakness is relieved by tetanic stimulation, thanks to the above-mentioned facilitatory processes. This is unlike classical *myasthenia gravis*, because of the auto-antibodies against the acetylcholine receptor of the motor endplate; neurotransmitter release is normally active in the latter pathology, and may be sufficient to sustain neuromuscular transmission at rest, but not during prolonged tetanic stimulation, when synaptic depression becomes predominant and the quantity of neurotransmitter released following a nerve action potential may become insufficient, in the presence of a reduced number of receptors, to trigger excitation and contraction of the muscle fibre.

4.3. Subcellular distribution of calcium in the nerve terminal

The inflow of calcium through the transient opening of voltage–operated channels that accompanies an action potential may not produce relevant and/or sustained changes in bulk cytoplasmic concentration of free calcium. On the contrary computer simulations suggest that transient and localized increases occur in the submembrane domain facing the channels [23]. If calcium is directly responsible for the fusion of synaptic vesicles with the presynaptic plasmalemma, then exocytosis is expected to occur to significant extents only at specific locations where calcium channels are concentrated. It may be appropriate to recall here that two main populations of secretory organelles exist in nerve terminals, and often coexist within the same nerve terminal (see THESLEFF, this volume): small, clear synaptic vesicles (SSV, about 40 nm diameter), containing small biogenic amines and larger granules with an electrondense core (granules and large, dense-core vesicles, LDCV) [24]. Whereas single-action potentials preferentially elicit release of neurotransmitter from small synaptic vesicles, pharmacological treatments and specific patterns of repetitive stimulation appear to be able to induce exocytosis from granules and LDCVs [25,26]. This may be explained by the hypothesis that single stimuli induce rapid, massive, localized increases in calcium concentration near the plasma membrane, whereas repetitive stimulation may produce sustained, though smaller, increases in bulk cytoplasmic calcium concentration.

In fact, changes in subcellular levels and distribution of free calcium ions appear to be well correlated with these differentially regulated secretory processes. A recent study in synaptosomes (isolated presynaptic compartments) has shown that a uniform rise of cytoplasmic calcium to 400 nM, as obtained by means of a calcium ionophore such as ionomycin, will elicit release of peptides from LDCVs, but not of classical neurotransmitter (contained in small synaptic vesicles), unless [Ca^{2+}] rises above 1 μM. Conversely, depolarization by elevated extracellular concentration of K^+ elicits secretion from SSVs even with increases in bulk cytoplasmic calcium concentrations too small to induce peptide release [27]. These observations suggest that the two mechanisms of exocytosis are differentially regulated due to a markedly different sensitivity to calcium: a regulatory mechanism for exocytosis by small vesicles, with low affinity for calcium, would be activated by localized and rapid but massive elevations of calcium concentration, such as can only occur in restricted regions of the cytoplasm, near the presynaptic membrane where the calcium channels are located; a second mechanism with much higher affinity for calcium would necessarily be slower and require sustained increases in calcium concentration, but it would be activated by much smaller changes in calcium levels, as can occur in the bulk of the nerve terminal cytoplasm, and induce exocytosis from granules and LDCV.

4.3.1. Rapid, localized, massive increases in presynaptic calcium concentration

It may be observed that if the calcium channels are not homogeneously distributed on the presynaptic membrane, then release of neurotransmitter from small synaptic vesicles may be localized to the areas of the presynaptic membrane where calcium channels are concentrated. It has been suggested that this occurs at the active zones, specialized structures where synaptic vesicles are particularly concentrated [28]; this arrangement would help in restricting release of the neurotransmitter in direct correspondence with particular regions of the postsynaptic membrane, e.g. where specific receptors are concentrated. Measurements of changes in the subcellular distribution of calcium, using the low-affinity, very fast, calcium-sensitive dye J-aequorine, tend to support the view of localized, massive increases in very limited juxta-membrane domains with no increase in bulk cytoplasmic calcium concentration detectable by high-affinity, slower dyes [29]. Such a specialization appears to be important at the neuromuscular junction in particular, where very rapid and brief activation of the postsynaptic receptor is achieved thanks to the localized secretion of a huge amount of acetylcholine (the concentration in the cleft reaches well above the micromolar range) onto a low affinity receptor, in the presence of the very fast hydrolysing enzyme acetylcholinesterase.

4.3.2. Diffuse increases in presynaptic cytoplasmic calcium concentration

Multiple impulses reaching the nerve terminal can produce a gradual accumulation of cytoplasmic free calcium, if the extrusion and sequestration processes cannot cope with the sustained calcium influx through voltage–activated channels. However, this is not the only possibility. As demonstrated for the vertebrate retina, IP_3 receptor can be localized in synaptic terminals [30], and IP_3 sensitive stores may play a role in transducing activation of presynaptic receptors into a diffuse rise of calcium concentration through release from intracellular stores.

5.   Conclusive remarks

In general, the speed of a chemical reaction is inversely related to its affinity constant. This, combined with the slowed intracellular diffusion of calcium ions, allows for a wide variety of response patterns, ranging from very fast, localized responses to transient but marked increases of local free-calcium concentration (like SSV-mediated exocytosis and other very low-affinity processes), to slow, poorly localized responses to small increases of bulk-cytoplasmic free-calcium concentration (high-affinity processes).

The existence of multiple mechanisms for active and passive transport of Ca^{2+} across the plasma membrane allows for fine spatial/temporal regulation of intracellular Ca^{2+} concentration, as a function of intra- and extracellular modifications: transmembrane potential and its past history, extracellular chemical mediators and hormones and intracellular second messengers (among which is Ca^{2+} itself).

The presence and complex regulation of intracellular calcium stores can introduce positive or negative feedback phenomena, thus leading to either transient increases and oscillations or damping of free calcium concentration. Thus, activation of fast (low-affinity) or slow (high-affinity) calcium-regulated processes can be selectively achieved. Alternatively, intermediate affinity proteins may be partially activated by rapid and transient calcium increases but remain in the activated state for a longer time, as seems to occur for Ca^{2+} calmodulin-dependent protein kinase. In response to repetitive calcium transients, the enzyme would be partially but virtually continuously activated, to a level proportional to the frequency of calcium spikes. Thus, selective activation of calcium-dependent enzymatic pathways, thanks to differential calcium affinity and/or spatial localization, and *frequency modulated* activation of specific calcium-dependent processes, can all be achieved by the properties of the dynamic regulation of intracellular calcium concentrations.

References

[1] S.W. KUFFLER, J.G. NICHOLLS and A.R. MARTIN, From Neuron to Brain, Sinauer: Sunderland (1984).

[2] B. KATZ, *Nerve, Muscle and Synapse,* McGraw-Hill, New York (1966).

[3] A. SCHWARTZ, E. MCKENNA and P.L. VAGHY, *Am. J. CARDIOL.,* **62**, 3G (1988).

[4] R.W. TSIEN, D. LIPSCOMBE, D.V. MADISON, K.R. BLEY and A.P. FOX, *Trends Neurosci.,* **11**, 431 (1988).

[5] H. JAHNSEN and R. LLINAS, *J. Phisiol.,* **349,** 227 (1984).

[6] R.R. LLINAS, *Science,* **242,** 1654 (1988).

[7] A.J. HUDSPETH, *Science,* **230,** 745 (1985).

[8] A.G. GILMAN, *Annu. Rev. Biochem.,* 56, 615 (1987).

[9] R.W. TSIEN and R.Y. TSIEN. *Annu. Rev. Cell Biol.,* **6,** 715 (1990).

[10] J. MELDOLESI, L. MADEDDU and T. POZZAN, *Bioch. Biophys. Acta,* **1055,** 130 (1990).

[11] P. VOLPE, T. POZZAN and J. MELDOLESI, *Semin. Cell Biol.,* **1,** 297 (1990).

[12] A. VILLA, P. PODINI, D.O. CLEGG, T. POZZAN and J. MELDOLESI, *J. Cell Biol.,* **4,** 779 (1991).

[13] W. MÜLLER and J.A. CONNOR, *Nature (London),* **354,** 73 (1991).

[14] G.L. COLLINGRIDGE and T.V.P. BLISS, *Trends Neurosci,* **10,** 288 (1987).

[15] G.L. COLLINGRIDGE and W. SINGER, *Trends Pharm. Sci.* **11,** 290 (1990).

[16] K.L. MAGLEBY, *Short-term changes in synaptic efficacy. In: Synaptic function.* (Editors) G.M. Edelman, W.E. Gall and W.M. Cowan. J. Wiley & Sons: New York, (1987), pp. 21 - 56.

[17] J.-W. LIN, M. SUGIMORI, R.R. LLINAS, T.L. MCGUINNESS and P. GREENGARD., *Proc. Natl. Acad. Sci. U.S.A.,* **87** (8257 (1987).

[18] R. ANWYL. *Trends Pharm. Sci.,* **10,** 236 (1989).

[19] P. FATT and B. KATZ., *J. Physiol.,* **115,** 320 (1951).

[20] B. KATZ and R. MILEDI. *Proc. Roy. Soc. Lond. B.* **161,** 496 (1965).

[21] R. LLINÁS, I.Z. STEINBERG and K. WALTON., *Proc. Natl. Acad. Sci. U.S.A.,* **73,** 2918 (1976).

[22] H. PARNAS and L.A. SEGEL, *Progr. Neurobiol.,* **32,** 1 (1989).

[23] R. LLINÁS., *Functional compartments in synaptic transmission.,* In: *Synaptic function.* G.M. EDELMAN, W.E. GALL and W.M. COWAN., (Editors), J. Wiley & Sons: New York (1987) pp 7-20.

[24] P. DE CAMILLI and R. JAHN, *Annu. Rev. Physiol.,* **52,** 000 (1990).

[25] J.M. LUNDBERG and T. HÖKFELT, *Trends Neurosci.* **6,** 325 (1983).

[26] M. MATTEOLI, C. HAIMANN, F. TORRI-TARELLI, J.M. POLAK, B. CECCARELLI and P. DE CAMILLI. *Proc. Natl. Acad. Sci. U.S.A.,* **85,** 7366 (1988).

[27] M. VERHAGE, H.T. MCMAHON, W.E.J.M. GHIJSEN, F. BOOMSMA, G. SCHOLTEN, V.M. WIEGANT and D.G. NICHOLLS, *Neuron,* **6,** 517 (1991).

[28] D.W. PUMPLIN, T.S. REESE and R. LLINÁS, *Proc. Natl. Acad. Sci. U.S.A.*, **78,** 7210 (1981).

[29] R. LLINÁS, M. SUGIMORI and R.B. SILVER, *Calcium microdomains in squid giant synapse.* Communication to the *XVIème Conférence en Neurobiologie de Gif.* Gif Sur Yvette, 5-6 Dec, 1991.

[30] Y.-W. PENG, A.H. SHARP, S.H. Snyder and K.-W. YAU, *Neuron,* **6,** 525 (1991).

BIOCHEMICAL ASPECTS OF PRESYNAPTIC FUNCTION

FLAVIA VALTORTA[1], FABIO BENFENATI[2], NUMA IEZZI[1], AND MARTIN BÄHLER[3]

[1] *"Bruno Ceccarelli" Centre, Department of Medical Pharmacology,
CNR Center of Cytopharmacology and S.Raffaele Institute, University of Milan, Italy;
[2]Institute of Human Physiology, University of Modena, Italy; [3]Friedrich-Miescher
Laboratorium der Max-Planck Gesellschaft, Tübingen, Germany*

Contents

Bioelectrochemistry IV
Edited by B.A. Melandri *et al.*, Plenum Press, New York, 1994

1.   The life cycle of synaptic vesicles

Neurons are highly polarized cells in which several compartments, such as cell body, dendrites, axons and nerve endings can be distinguished. The proteins of the presynaptic nerve terminal are synthesized in the cell body, from where they are targeted down the axon to their site of action. The sorting signal is not known, and there is some debate as to whether membrane proteins of synaptic vesicles leave the cell body assembled in mature synaptic vesicles or in pleiomorphic membranous structures from which synaptic vesicles are derived at later stages (DE CAMILLI and JAHN, [1]).

Synaptic vesicles (or their precursors) are transported along microtubules *via* the fast axonal transport. The mechanoenzyme responsible for their anterograde transport has been identified as kinesin, whereas cytoplasmic dynein seems to be responsible for the retrograde transport (VALLEE and SHPETNER, [2]).

Mature synaptic vesicles accumulate in the nerve terminal, where they cluster close to fusion sites (the *active zones*). Some of the vesicles are docked at the presynaptic plasma membrane, with which they fuse following the arrival of a nerve stimulus, thereby releasing their neurotransmitter content into the synaptic cleft. Fused vesicles are then retrieved *via* an endocytotic mechanism, they refill with neurotransmitter and can undergo subsequent cycles of fusion and retrieval (VALTORTA *et al.*, [3]).

In addition to the small clear vesicles which contain classical neurotransmitters, neurons also possess a population of larger vesicles, the large dense-core vesicles (LDCVs), which store and release neuropeptides. These vesicles are not specific for the nerve terminal and also undergo extra-synaptic exocytosis. After endocytosis, they are not locally recycled, but have to be transported back to the GOLGI apparatus for reloading with neuropeptides.

2.   Neurotransmitter uptake and storage

Synaptic vesicles possess uptake systems which are specific for the neurotransmitter that they store. Cholinergic synaptic vesicles from *Torpedo* electroplax were the first vesicles to be isolated and shown to store neurotransmitters (WHITTAKER *et al.*, [4]). The uptake of neurotransmitters into these vesicles was studied by using radioactively-labeled precursors of ACh and has been shown to require energy and to be dependent on the maintenance of an electrochemical proton gradient across the membrane (PARSONS and KOENIGSBERGER, [5]). The electrochemical gradient is generated by an ATP-dependent proton pump which has been shown to be present in the vesicle membrane. This proton pump (which belongs to the family of the so-called vacuolar type of proton pumps) is not specific for synaptic vesicles, since it can be found in more or less all intracellular organelles (NELSON and TAIZ, [6]; see below).

While the proton pump is probably shared by all synaptic vesicles, the neurotransmitter transporter molecules are specific for a certain neurotransmitter. The various neurotransmitter transporters have different dependencies on the membrane potential and proton gradient; for example, the glutamate transporter simply depends on the membrane potential, whereas the transport of dopamine depends exclusively on the proton gradient.

The neurotransmitter transporters present in the synaptic vesicle membrane have different properties with respect to those found in the plasma membrane, which are responsible for the re-uptake of the secreted transmitter. For example, the plasma membrane transporters are rapid and exhibit high affinity for the transmitter; the vesicle transporters are of lower affinity, but show a higher specificity for the transported transmitter (MAYCOX *et al.*[7],).

3.   Fusion and recycling

The fusion between intracellular membranes is a common event in all cells. Membrane fusion does not occur at random, but is probably preceded by a recognition step between the two membranes. The selectivity of fusion is fundamental to the ability of the cells to maintain the segregation of the various compartments.

The fusion of synaptic vesicles with the plasma membrane is a regulated event triggered by the influx of extracellular calcium. When dealing with the possible mechanisms of synaptic vesicle fusion, one has to take into account the enormous speed at which the event occurs. The synaptic delay is about 1 ms, and most of this time is consumed by the opening of the voltage-dependent calcium channels (LLINAS *et. al.,* [8]). This leaves not more than 200 μs for all the events occurring from calcium entry to the release of neurotransmitter. Pure phospholipid vesicles can be made to fuse, but at about a 100-fold lower rate. Also in the case of viral fusogenic proteins the fusion with the plasma membrane appears to be very slow (ALMERS and TSE, [9]).

One model which has been put forward to account for the rapidity of the fusion of synaptic vesicles with the plasma membrane is the *fusion pore* model. This model holds that oligomeric proteins from the plasma and vesicle membranes interact with each other to form a channel. The influx of calcium causes the opening of the preassembled proteinaceous fusion pore and the release of neurotransmitter. As stimulation proceeds, the subunits of the fusion pore may lose affinity and dissociate from each other, leading to widening of the pore and collapsing of the vesicle membrane into the presynaptic plasma membrane.

Experimental evidence supporting this model comes from electrophysiological recordings performed with mast cells of the beige mouse, which are characterized by the presence of enormously large granules. When fusion of the granules is stimulated, the first event to be recorded is the appearance of a small conductance, which can then flicker and finally develop into a large conductance. These observations suggest that fusion of the granules begins with the formation of a narrow pore which then dilates. The flickering indicates that the early phases of the event are reversible (BRECKENRIDGE and ALMERS, [10]). Two synaptic vesicle proteins have been proposed as components of the fusion pore and therefore as directly involved in exocytosis: synaptophysin and synaptoporin (see below).

It is possible that under physiological conditions exocytosis involves only the opening and closing of the fusion pore, with an immediate retrieval of the fused vesicle. This hypothesis has been tested by using antibodies against synaptic vesicle-specific proteins to trace the fate of the vesicle membrane during the exo-endocytotic cycle. The results indicate that the retrieval process is very rapid and selective for the vesicle membrane components even during high rates of stimulation. Flattening of the vesicle membrane into the plasma membrane occurs only under extreme conditions, when endocytosis is blocked. In this case, it is possible that molecular intermixing of the vesicle and plasma membrane components occurs and that recovery involves endocytosis through coated vesicles (VALTORTA *et al.*, [11]).

The protein responsible for the formation of the coat around endocytotic vesicles is clathrin. Coated vesicles isolated from the brain contain synaptic vesicle proteins (PFEFFER and KELLY; WIEDENMANN *et al.*, [12]).

A protein involved in endocytosis of synaptic vesicles has been identified by a genetic approach. The *shibire* mutant of the fruit fly *Drosophila melanogaster* exhibits a temperature-dependent inhibition of endocytosis. The protein encoded by *shibire* is dynamin, a protein initially discovered as a nucleotide-dependent microtubule binding protein (SHPETNER and VALLEE; CHEN *et. al.*; VAN DER BLIEK and MEYEROWITZ, [13]). Dynamin is highly enriched in the brain and its expression correlates with synaptogenesis (NAKATA *et al.*, [14]). However, it does not copurify with synaptic vesicles. Immunofluorescence experiments carried out on PC12 cells indicate that dynamin is predominantly localized in the cell body and in the tips of the neuritic processes. The staining appears punctate, suggesting that *in vivo* dynamin is associated with organelles (SCAIFE and MARGOLIS [15],).

4.   Synaptic vesicle proteins

The development of procedures for the preparation of highly purified synaptic vesicles from *Torpedo* electric organ and from rat brain has allowed the analysis of their protein content. Thus, several proteins specific for synaptic vesicles have been identified and, in many cases, purified, cloned and sequenced (for recent reviews, see DE CAMILLI and JAHN, 1990; SÜDHOF and JAHN [16], 1991). There are an enormous num-

ber of synaptic vesicles in the cortex of rat brain, as indicated by the relatively low purification factor of synaptic vesicle proteins (about 20 fold).

In the following paragraphs, the major families of synaptic vesicle proteins
are described.

4.1. Synapsins

The synapsins were discovered as endogenous substrates for cAMP- and
Ca²⁺-dependent phosphorylation in mammalian brain, and were subsequently found
to be specifically associated with the cytoplasmic surface of synaptic vesicles (JOHNSON
et al.; UEDA *et. al.*; SCHULMAN and GREENGARD; DE CAMILLI *et. al.*; HUTTNER *et al.* [17]).
The family of the synapsins comprises four homologous proteins, synapsins I*a* and I*b*
(collectively referred to as synapsin I) and synapsins II*a* and II*b* (collectively referred
to as synapsin II). Synapsins I and II are encoded by two distinct genes whose primary transcripts are differentially spliced to generate the *a* and *b* isoforms of each
protein. The four synapsins have an extensive sequence homology in the N-terminal
region, but they diverge in the C-terminal region (SÜDHOF *et al.* [18]). Synapsin I is a
very basic (isoelectric point higher than 10) and acid-soluble protein. The synapsin I
molecule is highly asymmetric, being composed of a globular NH₂-terminal region
(the *head* region) and of an elongated, proline-rich COOH-terminal region (the *tail*
region). The head region contains most of the hydrophobic aminoacid residues
found in the molecule, whereas the tail region is responsible for the basicity of the
protein (UEDA and GREENGARD [19]).   Synapsin IIa and IIb, which lack the tail region,
have a less elongated structure and are not basic (BROWNING *et al.* [20]; for recent reviews on the structure of the synapsins, see DE CAMILLI *et al.* and VALTORTA *et al.* [21]).

Cyclic AMP-dependent protein kinase phosphorylates synapsin I on a serine
residue located in the head region (Ser 9, site 1). The same residue is phosphorylated
by Ca²⁺/calmodulin-dependent protein kinase I; Ca²⁺/calmodulin-dependent protein
kinase II catalyses the incorporation of phosphate into two additional serine residues
located in the tail region (Ser 566, site 2 and Ser 603, site 3; CZERNIK *et al.*[22]; SÜDHOF
et al.[18]). Phosphorylation site 1 is present also in synapsin II, whereas phosphorylation sites 2 and 3 are specific for synapsin I.

The synapsins are neuron-specific proteins. In the brain cortex, they constitute about 0.6 % of the total protein (GOELZ *et al.*, [23]; BROWNING *et al.* [20]). They are
present in the central as well as in the peripheral nervous system and are expressed
by virtually all neurons irrespective of the neurotransmitter released. However, the
relative expression of the four isoforms may vary among the various types of neurons.
Immunoelectron microscopy as well as subcellular fractionation studies have shown
that synapsin I is specifically associated with the cytoplasmic side of the membrane of
small synaptic vesicles, constituting about 6 % of the total protein present in a highly

purified preparation of synaptic vesicles (DE CAMILLI *et al*; HUTTNER *et al*. [17]).

Synapsin I is able to penetrate into the hydrophobic core of the synaptic vesicle membrane (BENFENATI *et al*. [25]). However, it is not an integral membrane protein and can be quantitatively removed from the vesicle membrane by lowering the pH or increasing the ionic strength of the medium (HUTTNER *et al*. [17]). This property has allowed the study of the binding of exogenous synapsin I to synaptic vesicles depleted of endogenous synapsin I. This binding is characterized by high affinity and saturability and is modulated by phosphorylation, the binding affinity being reduced 5 fold upon phosphorylation of synapsin I by Ca^{2+}/calmodulin-dependent protein kinase II (SCHIEBLER *et al*. [26]). Synapsin I binds with high affinity to both protein and phospholipid components of synaptic vesicles, suggesting the presence of multiple binding sites in the vesicle membrane. The hydrophobic head region is responsible for the interaction with the acidic phospholipids of the cytoplasmic leaflet of the vesicle membrane and for the partial penetration of synapsin I into the core of the membrane. The hydrophilic tail region binds to a protein component of the vesicles and its binding is virtually abolished by phosphorylation of sites 2 and 3 (BENFENATI *et al*. [25, 27]). Preliminary evidence indicates that the binding protein for synapsin I is the vesicle associated form of Ca^{2+}/calmodulin-dependent protein kinase II (BENFENATI *et al*. [28]).   The high degree of homology with the head region of synapsin I suggests that synapsin II might also be able to interact with vesicle phospholipids and, possibly, to compete with synapsin I at this binding site.

Synapsin I is also able to interact with various cytoskeletal elements (microtubules, spectrin, neurofilaments, actin (BAINES and BENNETT; GOLDENRING *et al*.; BÄHLER and GREENGARD,; PETRUCCI and MORROW; STEINER *et al*. [29]). Its interaction with actin filaments is characterized by the formation of actin bundles. Phosphorylation of synapsin I by Ca^{2+}/calmodulin-dependent protein kinase II abolishes the actin-bundling activity (BÄHLER and GREENGARD; PETRUCCI and MORROW [29]) and markedly reduces actin binding (BÄHLER and GREENGARD [29]).

Since structure-function analysis has shown that the binding sites for synaptic vesicles and for actin are located in different parts of the synapsin I molecule (BÄHLE *et al*.,[30]; BENFENATI *et al*. [25, 27]), the possibility exists that synapsin I reversibly tethers synaptic vesicles to the F-actin-based cytoskeleton of the nerve terminal and that this linkage is impaired when synapsin I is phosphorylated by Ca^{2+}/calmodulin-dependent protein kinase II.

Synapsin I also affects the dynamics of actin filament assembly. Synapsin I triggers the condensation of actin monomers to form nuclei, thereby accelerating actin polymerization and promoting the formation of filaments also under conditions in which actin polymerization is not favoured. The actin-nucleating activity is decreased or virtually abolished by site-specific phosphorylation of synapsin I (BENFENATI *et al*.; VALTORTA *et al*.; FESCE *et al*. [31]). It is therefore possible that dephosphorylated synapsin I bound to synaptic vesicles induces the growth of actin filaments, rapidly embedding the vesicles in a cytoskeletal meshwork and making them unavailable for exocytosis. Such a mechanism could also play a role in the endocytotic retrieval of the vesicles after exocytosis. By inhibiting this activity, synapsin I phosphorylation would facilitate the release of the vesicles from the cytoskeleton and maintain their availability during the exo-endocytotic cycle (BÄHLER *et al.* [32]).

Phosphorylation of the synapsins is increased by virtually all conditions inducing or facilitating Ca^{2+}-dependent neurotransmitter release. These conditions include electrical tetanic stimulation of nerve cells, veratridine- or K^+-induced depolarization, and activation of certain classes of presynaptic receptors. In particular, the phosphorylation of synapsin I by Ca^{2+}/calmodulin-dependent protein kinase II occurs promptly and to high stoichiometry during stimulation of the nerve terminal (NESTLER and GREENGARD [33]).

Experimental support for a role of synapsin I in the modulation of neurotransmitter release has been obtained in *in vivo* experiments where synapsin I has been microinjected into the preterminal digit of the squid giant synapse. The injection of dephosphorylated synapsin I induces a depression in the amplitude and rate of rise of the postsynaptic potential-values evoked by presynaptic depolarizing steps. On the contrary, the injection of phosphorylated or heat-denatured synapsin I is totally ineffective. These effects on the postsynaptic response are not accompanied by detectable changes in the presynaptic inward calcium current, indicating an action on the number of quanta of neurotransmitters released in response to a given Ca^{2+} influx (LLINAS *et al.* [34],).

These data strongly suggest that dephosphorylated synapsin I may provide an inhibitory constraint to the release of neurotransmitters and that this inhibitory action is abolished by phosphorylation.

Simulation experiments demonstrated that phosphorylation of synapsin I by Ca^{2+}/calmodulin-dependent protein kinase II decreases the number of cross-bridges between synaptic vesicles and actin filaments by enhancing the dissociation of synapsin I from the actin sites and, to a much lesser extent, from the vesicle sites (BENFENATI *et al.* [35]). Experimental results supporting this model were obtained using rat brain synaptosomes, where it was found that the release of neurotransmitters evoked by K^+-induced depolarization was accompanied by an increase in the amount of synapsin I in the cytosolic fraction (SIHRA *et al.* [36]). Immunogold labeling of frog neuromuscular junctions fixed under conditions of enhanced neurotransmitter release demonstrated that the complete dissociation of synapsin I from the vesicle membrane is not necessary in order for the vesicle to undergo fusion and recycling (TORRI-TARELLI *et al.* [37]).[1]

ADDED IN PROOF.

[1]Morphometric analysis of nerve terminals electrically stimulated to release neurotransmitter indicates that synapsin I partially dissociates from synaptic vesicles upon exocytosis and reassociates with vesicle membrane after endocytosis (F. TORRI TARELLI, M. BOSSI, R. FESCE, P. GREENGARD and F. VALTORTA, *Neuron*, **9**, 1143 (1992).

It is therefore possible that synapsin I, by undergoing cycles of phosphorylation and dephosphorylation, is implicated in the transition of synaptic vesicles from a reserve pool (vesicles bound to the cytoskeleton) to a pool of vesicles released from the cytoskeleton and available for exocytosis. This mechanism could provide a biochemical basis for the changes in the functional properties of the nerve terminal in response to previous activity or presynaptic inputs.

Recent data suggest that the synapsins may also play a role in the morphological and functional rearrangements which occur during synaptogenesis. Overexpression of synapsin IIb in neuroblastoma x glioma hybrid cell lines induces, upon differentiation, dramatic increases in the number of varicosities and of synaptic vesicles per varicosity. Moreover, the levels of synaptic vesicle proteins are greatly increased (HAN *et al.* [38]). Microinjection of synapsin I into *Xenopus* embryos at the 2-8 cell stage induces a marked acceleration in the maturation of quantal secretion mechanisms in developing neuromuscular synapses in culture, as measured by both spontaneous and evoked neurotransmitter release (LU *et al.* [39]).

4.2. *The family of synaptophysin and synaptoporin*

Synaptophysin (previously referred to as p38) is a major integral membrane protein of synaptic vesicles, where it represents about 6 % of the total protein content. It is an acidic glycoprotein with an apparent molecular weight of 38,000 (JAHN *et al.*; [40]; WIEDENMANN *et al.* [40]). Synaptophysin can be endogenously phosphorylated on tyrosine and serine residues in both purified synaptic vesicles and synaptosomes (PANG *et al.* [41]). Subcellular fractionation as well as immunocytochemical studies indicate that within the nervous system synaptophysin is esclusively localized on synaptic vesicles from virtually all synapses studied so far (Fig.1; JAHN *et al.*; WIEDENMANN *et al.* [40]; NAVONE *et al.* [42]).

Synaptoporin (molecular weight, 37,000) was identified by low stringency hybridization using a synaptophysin probe (KNAUS *et al.* [43]). Primary structure analysis reveals a relatively high degree of homology between the two proteins. Hydrophobicity plots of synaptophysin and synaptoporin suggest the presence of four transmembrane spanning regions with both the N- and C- termini located in the cytoplasm. The hydropathy plots resemble those of connexins, the protein subunits of gap junctions, in spite of the absence of a real sequence similarity (LEUBE *et al.*; SÜDHOF *et al.* [44]; KNAUS *et al.* [43]).

When reconstituted in lipid bilayers, synaptophysin has been reported to form voltage-dependent channels (THOMAS *et al.* [45]). Synaptophysin has also been found to form a multimeric complex, although it is not yet clear whether the complex is a tetramer or a hexamer (REHM *et al.* [46]). A low molecular weight subunit (as yet unidentified) seems also to be part of the complex (JOHNSTON and SÜDHOF [47]).

Because of all these properties, it has been suggested that synaptophysin may be part of the putative fusion pore which leads to the release of neurotransmitters. The opening of the fusion pore would require the interaction of synaptophysin with its plasma membrane counterpart. A presynaptic membrane protein to which synaptophysin binds has indeed been identified and named physophylin (THOMAS and BETZ [48])[2].

4.3.   The family of synaptobrevins (VAMPs)

Synaptobrevin I and II have been identified in synaptic vesicles from rat brain and represent two differentially spliced products of the same gene (BAUMERT *et al.*; SÜDHOF *et al.* [49]). They are the mammalian homologues of VAMP-1 and VAMP-2, respectively, which were initially characterized in *Torpedo* (TRIMBLE *et al.*; EFFERINK *et al.* [50]). They are integral membrane proteins with a molecular weight of approximately 18,000 and have a single membrane-spanning domain. The distribution of synaptobrevins is similar to that of synaptophysin. The presence of sequence homology among *Torpedo*, *Drosophila* and humans indicates that they have been highly conserved during evolution, suggesting that they may play an as yet unidentified important role in synaptic vesicle function[3].

4.4.   The Rab proteins

Rab3A and Rab3B are small GTP-binding proteins, related to the p21/ras oncogene family (MATSUI *et al.*; ZAHRAOUI *et al.*; FISCHER VON MOLLARD *et al.* [51]). Small GTP-binding proteins are enzymes (GTPases) which are able to bind and hydrolyse GTP. The hydrolysis of the bound GTP to GDP produces a flow of information, causing the switching of the protein from an *on* to an *off* state. A large variety of small GTP-binding proteins have been identified. All of them seem to be involved in regulating vesicular traffic (from the endoplasmic reticulum to the GOLGI complex, within the GOLGI complex, *etc.*), possibly mediating the specific recognition and docking of a given vesicle with a subcellular compartment.

It has been hypothesized that the cycling of small GTP-binding proteins between a membrane-bound and a soluble state regulates membrane fusion events (BOURNE [52]). According to this model, in their inactive form (GDP bound) small GTP-binding proteins are in the soluble phase. The switching to the active state

ADDED IN PROOF.

[2]Microinjection experiments with antibodies and antisense oligonucleotides have demonstrated that the presence of synaptophysin is essential for vesicle fusion and release of neurotransmitter to occur (J. ALDER, B. LU, F. VALTORTA, P. GREENGARD and M.-M. POO, *Science*, **257**, 657 (1992); J. ALDER, Z.-P. XIE, F. VALTORTA, P. GREENGARD and M.-M. POO, *Neuron*, **9**, 759 (1992))

[3]Recently, it has been reported that tetanus and botulinum-B neurotoxins block neurotransmitter release by specific cleavage of synaptobrevin II (Vamp II) (G. SCHIAVO, F. BENFENATI, B. POULAIN, O. ROSSETTO, P. POLVERINO DE LAURETO, B. R. DASGUPTA and C. MONTECUCCO, *Nature*, **359**, 832 (1992))

causes their association with the donor membrane and its subsequent docking with the acceptor membrane. The hydrolysis of GTP leads to the dissociation of the GTP-binding protein from both membranes.

Within the nervous system, the majority (about 80 %) of Rab3A is membrane-bound, being specifically associated with synaptic vesicles. A small pool (about 20 %) of soluble Rab3A also exists (FISCHER VON MOLLARD *et al.* [51]). When nerve terminal preparations are stimulated to release neurotransmitters by depolarization, the amount of membrane-associated Rab3A decreases. However, during recovery after stimulation the re-association of Rab3A with the membrane appears to be slow and incomplete (FISCHER VON MOLLARD *et al.* [53]).

4.5. *Synaptotagmin*

Synaptotagmin (p65) is an integral membrane protein of synaptic vesicles with an apparent molecular weight of 65,000 (MATTHEW *et al.*; PERIN *et al.* [54]). It is evolutionarily highly conserved, as indicated by the comparison of the primary structures of the human and *Drosophila* proteins (PERIN *et al.* [55]). The protein possesses one transmembrane region, a large cytoplasmic domain and a small intravesicular domain. The cytoplasmic domain has two internal repeats which are homologous to each other and to the C_2 domain of the regulatory region of protein kinase C (PERIN *et al.* [56]). The C_2 domain of the kinase is the domain that confers Ca^{2+}-dependence to the enzyme (KIKKAWA *et al.* [57]). This led to the hypothesis that p65 binds Ca^{2+} and may be the Ca^{2+}-acceptor for the fusion event.

Synaptotagmin has the unusual property of being able to agglutinate erythrocytes (POPOLI and MENGANO [58]). This property can be ascribed to the ability of the two C_2-homologous repeats to bind with high affinity to negatively charged phospholipids. The ability of synaptotagmin to bind to phospholipids has allowed to speculation that it might be the protein responsible for the docking of synaptic vesicles to the presynaptic plasma membrane. Further support for this hypothesis has come from the recent demonstration that synaptotagmin is able to interact with the α-latrotoxin receptor (PETRENKO *et al.* [59]), an integral membrane protein of the presynaptic plasma membrane (VALTORTA *et al.* [60]).

4.6. *The vesicular proton pump*

The vesicular proton pump is a hetero-oligomer consisting of an undefined number of subunits. Most of the subunits seem to be shared by the vacuolar proton pumps of the various organelles, whereas some subunits may be specific for synaptic vesicles. Most of the subunits have been cloned. The 70 kDa subunit, which seems to contain the ATP-binding site, and the 17 kDa subunit, which is a lipoprotein and forms the proton pore, resemble the corresponding subunits of the mitochondrial and bacterial proton pumps (for review, see STONE *et al.* [61]).

4.7. p29 – protein

p29 – protein has been identified by the use of antibodies (BAUMERT *et al.* [62]). It is an integral membrane protein with an apparent molecular weight of 29,000. p29 is probably related to synaptophysin, since some antibodies raised against p29 cross-react with synaptophysin. In addition, p29, like synaptophysin, is phosphorylated on tyrosine residues.

4.8. Proteoglycan and SV2

These proteins were originally identified in synaptic vesicles purified from *Torpedo* and subsequently found also in mammalian synaptic vesicles (CARLSON *et al.*; BUCKLEY and KELLY [63]). The synaptic vesicle proteoglycan appears to be completely intravesicular, and it has been reported to be at least partially secreted during exocytosis. A structurally similar proteoglycan has been reported to be present in the extracellular matrix, although it is not clear whether this molecule is of vesicular origin (CARONI *et al.* [64]). SV2 is an integral membrane protein of approximately 100,000 molecular weight which appears to be heavily glycosylated.[4] No function for either of these proteins has been hypothesized so far.

5.   Conclusions

Physiological studies in a variety of systems have provided important clues as to the process of synaptic vesicle exo-endocytosis and its regulation. In addition, the application of protein purification as well as molecular cloning techniques to the study of synaptic vesicle proteins has led to the identification and characterization of the major components of these organelles. For several of the synaptic vesicle proteins studied so far a hypothetical function has been proposed. It is therefore possible that in the next few years a detailed characterization of the specific function of each of the proteins present on synaptic vesicles will be achieved, leading to major advancements in our understanding of the process of neurotransmitter release and of membrane fusion in general.

ADDED IN PROOF.

[4]Two groups have reported the cloning and sequencing of SV2 (M. B. FEANY, S. LEE, R. H. EDWARDS and K. M. BUCKLEY, *Cell*, **70**, 861 (1992); S. M. BAJJALIEH, K. PETERSON, R. SHINGHAL and R. SCHELLER, *Science*, **257**, 1271 (1992)). From these studies, it appears that SV2 is a novel chimeric transporter, having two domains homologous to two different families of transporters. The SV2 molecule contains 12 predicted transmembrane domains. The 6 most N-terminal of these domains are homologous to a family of transporters which includes bacterial sugar/proton contransporters and mammalian facilitated sugar transporters, whereas the 6 most C-terminal domains are homologous to the C-terminal domains of the plasma membrane transporters for neurotransmitters

References

[1] P. De Camillli and R. Jahn, *Annu. Rev. Physiol.*, **52**, 625 (1990).

[2] R.B. Vallee and H.S. Shpetner, *Annu. Rev. Biochem.*, **59**, 909 (1990).

[3] F. Valtorta, R. Fesce, F. Groovaz, C. Haimann, W.P. Hurlbut, N. Iezzi, F. Torri-Tarelli, A. Villa and B. Ceccarelli, *Neuroscience*, **35**, 477 (1990).

[4] V.P. Whittaker, I.A. Michaelson and R.J. Kirkland, *Biochem. J.*, **90**, 293 (1964).

[5] S.M. Parsons and R. Königsberger, *Proc. Natl. Acad. Sci. U.S.A.*, **77**, 6234 (1980).

[6] N. Nelson and L. Taiz, *Trends Biochem. Sci.*, **14**, 113 (1989).

[7] P.R. Maycox, J.W. Hell and R. Jahn, *Trends Neurosci.*, **13**, 83 (1990).

[8] R. Llinas, Z. Steinberg and K. Walton, *Biophys. J.*, **33**, 323 (1981).

[9] W. Almers and F.W. Tse, *Neuron*, **4**, 813 (1990).

[10] L.J. Breckenridge and W. Almers, *Nature (London)*, **328**, 814 (1987).

[11] F. Valtorta, R. Jahn, R. Fesce, R. Greengard and B. Ceccarelli, *J. Cell Biol.*, **107**, 2719 (1988).

[12] S.R. Pfeffer and R.B. Kelly, *Cell*, **40**, 949 (1985); B. Wiedenmann, K. Lawley, C. Grund and D. Branton, *J. Cell. Biol.*, **101**, 12 (1985).

[13] H.S. Shpetner and R.B. Vallee, *Cell*, **59**, 421 (1989); M.S. Chen, R.A. Obar, C.C. Schroeder, T.W. Austin, C.A. Poodry, S.C. Wadsworth and R.B. Vallee, *Nature (London)*, **351**, 583 (1991); A.M. van Der Bliek and E.M. Meyerowitz, *Nature (London)*, **351**, 411 (1991).

[14] T. Nakata, A. Iwamoto, Y. Noda, R. Takemura, H. Yoshikura and N. Hirokawa, *Neuron*, **7**, 461 (1991).

[15] R. Scaife and R.L. Margolis, *J. Cell. Biol.*, **111**, 3023 (1990).

[16] T.C. Südhof and R. Jahn, *Neuron*, **6**, 665 (1991).

[17] E.M. Johnson, E.E. Ueda, T. Maeno and P. Grengart, *J. Biol. Chem.*, **247**, 5650 (1972); T. Laeda, H. Maena and P. Greengard, *J. Biol. Chem.*, **248**, 8295 (1973); H. Schulman and P. Greengard, *Nature (London)*, **271**, 478 (1978); P. De Camilli, S.M.Harris, W.B. Huttner and P. Greencard, *J. Cell Biol.*, **96**, 1355 (1983); W.B. Huttner, W. Schiebler, P. Greengard and P. De Camilli, *J. Cell. Biol.*, **96**, 1374 (1983).

[18] T.C. Südhof, A.J. Czernik, K. Kao, P.A. Takei, P.A. Johnston, A. Horiuchi, S.D. Kanazir, M.A. Wagner, M.S. Perin, P. De Camilli and P. Greengard, *Science*, **245**, 1474 (1989).

[19] T. Ueda and P. Greengard, *J. Biol. Chem.*, **252**, 5155 (1977).

[20] M.D. Browning, C.K. Huang and P. Greengard, *J. Neurosci.*, **7**, 847 (1987).

[21] F. Valtorta and F Benfenati P. Greengard, *J. Biol. Chem.*, **267**, 7195 (1992).

[22] A. J. Czernik, D.T. Pang and P. Greengard, *Proc. Natl. Acad. Sci. U.S.A.*, **84**, 7518 (1987).

[23] S.E. Goelz, E.J. Nestler, B. Chehrazi and P. Greengard, *Proc. Natl. Acad. Sci. U.S.A.*, **78**, 2130 (1981).

[25] F. BENFENATI, M. BÄHLER, R. JAHN and P. GREENGARD, *J. Cell. Biol.*, **108**, 1863 (1989).

[26] W. SCHIEBLER, R. JAHN, J.P. DOUCET, J. ROTHLEIN and P. GREENGARD, *J. Biol. Chem.*, **261**, 8383 (1986).

[27] F. BENFENATI, P. GREENGARD, J. BRUNNER, M. BÄHLER, *J. Cell Biol.*, **108**, 1851 (1989).

[28] F. BENFENATI, F. VALTORTA, J.R. RUBENSTEIN, F.G. GORELICK, P. GREENGARD and A.J. CZERNIK, *Nature (London)*, **359**, 417 (1992).

[29] A. J. BAINES and V. BENNETT, *Nature (London)*, **315**, 410 (1985); *Nature (London)*, **319**, 145 (1986); GOLDENRING, R.S. LASHER, M.L. VALLANO, T. UEDA, S. NAITO, N. STERNBERGER and A. DE LORENZO, *J. Biol. Chem.*, **261**, 8495 (1986); M. BÄHLER and P. GREENGARD, *Nature (London)*, **326**, 204 (1987); T.C. PETRUCCI and J.S. MORROW, *J. Cell Biol.*, **105**, 1355 (1987); J.P. STEINER, E. LING and V. BENNETT, **262**, 905 (1987).

[30] M. BÄHLER, F. BENFENATI, F. VALTORTA, A. J. CZERNIK and P. GREENGARD, *J. Cell Biol.*, **108,** 1841 (1989).

[31] F. BENFENATI, F. VALTORTA, E. CHIEREGATTI and P. GREENGARD, *Neuron*, **8**, 377 (1992); F. VALTORTA, P. GREENGARD, R. FESCE, E. CHIEREGATTI and F. BENFENATI, *J. Biol. Chem.*, **267**, 11281 (1992); R. FESCE, F. BENFENATI, P. GREENGARD, and F. VALTORTA, *J. Biol. Chem.*, **267**, 11289 (1992).

[32] M. BÄHLER, F. BENFENATI, F. VALTORTA and P. GREENGARD, *Bio Essays*, **12**, 259 (1990).

[33] E.J. NESTLER and P. GREENGARD, *Protein Phosphorylation in the Nervous System*, J. Wiley & Sons, New York (1984).

[34] R. LLINAS, M. SUGIMORI, T.L. MC GUINNESS, C.S. LEONARD, and P. GREENGARD, *Proc. Natl. Acad. Sci. U.S.A.*, **82**, 3035 (1983); R. LLINAS, J.A. GRUNER, M. SUGIMORI, T.L. MC GUINNESS and P. GREENGARD, *J. Physiol.*, **436**, 257 (1991).

[35] F. BENFENATI, F. VALTORTA and P. GREENGARD, *Proc. Natl. Acad. Sci. U.S.A.*, **88**, 575 (1991).

[36] T.S. SIHRA, J.K.T. WANG, F.S. GORELICK and P. GREENGARD, *Proc. Natl. Acad. Sci. U.S.A.*, **86**, 8108 (1989).

[37] F. TORRI-TARELLI, A. VILLA, F. VALTORTA, P. DE CAMILLI, P. GREENGARD and B. CECCARELLI, *J. Cell Biol.*, **110**, 449 (1990).

[38] H.Q. HAN, R.A. NICHOLS, M.R. RUBIN, M. BÄHLER and P. GREENGARD, *Nature (London)*, **349**, 697 (1991).

[39] B. LU, P. GREENGARD and M.M. POO, *Neuron*, **8**, 521 (1992).

[40] R. JAHN, W. SCHIEBLER, C. OUIMET and P. GREENGARD, *Proc. Natl. Acad. Sci. U.S.A.*, **82**, 4137 (1985); B. WIEDENMANN and W. FRANKE, *Cell.*, **41**, 1017 (1985).

[41] D. PANG, J. WANG, F. VALTORTA, F. BENFENATI and P. GREENGARD, *Proc. Natl. Acad. Sci. U.S.A.*, **85**, 762 (1988).

[42] F. NAVONE R. JAHN, G. DI GIOIA, STUKENBROK, P. GREENGARD and P. DE CAMILLI, *J. Cell Biol.*, **103**, 2511 (1986).

[43] P. KNAUS, B. MARQUEZE-POUEY, H. SCHERER and H BETZ, *Neuron*, **5**, 453 (1990).

[44] R.E. LEUBE, P. KAISER, A. SEITER, R. ZIMBELMANN, W.A. FRANKE, N.P. REHM, P. KNAUS,

P. PRIOR, H. BETZ, H. REINKE, K. BEYREUTHER and B. WIEDENMANN, *EMBO J.*, **6**, 3261 (1987); T.C. SÜDHOF, F. LOTTSPEICH, P. GREENGARD, E. MEHL and R. JAHN, *Science*, **238**, 1142 (1987).

[45] L. THOMAS, K. HARTUNG, D. LANGOSCH, H. REHM, E. BAMBERG, W.W. FRANKE and H. BETZ, *Science*, **242**, 1050 (1988).

[46] H. REHM, B. WIEDENMANN and H. BETZ, *EMBO J.*, **5**, 535 (1986).

[47] P.A. JOHNSTON and T.C. SÜDHOF, *J. Biol. Chem.*, **265**, 7849 (1990).

[48] L. THOMAS and H. BETZ, *J. Cell Biol.*, **111**, 2041 (1990).

[49] M. BAUMERT, P.R. MAYCOX, F. NAVONE, P. DE CAMILLI and R. JAHN, *EMBO J.*, **8**, 379 (1989); T.C. SÜDHOF, M. BAUMERT, M.S. PERIN and R. JAHN, *Neuron*, **2**, 1475 (1989).

[50] W.S. TRIMBLE, D. M. COWAN and R. H. SCHELLER, *Proc. Natl. Acad. Sci. U.S.A.*, **85**, 4538 (1988); L.A. EFFERINK, W. S. TRIMBLE and R.H. SCHELLER, *J. Biol. Chem.*, **264**, 11061 (1989).

[51] Y. MATSUI, A. KIKUCHI, J. KONDO, T. HISHIDA, Y. TERANISHI and Y. TAKAI, *J. Biol. Chem.*, **263**, 11071 (1988); A. ZAHRAOUI, N. TOUCHOT, P. CHARDIN and A. TAVITIAN, *J. Biol. Chem.*, **264,** 12394 (1989); G. FISCHER VON MOLLARD, G.R. MIGNEY, M. BAUMERT, M.S. PERIN, T.J. HANSON, P.M. BURGER, R. JAHN and T.C. SÜDHOF, *Proc. Natl. Acad. Sci. U.S.A.*, **87**, 1988 (1990).

[52] H.R. BOURNE, *Cell*, **53**, 669 (1988).

[53] G. FISCHER VON MOLLARD, T.C. SÜDHOF and R. JAHN, *Nature (London)*, **349**, 79 (1991).

[54] W.D. MATTHEW, L. TSAVALER and L.F. REICHARDT, *J. Cell Biol.*, **91**, 257 (1981).; M. S. PERIN, V.A. FRIED, G.A. MIGNERY, R. JAHN and T.C. SÜDHOF, *Nature (London)*, **345**, 260 (1990).

[55] M.S. PERIN, P.A. JOHNSTON, T. ÖZCELIK, R. JAHN, U. FRANKE and T.C. SÜDHOF, *J. Biol. Chem.*, **266**, 615 (1991).

[56] M.S. PERIN, N. BROSE, R. JAHN and T.C. SÜDHOF, *J. Biol. Chem.*, **266**, 623 (1991).

[57] U. KIKKAWA, A. KIEHIMOTO and Y.A. NISHIZUKA, *Annu. Rev. Biochem.*, **58**, 31 (1989).

[58] M. POPOLI and A. MENGANO, *Neurochem. Res.*, **13**, 63 (1988).

[59] A.G. PETRENKO, M.S. PERIN, B.A. YAVLETOV, Y.A. USHKARYOV, M. GEPPERT and T.C. SÜDHOF, *Nature (London)*, **353**, 65 (1991).

[60] F. VALTORTA, L. MADEDDU, J. MELDOLESI and B. CECCARELLI, *J. Cell Biol.*, **99,** 124 (1984).

[61] D.K. STONE, B.P. CIDER, T.C. SÜDHOF and X.S. XIE, *J. Bioenerget. Biomembr.*, **21**, 605 (1989).

[62] M. BAUMERT, K. TAKEI, N. HARTINGER, P.M. BURGER, G. FISCHER VON MOLLARD, P.R. MAYCOX, P. DE CAMILLI and R. JAHN, *J. Cell. Biol.*, **110**, 1285 (1990).

[63] S.S. CARLSON, P. CARONI and R.B. KELLY, *J. Cell Biol.*, **103**, 509 (1986); K. BUCKLEY and R. R. KELLY, *J. Cell Biol.*, **100**, 1284 (1985).

[64] P. CARONI, S.S. CARLSON, E. SCHWEITZER and R.B. KELLY, *Nature (London)*, **314**, 441 (1985).

VARIOUS TYPES OF ACETYLCHOLINE RELEASE
FROM THE MOTOR NERVE

STEPHEN THESLEFF

Deparment of Pharmacology, University of Lund,
S- 22362, Lund, Sweden

Content

Bioelectrochemistry IV
Edited by B.A. Melandri *et al.*, Plenum Press, New York, 1994

1. Introduction

This presentation discusses different kinds of acetylcholine (ACh) release at the neuromuscular junction and their possible physiological role. Emphasis is given to recent studies and to musculotrophic influences (trophic effects on muscle).

2. Various quantal releases of ACh.

The best known type of ACh release is the quantal vesicular release, extensively studied and described by KATZ and his coworkers (see a review by KATZ [1]). This type of transmitter release, illustrated in Fig. 1A, is responsible for nerve-impulses-evoking synchronous release of quanta of ACh, which in the underlying muscle membrane gives rise to large potential changes, the endplate potential (e.p.p.), and therby to a propagated action potential and a muscle twitch. A closely related type of release is the spontaneous secretion of a single quantum of ACh resulting in the appearence in the muscle endplate of small, uniformly sized potentials, the socalled miniature end plate potentials (m.e.p.p.s.) illustrated in Fig. 1B. As originally suggested by

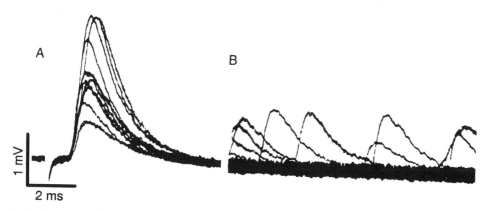

FIG. 1. Superimposed nerve impulse evoked quantal e. p. p.s (A) in a normal rat extensor digitorum longus muscle. The e.p. p.s were recorded in KREBS-RINGER solution containing 15 mM magnesium to reduce the quantal content. Record (B) shows spontaneous m. e.p.p.s in the same fibre.

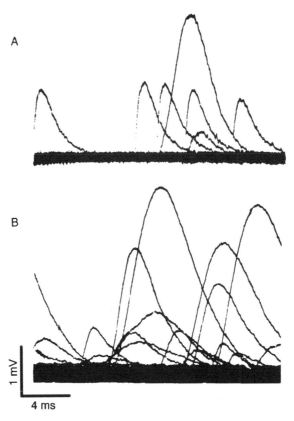

FIG. 2. (A) Superimposed traces showing spontaneous m.e.p.p.s in a normal rat extensor digitorum longus muscle. Note the appearance of one giant m.e.p.p. and one subm.e.p.p. (B) The frequencies of spontaneous giant m.e.p.p.s and of subm. e.p.p.s are greatly increased following prolonged blockade of neuromuscular transmission. The record is from a rat extensor digitorum longus muscle paralysed for 16 days by the use of botulinum toxin type A.

DEL CASTILLO and KATZ [2] each quantum of ACh, whether released by a nerve impulse or spontaneous, is preformed within synaptic vesicles in the nerve terminal. It has been suggested that Ca^{2+} ions inside the nerve terminal induce fusion of the synaptic vesicle with the nerve plasmalemma at morphologically distinct sites, the *active zones*. Upon fusion, the ACh content of the vesicle is exocytosed into the synaptic cleft and, after contact with postsynaptic ACh-receptors, gives rise to the potential changes mentioned above. Both types of ACh release are strictly dependent on the level of intracellular calcium ions and are treated in more detail in other chapters.

The obvious physiological role of nerve-impulse-evoked ACh release is to induce muscle activity and tone. Since the pattern of electrical and mechanical activity in muscle influences a number of membrane and contractile properties, this type of

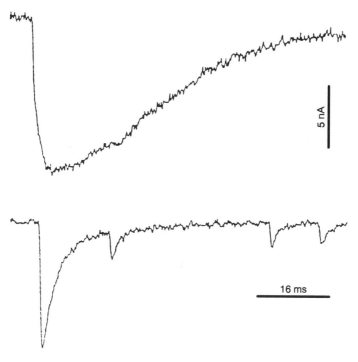

FIG. 3. Spontaneous miniature (lower tracing) and giant miniature (upper tracing) endplate currents recorded on the same frog neuromuscular junction at a holding membrane potential of -90 mV. The normal RINGER solution contained 20 μM Tacrine and 1 μM tetrodotoxin. Note, the difference in time course between *normal* and *giant* miniature endplate currents. By courtesy of J. MOLGO'.

ACh release also constitutes a major musculotrophic influence, (see reviews by PETTE and VRBOVA [3] and by LÖMO and GUNDERSEN [4]). No one has so far been able to demonstrate a physiological role for spontaneous m.e.p.p.s. although it has been speculated that they might have trophic influences on muscle [5].

In 1957 LILEY [6] described a third kind of quantal transmitter release. It is a spontaneous release of large amounts of ACh which are recorded as big postsynaptic potentials, socalled giant m.e.p.p.s. The potentials are produced by ACh, since they are blocked by curare, and their amplitude and time course are increased by inhibition of cholinesterase. Their frequency varies among fibres, from zero to 20 % of all m.e.p.p.s recorded, and their number is greatly increased at newly formed endplates [7,8] and following long-term block of neuromuscular transmission. In the latter condition, frequencies approaching one Hz have been recorded [9–11]. Unlike normal m.e.p.p.s their frequency is unaltered by nerve stimulation, except for prolonged nerve stimulation which enhances their frequency [12].

Changes in extracellular calcium ion concentrations do not affect their frequency, indicating that their dependency on intraterminal calcium ion concentrations is less than that of normal m.e.p.p.s [13]. Figure 2 illustrates such giant m.e.p.ps in a normal rat muscle (graph A) and in a rat muscle paralysed for 16 days by the use of botulinum toxin type A (graph B) . As shown, giant m.e.p.p.s are not only larger than normal but frequently also have a slower time-to-peak. Figure 3 shows recordings of endplate currents resulting from giant m.e.p.p.s and normal m.e.p.p.s respectively. It should be noted that the total charge transfer during a giant m.e.p.p. may be one order of magnitude larger than that of normal m.e.p.p.s. Hence, giant m.e.p.p.s. signal the spontaneous release of huge amounts of ACh.

The ACh responsible for giant m.e.p.p.s comes from the same pool of transmitter as that released by nerve impulses, since the drug hemicholinium-3 (5 μM) which blocks ACh synthesis, reduces the amplitude of all synaptic potentials including giant m.e.p.p.s but only if the nerve is stimulated and the preformed store depleted [14]. Furthermore, the ACh apparently is of synaptic vesicular origin, since blockade of the specific ACh uptake system into vesicles by 1 μM 2-(4-phenylpiperidino)cyclohexanol (AH-5183) abolishes giant m.e.p.p.s [14].

Lowering of temperature from 30 to 14 °C reduces giant m.e.p.p. frequency with a Q_{10} of about 12 (where Q_{10} stands for the frequency change due to a change of temperature of 10 °C) while normal m.e.p.p.s are affected by a Q_{10} of 2 to 3 [9]. Inhibitors of oxidative metabolism block the appearence of giant m.e.p.p.s, but increase the frequency of normal m.e.p.p.s [15]. Colchicine (5 mM), which aggregates microtubuli and thereby blocks proximo-distal axoplasmic transport, prevents giant m.e.p.p.s, but not normal m.e.p.p.s from appearing [16]. Hence, temperature sensitive metabolic processes and transport by microtubuli are necessary for the ACh release which gives rise to giant m.e.p.p.s.

Which synaptic structures are responsible for the release of the large amounts of ACh causing giant m.e.p.p.s? Nerve terminals have been examined for the presence of unusually large synaptic vesicles, cisternae or clusters of vesicles that might explain the release of large amounts of ACh, but with negative results [17]. However, one observes so-called large dense-core synaptic vesicles with a volume about ten-times that of the smaller clear synaptic vesicles whose content of ACh gives rise to normal m.e.p.p.s. Large dense-core synaptic vesicles contain various neuropeptides but presumably also the classical transmitter *i.e.* in motor nerves ACh (for a recent review of the subject see DE CAMILLI and JAHN [18]). The large dense-core synaptic vesicles exocytose their content differently from that of small clear synaptic vesicles. They have a much lower dependency on calcium ions and only high frequency nerve stimulation induces their exocytosis.

Furthermore, they do not seem to release their content at *active zones* in the terminal, but at sites outside those regions [19]. These kinds of behaviour correspond well with that observed for the release of the ACh giving rise to giant m.e.p.p.s. In contrast to clear synaptic vesicles , the large dense-core ones are carried by proximo-

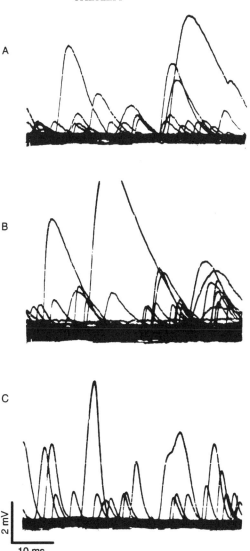

FIG. 4. Examples of giant m.e.p.p.s induced in the mouse diaphragm muscle by (A) 4-aminoquinoline (200 μM), (B) tetrahydroaminoacridine (Tacrine, 20 μM) and (C) emetine (20 μM).

distal axoplasmic transport to the nerve terminal. Large dense-core synaptic vesicles are particularly numerous in newly formed nerve terminals and in nerve terminals following prolonged neuromuscular blockade, *i.e.* in conditions with increased giant m.e.p.p. frequency [20—22]. Thus, it appears that the release mechanism of the large dense-core vesicle content has all the characteristics that have been observed for the release of the ACh causing giant m.e.p.p.s.

In connection with the possibility that giant m.e.p.p.s. may signal the release of

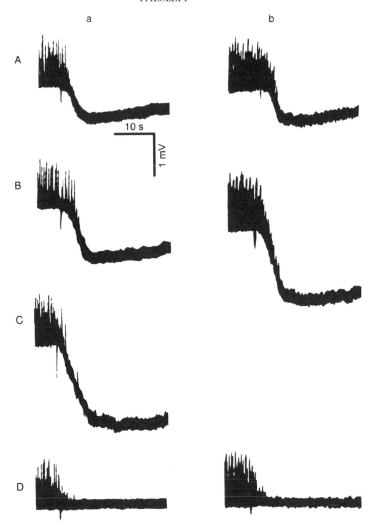

FIG. 5. Application of curare to the endplate in a mouse diaphragm muscle causes, in the presence of cholinesterase inhibition by 6 μM prostigmine (a) or by pretreatment with soman (b), a hyperpolarizing response (A). Inhibition of the Na^+-K^+-activated ATPase by 20 μM ouabain (B) or a K^+-free solution (C) enhances
the hyperpolarizing response, while activation of ATPase by readmission of K^+ (D) blocks the response.
From Ref. 49.

the content of large dense-core synaptic vesicles, it is of interest that certain drugs have
been shown to selectively enhance the frequency of giant m.e.p.p.s. Fig. 4 shows that
4-aminoquinoline (200 μM), tetrahydroaminoacridine (Tacrine 20 μM) and emetine
(20 μM) enhance the frequency of giant m.e.p.p.s. The frequency of normal m.e.p.p.s.
is not altered . The drug-induced giant m.e.p.p.s. have all the characteristics previously
described for the spontaneous giant m.e.p.p.s and those induced by long-term muscle

paralysis. As shown by Fig. 4, the time-to-peak of emetine-induced giant m.e.p.p.s. is faster than those induced by the other drugs or by paralysis . 4 – Aminoquinoline and Tacrine are potent cholinesterase inhibitors and long-term paralysis reduces the amount of cholinesterase in the junction. Emetine has no cholinesterase blocking activity, and this might explain the faster times-to-peak of the potentials induced by that drug. When emetine is administred together with the cholinesterase inhibitor edrophonium (50 μM) the time-to-peak of the potentials is lengthened, as observed with the other drugs.

The prevalence of giant m.e.p.p.s during embryogenesis, synapse formation and long-term transmission block indicate that the potentials are related, in some way, to synapse formation and maintenance. Neuropeptides are believed to have a regulatory influence on the appearance and number of postsynaptic nicotinic ACh receptors and to regulate the aggregation of receptors and cholinesterase at the endplate region [23–25]. If ACh is co-released with neuropeptides one might ask whether its role is to facilitate the uptake or to modulate the action of neuropeptides in the muscle cell. Frequently, the giant m.e.p.p.s are large enough to induce a muscle action potential and may therefore also contribute to the musculotrophic effects of muscle activity. The possibility that certain drugs selectively stimulate the release of neuropeptides with trophic actions might be of therapeutic value in neurodegenerative diseases.

In many instances unusually small m.e.p.p.s. are recorded (see for instance Fig. 1B). Such subm.e.p.p.s. have been extensively studied by KRIEBEL and his coworkers (for reviews see Ref. 26 and 27). KRIEBEL has proposed that normal sized m.e.p.p.s are generated by the almost simultaneous release of a number of subm.e.p.p.s; a view that has been criticised by a number of scientists [28–31]. Recently KRIEBEL *at al.* [32] have proposed a transmitter release process at the neuromuscular junction whose characteristics are similar to those of droplet formation. It however lacks experimental proof and can, at present, only be regarded as a speculation to explain some of the peculiarities seen in the size variation of transmitter quanta.

3.   Non-quantal release of ACh

There has long been a large discrepancy between the amount of ACh released from a neuromuscular preparation, as determined biochemically, and the amount calculated from the release of preformed quanta of ACh, as measured electrophysiologically. This difference is explained by the observation that in addition to the intermittent quantal releases of ACh mentioned above there is a continuous molecular leakage of ACh from the nerve terminal that quantitatively is almost two orders of magnitude larger than quantal releases. Electrophysiologically, this type of ACh release is detected when curare is added to the nerve-muscle reparation in the presence of a cholinesterase inhibitor. The resulting blockade of cholinergic muscle receptors stimulated by the leakage of ACh causes a local hyperpolarization of the endplate region amounting to 0.04-9 mV, depending upon species and experimental conditions [33,34]. This hyperpolarization called the H-response, is illustrated in Fig. 5 from the endplate region of a mouse, together with some of the experimental conditions that influence the release. The H-response is enhanced by nerve stimulation. EDWARDS *at al.* [35] and YU and VAN DER KLOOT [36] have suggested that the non-quantal, continuous leakage of ACh from nerve terminals is the result of the incorporation of the synaptic vesicle membrane into the axolemma, following vesicle fusion and exocytosis of its content. As mentioned, the vesicle membrane contains a specific ACh transport system that carries ACh from the axoplasm into the vesicle. If this transport continues to operate, following vesicle fusion with axolemma, it would transport ACh from the axoplasm into the synaptic cleft. In support of this idea, it has been observed that nerve stimulation with accelerated vesicle fusion enhances leakage, while magnesium ions, that oppose fusion, reduces it [37]. Furthermore, blockage of the vesicular transport system by 2-(4-phenylpiperidino)cyclohexanol abolishes leakage, as does blockade of ACh synthesis by hemicholinium-3 [38].

Growth cones of cholinergic neurones have this type of ACh secretion [39,40]. POO *et al.* [41,42] and LIN-LIU *et al.* [43] have demonstrated that a uniform electric field across the membrane of a myotube causes the accumulation of ACh receptors at the cathodal pole. It is therefore possible that the continuous depolarization caused by ACh leakage contributes to the aggregation of ACh receptors at the endplate. It has been proposed that desensitization of nicotinic cholinergic receptors is a mechanism for short-term regulation of synaptic efficacy [44 – 47]. Continuous leakage of ACh may cause receptor desensitization and thereby regulate the chemical sensitivity of the endplate [48].

4.   Comments

Of the different kinds of ACh release at the neuromuscular junction, only one type, nerve-impulse-evoked synchronous quantal ACh release, has been shown to have a physiological function. I have indicated that the other types of ACh secretion may have trophic effects on muscle. Experimental proof is, however, generally lacking. Future studies should clarify whether the various release processes are simply rudimentary leftovers from previous evolutionary stages or have still unknown physiological functions.

References

[1]   B. KATZ, *The Release of Neural Transmitter Substances*, in *The Sherrington Lectures* X, Vol. **10**, Liverpool Univ. Press, Liverpool, (1969).

[2]   J. DEL CASTILLO and B. KATZ, *J. Physiol. (London)*, **128**, 396 (1955).

[3]   D. PETTE and G. VRBOVA', *Muscle Nerve*, **8**, 676 (1985).

[4]   T. LÖMO and K. GUNDERSEN, in *Nerve-Muscle Cell Trophic Communications* H.L. FERNANDZEZ (Editor), CRC Press, Boca Raton, (1988), p.61.

[5]   S.THESLEFF, *J. Physiol. (London)*, **151**, 598 (1960).

[6]   A.W.LILEY, *J. Physiol. (London)*, **136**, 595 (1957).

[7]   M.R. BENNETT, E.M. MCLACHLAN and R.S. TAYLOR, *J. Physiol. (London)*, **233**, 481 (1973).

[8]   C.COLMÉUS, S. GOMEZ, J. MOLGO' and S. THESLEFF, *Proc. R. Soc. Lond. B, Biol. Sci.*, **215**, 63 (1982).

[9]   S. THESLEFF, J. MOLGO' and H. LUNDH, *Brain Res.*, **264,** 89 (1982).

[10]   R. DING, J. K. S. JANSEN, N. G. LAING and H. TÖNNESEN, *J. Neurocytol.*, **12,** 887 (1983).

[11]   K. GUNDERSEN *J. Physiol. (London)*, **430,** 399, (1990).

[12]   J. E. HEUSER, *J. Physiol. (London)*, **239**, 106P (1974).

[13]   S.THESLEFF, and J. MOLGO', *Neuroscience*, **9**, 1 (1983).

[14]   M. T. LUPA, N. TABTI, S. THESLEFF F. VYSKOCIL and S-P. YU, *J. Physiol. (London)*, **381**, 607 (1986).

[15]   N. TABTI, M. T. LUPA, S. -P. YU and S. THESLEFF, *Acta Physiol. Scand.*, **128**, 423 (1986).

[16]   S. THESLEFF, L. C. SELLIN and S. TAGERUD, *Br. J. Pharmacol.*, **100,** 487, (1990).

[17]   M. PÉCOT-DECHAVASSINE and J. MOLGO', *Biol. Cell.*, **46,** 93 (1982).

[18]   P. DE CAMILLI and R. JAHN, *Ann. Rev. Physiol.*, **52**, 625 (1990).

[19]   M. VERHAGE, H. T. MCMAHON, W. E. J. M. GHIJSEN, F. BOOMSMA, G. SCHOLTEN, V.M. WIEGANT and D. G. NICHOLLS *Neuron*, **6**, 517 (1991).

[20]   I. JIRMANOVA' and S. THESLEFF, *Neuroscience*, **1**, 345 (1976).

[21]   R. LÜLLMANN-RAUCH, *Z. Zellforsch.*, **121**, 593 (1971).

[22]   M. PÉCOT-DECHAVASSINE, J. MOLGO' and S. THESLEFF, *Neurosci. Lett.*, **130**, 5 (1991).

[23]   R. LAUFER, and J. P. CHANGEUX, *EMBO J.*, **6,** 901 (1987).

[24]   B. FONTAINE, A. KLARSFELD, T. HÖKFELT and J. P. CHANGEUX.,*Neurosci. Lett.*, **71**, 59 (1986).

[25]   G. D. FISCHBACH, D. A. HARRIS, D. I. FALLS, J.M. DUBINSKY, K. MORGAN, K.L. ENGLISH and F.A. JOHNSON in *Neuromuscular Junction*, , L. C. SELLIN, R. LIBELIUS and S. THESLEFF (Editors) Elsevier, Amsterdam (1989), p. 515.

[26]   M. E. KRIEBEL and C. ERXLEBEN, in *Calcium, Neuronal Function* and *Transmitter Release,* R. RAHAMIMOFF and B. KATZ (Editors), Nijhoff, Boston (1986), p. 299.

[27]   M. E. KRIEBEL and I. MONTELICA-HEINO, *Neuroscience, 23,* 757 (1987).

[28]   D. C. MILLER, M.M. WEINSTOCK and K. L. MAGLEBY, *Nature (London),* **274**, 388 (1978).

[29]   K. L. MAGLEBY and D.C. MILLER, *J. Physiol.,***311**, 267 (1981).

[30]   W. VAN DER KLOOT, *J. Neurosci.,* **27**, 81 (1989).

[31]   W. VAN DER KLOOT, *Progr. in Neurobiol.,* **36**, 93 (1991).

[32]   M. E. KRIEBEL, J. VAUTRIN and J. HOLSAPPLE, *Brain Res. Revs.,* **15**, 167 (1990).

[33]   B. KATZ and R. MILEDI,*Proc. R. Soc. Lond. B. Biol. Sci.* **196**, 59 (1977).

[34]   F. VYSKOCIL and P. ILLÉS, *Pflügers Arch.,* **370**, 295 (1977).

[35]   C. EDWARDS, V. DOLEZAL. S. TUCEK, H. ZEMKOVA' and F. VISKOCIL, *Proc. Natl. Acad. Sci U.S.A.,* **82**, 3514 (1985).

[36]   S-P. YU and W. VAN DER KLOOT, *Neurosci. Lett.,***117**, 111 (1990).

[37]   H. ZEMKOVA' and F. VYSKOCIL, *Neurosci. Lett.,* **103**, 293 (1989).

[38]   E. E. NIKOLSKY, V. A. VORONIN. T. I. ORANSKA and F. VYSKOCIL, *Pflügers Arch.,* **418**, 74 (1991).

[39]   M. POO and K. R. ROBINSON, *Nature (London),* **265**, 602 (1977).

[40]   Y. SUN and M. POO, *J. Neurosci.,* **5**, 634 (1985).

[41]   M. POO, W. H. POO and J. W. LAM, *J. Cell. Biol.,* **76**, 483 (1978).

[42]   M. POO, J.W. LAM and N. ORIDA, *Biophysics, 26,* 1 (1979).

[43]   S. LIN-LIU, W. R. ADEY and M. POO, *Biophys. J.,* **45**, 1211 (1984).

[44]   S. THESLEFF, *Physiol. Rev.* **40**, 734 (1960).

[45]   T. HEIDMANN and J. P. CHANGEUX, *C.R. Acad. Sci. (Paris),* **295**, 665 (1982).

[46]   J. P. CHANGEUX, F. BON, J. CARTAUD, A. DEVILLERS-THIÉRY, J. GIRAUDET, T. HEIDMANN, B. HOLTON, H. O. NGHIEM, J. L. POPOT, R. VAN RAPENBUSCH and S. TZARTOS, *Torpedo marmorata, Cold Spring Harbor Symp., Quant. Biol.,* **35**,...(1983).

[47]   J. P. CHANGEUX, *Prog. Brain Res.,* **68**, 373 (1986).

[48]   K. MILES and R. L. HUGANIR, *Mol. Neurobiol.,* **2**, 91 (1988).

[49]   F. VYSKOCIL and P. ILLÉS, *Physiol. Bohemoslov., 27,* 449 (1978).

BIOPHYSICAL ASPECTS OF PRESYNAPTIC ACTIVITY

RICCARDO FESCE

"Bruno Ceccarelli" Centre and C.N.R. Centre of Cytopharmacology.
Dept. of Medical Pharmacology. University of Milan
Dibit H. S. Raffaele, Milan, Italy

Contents

Bioelectrochemistry IV
Edited by B.A. Melandri *et al.*, Plenum Press, New York, 1994

1.   Introduction

In this lecture some aspects of presynaptic activity will be reviewed, focusing on the mechanisms of quantal secretion of neurotransmitter at the neuromuscular junction. The statistics of quantal release evoked by the nerve action potential and their relevance to the understanding of the secretory machinery will be discussed first. The second section will deal with the vesicle hypothesis of quantal release of neurotransmitter, and report findings on this issue from comparative morpho-functional studies. The general features and possible mechanisms of the modulation of quantal release will be discussed next, and the biophysical aspects of vesicle fusion to release neurotransmitter will be discussed in the final section.

2.   Statistics of quantal release of neurotransmitter.

The classical studies of FATT and KATZ in the early 1950s [1,2] demonstrated the occurrence of small transient changes in the membrane potential at the motor end plate. As these transients resembled the much higher deflections produced by stimulation of the nerve, which trigger the postsynaptic spike and muscle fibre contraction, they called these events miniature endplate potentials (MEPPs). MEPPs arise from spontaneous release of neurotransmitter from the resting motor nerve ending. Detailed studies of their amplitude, in the frog by DEL CASTILLO and KATZ [3], in mammals by BOYD and MARTIN [4] and LILEY [5], and in crayfish by DUDEL and KUFFLER [6], showed that they are elementary events with fairly uniform amplitudes, and that the summation of a certain number of such events gives rise to the endplate potential (EPP). This quantal theory of neurotransmitter release was further substantiated by many statistical studies on the amplitude of endplate and postsynaptic potentials in many preparations. The amplitude of the EPP is very sensitive to extracellular calcium concentration. When calcium concentration is reduced and/or magnesium concentration is raised in the bathing fluid, EPPs become smaller (a few mV) and the amplitudes of successive EPPs tend to cluster around integer multiples of the MEPP amplitude; as MEPPs do not have a fixed amplitude but a continuous distribution of amplitudes, smoothed peaks, rather than a discrete distribution of EPP amplitudes, are observed [3,4].

The first statistical fits to the data of evoked quantal secretion were in agreement with POISSON statistics, i.e. the distribution of quantal contents of EPPs (number of quanta constituting each EPP) suggested the existence of a great many release sites (or releasable quanta), each with a very low probability of release. POISSON statistics represent a limiting case of the more general binomial statistics, for probability of release very low, approaching zero. With intermediate calcium concentrations and a higher probability of release, binomial statistics, based on a finite number of release sites (or quanta), n, with probability of release, p, described better the observed distributions of quantal contents. The estimated number of release sites, n, was generally <1000, suggesting that this statistical parameter is not related to the very high number of synaptic vesicles observed in nerve terminals, but rather to a limited number of release sites, like the specialized regions of the presynaptic membrane called *active zones*, or to a small fraction of the quanta constituting a pool of readily available quanta [7]. The idea that binomial n and p may have precise correlates in nerve terminal cellular physiology, however, is far from proven.

The binomial model of quantal release appeared to be particularly sensible since it could explain the effect of different Ca^{2+} concentrations on evoked release in terms of probability of release, the number of release sites remaining unchanged; however, if n represents the number of available quanta or active release sites there is no reason to assume that it must remain constant during repetitive stimulation [8]. Furthermore, a series of problems arise in attempting this kind of correlation, especially when quantal content is not kept artificially low: nonlinear summation of EPPs [9,10]; erratic behaviour of the estimates of n when p is small; difficulty in estimating quantal size when release is not artificially lowered and receptor blocking drugs (curare) have to be used to reduce the size of EPPs and prevent the muscle fibre from firing and contracting. The main problem in the reliability of statistical studies of evoked quantal release, however, stems from the fact that the number of available quanta (or sites) may vary during repetitive stimulation and the probability of release may not be homogeneous among sites and in successive trials. These factors, if neglected, can introduce relevant distortions in EPP amplitude distributions and lead to substantial errors in the estimates of the binomial parameters; in fact under many conditions the simple binomial model (with the wrong parameters) will still fit very well the observed distributions and will not be rejected [11].

Several approaches have been proposed to analyze the statistics of quantal content during repetitive stimulation and to circumvent the problems of classical approaches:

(*i*) the use of higher moments of the distribution [12];

(*ii*) estimating the probability of release (and the size of the pool of readily available quanta) from the fractional reduction of the size of each successive EPP, in a train of stimuli at high frequency, assuming the fractional reduction to be proportional to fractional release by the preceding impulse [13];

(*iii*) obtaining independent estimates of spatial and temporal variations in the binomial parameters [14], and

(*iv*) fitting maximum-likelihood binomial models to the complete distribution of evoked responses with the help of deconvolution methods [15,16].

Notwithstanding the complexity of the problem, and the high sensitivity of approaches used so far to explicit and implicit unproven assumptions, the possibility of describing the process of evoked quantal release by a statistical model is still extremely attractive, because it may help in clarifying the mechanisms underlying the Ca^{2+}-dependence of quantal release and in understanding regulatory processes in synaptic activity.

3.   Morpho-functional correlations
The vesicle hypothesis of quantal release

The vesicle hypothesis of the release of quanta of neurotransmitter was proposed by DEL CASTILLO and KATZ in 1956 [17], following the electrophysiological studies demonstrating the quantal nature of synaptic transmission (for a review see Ref. 18) and the demonstration by electron microscopy that nerve terminals contained large numbers of small membrane-bound compartments, called synaptic vesicles [19-22].

This hypothesis was challenged because electron microscopic images of vesicles fused with the axolemma are rarely seen; it is difficult to deplete the population of synaptic vesicles by stimulating a synapse; most of the acetylcholine contained in nerve terminals is not confined into synaptic vesicles; newly synthesized acetylcholine is preferentially released, and biochemical estimates of the number of acetylcholine molecules in a quantum are not consistent with those of the number of molecules contained in a synaptic vesicle (*e.g.* Ref. 23).

After the development of the freeze-fracture technique, which allows wide plasmalemmal surface areas to be visualized, large numbers of synaptic vesicles were seen to fuse with the presynaptic membrane (Fig. 1) during intense quantal release [24-26]. Experiments with extracellular tracers showed that vesicles are recycled (Fig. 2) during intense synaptic activity [27,28], and recycled vesicles appear to be preferentially reloaded with freshly synthesized acetylcholine [29], yielding a possible explanation for the preferential release of newly synthesized ACh. Ionophoretic evaluations of the number of ACh molecules in a quantum [30] yielded an upper limit of 10^4, which is fairly consistent with the estimated content of a synaptic vesicle. However, all this evidence does not demonstrate that the release of a quantum of neurotransmitter coincides with the fusion of one vesicle with the presynaptic membrane to release its content. The issue is further complicated by the observation that most (> 90 %) of the ACh released by a resting motor nerve terminal exits the axolemma in molecular rather than quantal form [31,32].

Two approaches have yielded further support to the vesicle hypothesis of quantal release during the last 15 years: the investigation of the precise timing of vesicle fusion after the delivery of a single shock to the nerve, and the quantitative corre-

FIG. 1. Electron micrograph of a freeze-fractured motor nerve terminal (frog *cutaneus pectoris* muscle) stimulated by immersion in 20 mM K$^+$- containing Ringer solution for 1 min before chemical fixation. Note the active zones, characterized in the frog by double rows of particles aligned alongside a ridge that crosses the whole presynaptic membrane. Several pits, generated by synaptic vesicles in the process of fusing with the plasmalemma, are seen, mostly in association with the active zones (magnification: x 75,000).

lation between the number of quanta secreted and the number of vesicles fused with the presynaptic axolemma during intense synaptic activity.

3.1.   Timing of vesicle fusion

In the late 1970's HEUSER and his coworkers implemented a procedure for rapidly freezing biological specimens, by smashing the preparation on to a copper block cooled to about 15 K by liquid helium [33,34]. The superficial layer of the preparation (about 10 μm thickness) is frozen by this procedure within less than 1 ms, providing both a good preservation of unfixed tissue (the speed of freezing prevents the formation of ice crystals large enough to damage the ultrastructure of the specimen) and a reasonable time resolution for studying synaptic events. In quick-frozen, freeze-fractured nerve-muscle preparations, these authors observed many vesicle fusions with the presynaptic axolemma as soon as 4-6 ms after the delivery of a single shock to the motor nerve. Aminopyridine was present in these experiments to prolong the duration of the nerve action potential and increase release of ACh by severalfold. Considering the slowing of conduction in the nerve due to *precooling* of the preparation during the approach to the copper block, this was a reasonable delay. Using an improved quick-freezing apparatus, studied to minimize precooling, fusions of synaptic vesicles with the presynaptic axolemma could be observed in preparations quick-frozen 2.5 ms after the nerve stimulus, fixed by cryosubstitution and thin-sectioned [35] (see Fig. 3). In these preparations the end-plate potential occurred about 3 ms after the stimulus: about 2 ms conduction time and about 1 ms synaptic delay, most of which is spent waiting for the voltage-dependent calcium channel to open [36]. Thus, fusion of vesicles begins to occur exactly when release of ACh starts. This precise temporal coincidence of vesicle fusion with the release of quanta of ACh strongly supports the vesicle hypothesis of quantal release.

3.2.   Correlation between number of quanta released and loss of synaptic vesicles

The recycling of vesicles and quanta is relevant, and significant changes in vesicle population are observed, when quantal secretion is vigorous and prolonged. Accurate information on the time course of quantal secretion, combined with morphometry, permits evaluation of the extent of quantal turnover and a quantitative correlation with vesicle recycling under different experimental conditions.

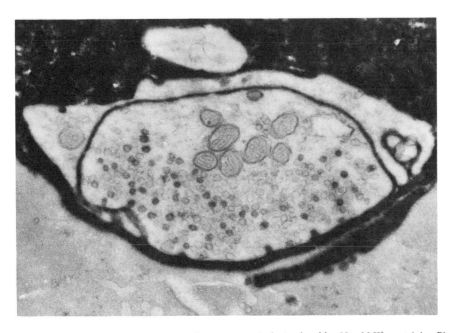

FIG. 2. Electron micrograph of a thin section of a nerve terminal stimulated by 20 mM K$^+$-containing Ringer solution, in the presence of the extracellular tracer horseradish peroxidase. Peroxidase reaction product stains the extracellular spaces. Many synaptic vesicles appear loaded with the tracer, indicating that they have fused with, and reformed from, the presynaptic axolemma. Images of vesicles fused with the axolemma can be seen (magnification x 45,000). Courtesy of B. CECCARELLI, F. GROHOVAZ and N. IEZZI [49 b].

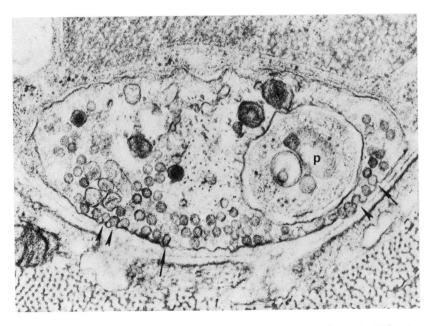

FIG. 3. Micrograph from a preparation quick-frozen 2.5 ms after a single stimulus in 1 mM 4-aminopyridine. Different degrees of association between vesicles and the prejunctional membrane are evident in correspondence with the active zones. Arrow-heads indicate clearcut openings whereas arrows indicate images suggestive of intermediate steps between fusion and fission of the two membranes. **p**. SCHWANN cell process. (magnification x 70,000). Reprinted from Ref. 35, Copyright The Rockefeller University Press, N.Y.

Intense synchronous secretion evoked by electrical stimulation of the nerve can be measured from the amplitude of the endplate potentials in curare. However, recycling is very active under these conditions and it is difficult to correlate vesicle loss with the number of quanta secreted. Depletion of vesicles and possibly block of vesicle recycling are observed under several conditions (see below) that vigorously activate asynchronous quantal release. To measure the high rates of asynchronous quantal release attained under these conditions, an original method of statistical analysis was developed and applied to electrophysiological recordings obtained from endplates. This procedure is based on the principles of noise analysis: the mean, the variance, and a series of other statistical parameters (semi-invariants and their spectral composition) add up linearly for independent random variables; thus, for a signal generated by the summation of randomly occurring events, these parameters equal those of the single event multiplied by the rate of occurrence of the elementary events. The waveform, amplitude and rate of occurrence of the elementary events constituting a composite signal (noise) can therefore be deduced from the power spectrum and a few semi-invariants (*e.g.* mean and variance) of the noise. In particular, for endplate recordings the power spectrum, variance and skew

(3rd order semi-invariant: average cube value of the departure from the mean) were used to derive amplitude, rate of occurrence and waveform of the MEPPs. Corrections were applied for nonlinear summation of MEPPs and for distributed MEPP amplitudes [37-40].

Early attempts at applying fluctuation analysis to endplate recordings were reported by HEUSER and MILEDI [41] and KATZ and MILEDI [42], but those approaches (using the mean depolarization and the variance of the fluctuations as indexes of MEPP rate and amplitude) could only be applied to short recordings where the membrane potential was not affected by factors other than the summation of MEPPs. Nonlinear summation and extraneous sources of depolarization invalidate the mean depolarization as an indicator of MEPP rate and amplitude, whereas they do not interfere markedly with the fluctuations of the membrane potential about its mean value, and therefore with the variance, skew and power spectrum. The procedure for noise analysis of quantal recordings has been further improved to follow rapid changes in quantal amplitude and rate of occurrence [40, 43-45].

Particularly interesting for these kinds of studies was the toxin purified from the black widow spider venom (BWSV), α-Latrotoxin [46]. This toxin induces massive release of neurotransmitter from all vertebrate synapses tested, and at high doses or in the absence of calcium it blocks vesicle recycling leading to complete depletion of vesicles and block of transmission within about 1 hour [47-48].

Table 1 illustrates the results of combined morphometric and noise analysis studies using several agents that vigorously activate asynchronous quantal release and in some cases block vesicle (and quantum) recycling. At the frog neuromuscular junction, under conditions where vesicle recycling is blocked, a number between 0.5 and 1 million quanta are released before secretion vanishes, which is in good agreement with the average number of vesicles present in nerve terminals at rest. Furthermore, after 1 hour of intense release, high rates of quantal release (in the order of 100/s) are sustained under conditions where vesicle recycling is active, as indicated by the persistence of a relevant number of vesicles in the nerve terminal and by the uptake of extracellular tracers, whereas release vanishes (<10/s) under conditions that lead to severe depletion of vesicles.

These data strongly support the idea of a one-to-one correspondence of vesicles and ACh quanta [37-39, 49]. Recently, a correlation has been attempted in the same preparation between the time course of quantal release and the number of vesicles remaining per μm of nerve terminal, during treatment with BWSV or α-Latrotoxin in the absence of calcium (with blocked vesicle recycling). Figure 4 illustrates these results, which confirm the one-to-one correlation between vesicle loss and quantal release and so yield further solid evidence in support to the vesicle hypothesis of quantal release [50].

TABLE 1. CORRELATION BETWEEN QUANTAL RELEASE AND VESICLE LOSS AT THE FROG ENDPLATE

Experimental condition		Quanta secreted	Vesicle counts			References
			Remaining	Lost‡	Recycled	
Electrical stimulation (2 Hz)	30 min	1330*	675 ± 225	206	1100	Ceccarelli et al., [27]
	1 hr	2500*	506 ± 131	375	2100	Torri-Tarelli et al., [72]
Ouabain 0.5-1 mM	0 mM Ca²⁺, 4 mM Mg²⁺	1230*	291 ± 170	590	640	Haimann et. al., [49]
	1.8 mM Ca²⁺, 4 mM Mg²⁺	1470*	147 ± 106	735	740	
Lanthanum 0.1 mM room temperature	no Ca²⁺	4500*	110 ± 39	718†	3800	Segal et. al., [37]
	1.8 mM Ca²⁺	5830*	227 ± 73	601†	5200	Fesce et. al., [39]
BWSV 5 µg/cm³ ●	0 mM Ca²⁺, 4 mM Mg²⁺	1170*	none	880†	290	Fesce et. al., [39]
	1.8 mM Ca²⁺, 4 mM Mg²⁺	2830*	650 ± 192	178†	2650	Torri-Tarelli et. al., [72]
α-LTx 0.2 µg/cm³ room temperature	0 mM Ca²⁺, 4 mM Mg²⁺	1380*	none	N.D.	N.D.	Valtorta et. al., [70]
	1.8 mM Ca²⁺, 4 mM Mg²⁺	3000*	N.D.	N.D.	N.D.	
α - LTx 0.5 µg/cm³ 0 mM Ca²⁺, 4 mM, Mg²⁺ room temperature	5 min	370	838 ± 738	812	<0	Hurlbut et. al., [50]
	10 min	1100	388 ± 438	1260	<0	
	15 min	1160	275 ± 250	1375	<0	
α-LTx 1-3 °C	0 mM Ca²⁺, 4 mM Mg²⁺	970*	390 ± 340	950	20	Ceccarelli et. al., [49]
2 µg/cm³	1.8 mM Ca²⁺, 4 mM Mg²⁺	1180*	350 ± 312	990	190	
α-LTx 9-10 °C 1 µg/cm³ 0 mM Ca²⁺, 4 mM Mg²⁺	10 min	120	1700 ± 950	650	<0	Hurlbut et. al., [50]
	20 min	680	1313 ± 613	1040	<0	
	40 min	1440	338 ± 175	2012	<0	
	75 min	1700	88 ± 75	2270	<0	

Notes: (‡) Vesicles lost = resting number - remaining. (†) Resting vesicle population not determined in this particular batch of frogs. (*) The length of the junction recorded from was not measured (secretion per active zone computed assuming 600 active zones per terminal). (●) Corresponding to ≈ 0.2 µg/cm³ α-LTx. N.D.: not determined.

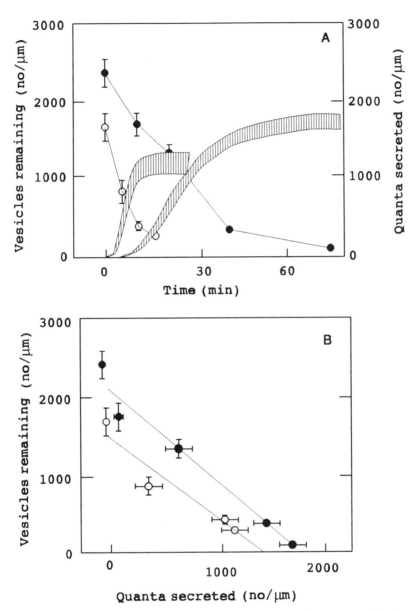

FIG. 4. Upper panel (A). Number of vesicles remaining per micrometre of nerve terminal and number of quanta secreted per micrometre at frog neuromuscular junctions stimulated by α-Latrotoxin in the absence of extracellular calcium (a condition where vesicle recycling is blocked); open circles and hatched curve to the left, 22-23 °C; filled circles and hatched curve to the right, 9-10 °C. The bars and hatchmarks indicate standard errors. Lower panel (B): correlation between remaining vesicles and quanta secreted (open circles, 22-23 °C; filled circles, 9-10 °C). Bars are standard errors and the lines are regression curves at the two temperatures (slopes= 1.11 and 1.21, respectively). (Modified from Ref. 50).

4.   Modulation of quantal release

The number of quanta released by a nerve terminal in response to the arrival of an action potential is not fixed: ionic conditions, neurotransmitters, drugs and previous activity may either facilitate or depress quantal release.

4.1. Modulation by previous activity

In the period that follows the delivery of a single stimulus to the nerve, quantal release is facilitated, and facilitation from successive stimuli adds up. This facilitation regards not only evoked quantal release, but also spontaneous, asynchronous quantal release. The facilitation of release decays with time and at least four different phases can be distinguished in this decay; the different phases have been interpreted as resulting from different regulatory phenomena and have been called first and second component of facilitation (with decay time constants of 50 and 300 ms, respectively), augmentation (7 s time constant) and potentiation (> 20 s) [51]. The longer-lasting facilitatory phenomena, and in particular potentiation, are not evident after single stimuli, as their magnitude is small, but they become relevant and dominate the facilitatory processes after prolonged repetitive stimulation, as the effects of many successive stimuli add up during the lifetime of augmentation, and even more so for potentiation. Thus, the latter phenomenon is often referred to as post-tetanic potentiation.

The mechanisms that sustain activity-induced facilitatory phenomena are not well understood. The first component of facilitation, and possibly the second one as well, are generally believed to be related to the lifetime of calcium ions that have entered the nerve terminal during the action potential (*residual calcium* mechanism). Thus, the time constant of decay of facilitation is likely to be related to the time constants of calcium diffusion from the sub-plasmalemmal region and of extrusion across the plasma membrane and sequestration into the intracellular calcium store compartments (see Ref. 52). Much less is known on the cellular basis for augmentation and post-tetanic potentiation: activation of calcium-dependent protein kinases and phosphorylation of proteins involved in ion fluxes, in cytoskeletal assembly, in the interactions of vesicle membrane with the cytoskeleton or the presynaptic membrane, have been proposed as possible mechanisms. In the squid giant synapse, electrophysiological studies, combined with fluorometric measurements of calcium concentration in the bulk of the nerve terminal cytoplasm, indicate that increases in the latter, produced by a tetanus, decay with the same time constant as augmentation does, and augmentation is abolished if slow calcium chelators (EGTA) are present intracellularly during the tetanus. Among the metabolic calcium-dependent changes that may be involved in augmentation and potentiation, the possible role of synapsin I (see VALTORTA *et. al.*, this volume) and the facilitatory effect of the calcium-sensitive protein-kinase C, which phorbol esters activate, mimicking diacyl-glycerol [53, 54], must be mentioned.

Electrical stimulation of the nerve elicits a negative modulation of quantal release as well (synaptic depression). The facilitatory phenomena are apparent in quantal release when its magnitude is low, as is the case in many central synapses or when neuromuscular junctions are bathed in low calcium and/or elevated magnesium concentration. At the neuromuscular junction in physiological ionic conditions, facilitatory phenomena are not very apparent as they are masked or even reversed by synaptic depression.

Facilitatory mechanisms are dependent on the extent of previous stimulation but not necessarily on the amount of neurotransmitter released. Indeed, post-tetanic potentiation is clearly elicited by repetitive stimulation even in the absence of calcium, provided high concentrations of magnesium are present; MEPP frequency increases noticeably although no evoked activity (EPP) is detectable, and on readmission of calcium evoked release is clearly potentiated [55]. Synaptic depression is instead particularly marked when quantal release is intense, and this has lead to the hypothesis that it arises from depletion of the store of readily releasable transmitter quanta [7]. However, statistical analysis of quantal content of EPPs, during repetitive stimulation that gives rise to synaptic depression, indicates that the probability of release is reduced as well [56]. This observation contributes to the difficulty in compelling current information on evoked quantal release into a fully consistent statistical model.

The simultaneous presence of facilitatory and depressory phenomena during repetitive stimulation of the nerve accounts for the marked differences in the responses of pathologically altered neuromuscular junctions to tetanic stimulation. Two forms of neuromuscular pathology, leading to muscular weakness (myasthenia), have been well characterized in man. At normal neuromuscular junctions the number of quanta of ACh released by nerve impulse is far in eccess with respect to the minimum quantity necessary to produce a current in the acetylcholine receptor (AChR) sufficient to trigger the muscle fibre action potential, and contraction. This *safety factor* in neuromuscular transmission is heavily reduced in classical *Miasthenia gravis* by auto-immune degradation of AChR at the endplate. Thus, although single nerve impulses elicit muscle action potentials and contraction, during tetanic stimulation synaptic depression lowers quantal release to a level that may not be sufficient to trigger action potentials and contraction in many muscle fibres, and the patient cannot sustain a strong, prolonged muscle contraction. In the LAMBERT-EATON myasthenic syndrome (LEMS), on the other hand, neuronal calcium channels are the target of an auto-immune reaction. Thus, less calcium enters the nerve terminal at the arrival of the action potential and quantal release may be reduced to the point that neuromuscular transmission is compromised. Under these conditions, as in isolated neuromuscular preparations exposed to low calcium, facilitatory effects predominate during repetitive stimulation, and efficient neuromuscular transmission may therefore be reestablished by tetanic stimulation. Consequently, in these patients the efficiency of neuromuscular transmission improves during a tetanus, unlike the case of *myasthenia gravis*.

Long-term changes (days or weeks) in synaptic efficiency are observed in specific synapses (*e.g.* long-term potentiation in hippocampal neurons); these changes are often mediated by the activation of a glutamate receptor subtype, the so-called N-methyl-D-aspartate (NMDA) receptor, and appear to involve presynaptic as well as postsynaptic changes. Though the question is still controversial [57, 58], presynaptic changes may occur following activation of metabolic processes triggered either by glutamatergic autoreceptors or by transmitter substances released back by the postsynaptic site (the most likely candidate appears to be nitrous oxide, NO, an activator of guanylyl cyclase and possibly other enzymes or processes, see for example Ref. 59).

4.2. Modulation by transmitters and drugs

An extensive series of mediators have been shown to interfere with acetylcholine release from the neuromuscular junction [60]. Among these are acetylcholine itself, through both muscarinic (facilitatory) and nicotinic (inhibitory) receptors [61]; substance P, a peptidic transmitter contained in large dense-core vesicles in motor nerve endings [62]; ATP, which is contained at high concentrations in synaptic vesicles at the neuromuscular junction and appears to potentiate spontaneous release whereas it inhibits nerve-stimulus evoked release [63]; various metabolites of arachidonic acid and several drugs that interact more or less specifically with calcium fluxes, kinases and phosphatases [66, 67], cyclic nucleotides [68] or calcium binding proteins [69]. The overall picture is complex and the relevance of the mechanisms involved in these effects to the normal function of the neuromuscular junction is not clear.

5.   The fusion of synaptic vesicles with the axolemma to release transmitters

Several extensive studies have been carried out in order to elucidate the mechanisms of exocytosis [64-65] and in particular of the fusion between biological membranes [66]. It is not clear, however, whether release of neurotransmitter by exocytosis involves a direct interaction between two biological membranes, as can be modelled and studied using artificial planar lipid bilayers, or rather a protein–mediated interaction, membrane proteins being responsible for reciprocal recognition and possibly for vesicle anchoring and formation of a hydrophilic pore. Furthermore it is not clear whether true fusion must always occur, with intermixing of the lipid matrix, or exocytosis may also occur through a transiently open pore in a proteic complex.

A nice approach to this problem was introduced by NEHER and MARTY [67], using the patch clamp technique to measure cell membrane capacitance (proportional to surface area), thus monitoring the fusion of granules in chromaffine or mast-cells. The addition of the membrane surface of a secretory granule to the plasmalemmal surface area, which accompanies the fusion of the granule to exocytose its content, is detected as a sudden increase in cell membrane capacitance. In mast-cells, BRECKEN-RIDGE and ALMERS [68] showed that the fusion event is accompanied by a complex series of electrical events, as depicted in Fig. 5: a transient of capacitive current occurs first, which is due to charging of the membrane of the granule to bring the potential drop across its membrane to the same potential difference as the rest of the plasmalemma (membrane potential); afterwards, a measurable increase is detected in the cell membrane capacitance. The time constant of decay of the capacitive transient is given by $\tau = RC$, where C is the granule membrane capacitance and R is the series resistance through which this capacitance is charged; in particular, the value of R indicates that the granule membrane is charged through a pore with electrical conductance ≈ 300 pS, a value similar to the conductance of a typical membrane ion channel with 7 nm internal diameter pore. Furthermore, the increase in membrane capacitance may initially be transiently reversible, giving rise to *flickering* of capacitance between the two levels. All this suggests that continuity between the extracellular fluid and the interior of the granule is initially established through a typical proteic pore, which may transiently close again, whereas intermixing of lipids, which leads to integration of the granule membrane into the plasmalemma, follows later.

These observations cannot be generalized in a straightforward manner to any exocytotic event, and in particular to exocytosis mediated by small synaptic vesicles, as the membrane of these organelles has a protein composition rather different from big granules [69]. However, these data do suggest that two phases can be identified in the exocytotic process:

(*a*) proteic interaction of the two membranes with opening of a water-permeable pore between the interior of the secretory organelle and the extracellular fluid and

(*b*) fusion of the lipid matrix of the two membranes. This aspect is of interest because controversial evidence has accumulated on the fate of the synaptic vesicle membrane after the exocytotic fusion. Whereas some experiments at the frog neuromuscular junction indicate that recycling of the vesicle membrane requires the formation of clathrin-coated pits and vesicles to pinch out fragments of plasmalemma, which are then accumulated into intracellular cisternae and large vesicular bodies, from which new synaptic vesicles form [28], other experiments in the same preparation suggest that the vesicle may recycle directly from the plasmalemma after exocytotic fusion [27,35,70]. It has been proposed that the two mechanisms of recycling coexist, i.e. that vesicle fusion may be reversible under conditions of mild to mo-

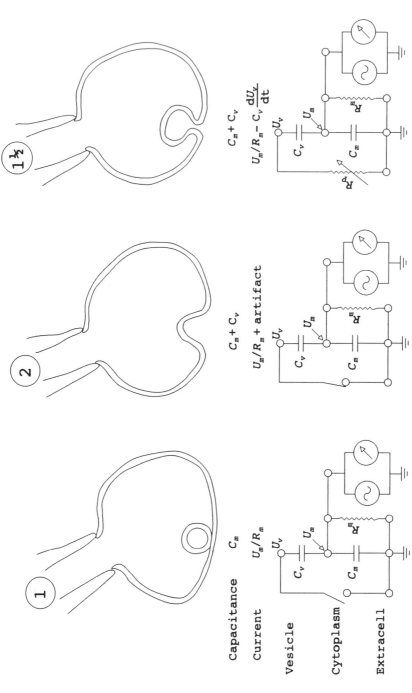

Fig. 5. Diagrams representing a patch-clamp pipette attached to a secretory cell, containing a vesicle or granule before (1), after (2) and during (1 1/2) fusion with the plasmalemma. The corresponding equations for membrane capacitance and current, and the simplified equivalent circuits for the three conditions are reported below each panel. U_v = potential difference between the interior of the vesicle and the bathing solution (grounded); U_m = membrane potential, C_v and C_m = capacitance of the vesicle and cell membrane, respectively; R_m and R_p = membrane and pore resistance. The generator with galvanometer depicted to the right represents the patch-clamp pipette and amplifier. For explanation, see text.

derate activation of release, but their membrane gets mixed with the plasmalemma under conditions of strong stimulation [71,72], thereby requiring sorting of specific proteins into small areas of plasmalemma and retrieval of these patches (sorting and retrieval of membrane proteins are the typical functions of coated pits and vesicles). This interpretation is in agreement with the observations that typical *omega-shaped* images of vesicles collapsing into the axolemma are not observed in motor nerve terminals quick frozen 2 to 10 ms after the delivery of single stimuli to the nerve [35], and that no relevant presence of vesicle-specific membrane proteins is detected in the presynaptic membrane of preparations stimulated by black widow toxin so that prolonged and intense release is induced without marked depletion of the vesicle population of the nerve terminal [70,73].

Thus, it appears that a rich biochemical machinery is involved in exocytotic release of neurotransmitters: a structural specialization is present at the *active zones*, where vesicle fusion preferentially occurs, and specific proteins have been characterized in the vesicle membrane (see VALTORTA and BÄHLER, this volume) that reversibly link vesicles to the cytoskeleton (synapsins), or that may play the role of *docking* the vesicle membrane to the plasmalemma (synaptotagmin), or that are capable of constituting oligomeric complexes behaving as large ion channels (synaptophysin). As opposed to the classical study of model systems for membrane fusion, the view that the approach of the vesicle to the presynaptic membrane is under the control of a complex biochemical system, and that protein-mediated, transient opening of the interior of the synaptic vesicle to the synaptic cleft may occur under physiological operation of the synapse, offers an energetically more favourable interpretation of the process, as extensive rearrangement of the lipid matrix of the vesicle membrane, and the work to retrieve and reconstitute the vesicle, would be saved under conditions of moderate activation of release. The study of the fast events that accompany neurotransmitter release, which was pioneered by biophysicists and physiologists, has now become a multidisciplinary field of research. Many issues still need to be clarified, and we look forward to what modern biochemistry and molecular biology will be able to tell us in the near future about the mechanisms of neurotransmitter release and its regulation.

References

[1]   P. FATT and B. KATZ, *Nature (London)*, **166,** 597 (1950).

[2]   P. FATT and B. KATZ, *J. Physiol.*, **115,** 320 (1951).

[3]   J. DEL CASTILLO and B. KATZ, *J. Physiol.*, **124,** 560 (1954).

[4]   I.A. BOYD and A.R. MARTIN, *J. Physiol.*, **132,** 74 (1956).

[5]   A.W. LILEY, *J. Physiol.*, **132,** 650 (1956).

[6]   J. DUDEL and S.W. KUFFLER, *J. Physiol.*, **155,** 514 (1961).

[7]   A.W. LILEY and K.A.K. NORTH, *J. Neurophysiol.*, **16,** 509 (1963).

[8] R.S. ZUCKER, *J. Physiol.*, **229**, 787 (1972).

[9] A.R. MARTIN, *J. Physiol.*, **130**, 114 (1955).

[10] E.M. McLACHLAN and A.R. MARTIN, *J. Physiol.*, **311**, 307 (1981).

[11] T.H. BROWN, D.H. PERKEL and M.W. FELDMAN, *Proc. Natl. Acad. Sci. U.S.A.*, **73**, 2913 (1976).

[12] K.R. COURTNEY, *J. Theor. Biol.*, **73**, 285 (1978).

[13] D. ELMQUIST and D.M.J. QUASTEL, *J. Physiol.*, **178**, 505 (1965).

[14] D.H. PERKEL and M.W. FELDMAN, *J. Mathemat. Biol.*, **7**, 31 (1979).

[15] B. WALMSLEY, F.R. EDWARDS and D.J. TRACEY, *J. Neurosci.*, **7**, 1037 (1987).

[16] H. KORN and D.S. FABER, *Synaptic Function*, G.M. EDELMAN, W.E. GALL and W.M. COWAN (Editors), JOHN WILEY and SONS (Editors), New York (1987) p. 57.

[17] J. DEL CASTILLO and B. KATZ, *J. Physiol.*, **132**, 630 (1956).

[18] A.R. MARTIN, *Physiol. Rev.*, **46**, 51 (1966).

[19] E.D.P. DE ROBERTIS, and H.S. BENNET, *Federation Proc.*, **13**, 35 (1954).

[20] G.E. PALADE, *Anat. Rec.*, **118**, 335 (1954).

[21] S.L. PALAY, *Anat. Rec.*, **118**, 336 (1954).

[22] F.S. SJÖSTRAND, *J. Appl. Phys.*, **24**, 1422 (1953).

[23] R.M. MARCHBANKS, *Neurosci*, **1**, 83 (1978).

[24] J.E. HEUSER, T.S. REESE and D.M.D. LANDIS, *J. Neurocytol.*, **3**, 109 (1974).

[25] B. CECCARELLI, F. GROHOVAZ and W.P. HURLBUT, *J. Cell Biol.*, **81**, 163 (1979).

[26] B. CECCARELLI, F. GROHOVAZ and W.P. HURLBUT, *J. Cell Biol.*, **81**, 178 (1979).

[27] B. CECCARELLI, W.P. HURLBUT and A. MAURO, *J. Cell Biol.*, **57**, 499 (1973).

[28] J.E. HEUSER and T.S. REESE, *J. Cell Biol.*, **57**, 315 (1973).

[29] H. ZIMMERMANN and C.R. DENSTON, *Neurosci.*, **2**, 715 (1977).

[30] S.W. KUFFLER and D. YOSHIKAMI, *J. Physiol.*, **240**, 465 (1975).

[31] B. KATZ and R. MILEDI, *Proc. Roy. Soc. Lond. B*, **196**, 59 (1977).

[32] A. GORIO, W.P. HURLBUT and B. CECCARELLI, *J. Cell Biol.*, **78**, 716 (1978).

[33] J.E. HEUSER, T.S. REESE. and D.M.D. LANDIS, *Cold Spring Harbor Quant. Biol.*, **40**, 17 (1976).

[34] J.E. HEUSER, T.S. REESE, M.J. DENNIS, Y. JAN, L. JAN and L. EVANS, *J. Cell Biol.*, **81**, 275 (1979).

[35] F. TORRI-TARELLI, F. GROHOVAZ, R. FESCE, and B. CECCARELLI, *J. Cell Biol.*, **101**, 1386 (1985).

[36] R. LLINÁS, I.Z. STEINBERG and K. WALTON, *Biophys. J.*, **33**, 323 (1981).

[37] J.R. SEGAL, B. CECCARELLI, R. FESCE and W.P. HURLBUT, *Biophys. J.*, **47**, 183 (1985).

[38] R. FESCE, J.R. SEGAL and W.P. HURLBUT, *J. Gen. Physiol.*, **88**, 25 (1986).

[39] R. FESCE, J.R. SEGAL, B. CECCARELLI and W.P. HURLBUT, *J. Gen. Physiol.*, **88**, 59 (1986).

[40] R. FESCE, *Progr. Neurobiol.*, **35**, 85 (1990).

[41] J.E. HEUSER and R. MILEDI, *Proc. R. Soc. London B*, **179**, 247 (1971).

[42] B. KATZ and R. MILEDI, *J. Physiol.*, **224**, 665 (1972).

[43] B. CECCARELLI, R. FESCE, F. GROHOVAZ and C. HAIMANN, *J. Physiol.*, **41**, 163 (1988).

[44] M.L. ROSSI, C. BONIFAZZI, M. MARTINI and R. FESCE, *J. Gen. Physiol.*, **94**, 303 (1989).

[45] M.L. Rossi, C. Bonifazzi, M. Martini and R. Fesce, *J. Gen. Physiol.*, in press. (1993).

[46] N. Frontali, B. Ceccarelli, A. Gorio, A. Mauro, P. Siekevitz, M.C. Tzeng and W.P. Hurlbut, *J. Cell Biol.*, **68**, 462 (1976).

[47] W.P. Hurlbut and C. Ceccarelli, *Adv. Cytopharm.*, **3**, 87 (1979).

[48] B. Ceccarelli and W.P. Hurlbut, *J. Cell Biol.*, **87**, 297 (1980).

[49] C. Haimann, F. Torri-Tarelli, R. Fesce and B. Ceccarelli, *J. Cell Biol.*, **101**, 1953 (1985).

[49 b] B. Ceccarelli, W.P. Hurlbut and N. Iezzi, *J. Physiol.*, **402**, 195 (1988).

[50] W.P. Hurlbut, N. Iezzi, R. Fesce and B. Ceccarelli, *J. Physiol.*, **425**, 501 (1990).

[51] K.L. Magleby, in *The Cholinergic Synapse. Progr. Brain Res.*, **49**, 175 (1979) S. Tucek (Editor) Elsevier Amsterdam.

[52] H. Parnas and L.A. Segel, *Progr. Neurobiol.*, **32**, 1 (1989).

[53] C. Haimann, J. Meldolesi and B. Ceccarelli, *Pflügers Arch.* **408**, 27 (1981).

[54] R. Shapira, S.D. Silberberg, S. Ginsburg and R. Rahamimoff, *Nature (London)*, **325**, 58 (1987).

[55] W.P. Hurlbut, H.B. Longenecker and A. Mauro, *J. Physiol.*, **219**, 17 (1971).

[56] B.N. Christensen and A.R. Martin, *J. Physiol.*, **210**, 933 (1970).

[57] G.L. Collingridge and W. Singer, *Trends Pharm. Sci.*, **11**, 290 (1990).

[58] F.A. Edwards, *Nature (London)*, **355**, 21 (1990).

[58 b] A.M. Malgaroli and R.W. Tsien, *Nature (London)*, **357**, 134 (1992).

[59] T. McCall and P. Vallance, *Trends Pharm. Sci.*, **13**, 1 (1992).

[60] B.L. Ginsborg and D.H. Jenkinson, *Neuromuscular Junction. Handb. Exper. Pharmacol.*, **42**, 229 (1976) E. Zaimis, (Editor), Springer-Verlag.

[61] W.C. Bowman, I.G. Marshall, A.J. Gibb and A.J. Harborne, *Trends Pharm. Sci.*, **9**, 16 (1988).

[62] A. Steinacker, *Nature (London)*, **267**, 268 (1977).

[63] W. Fu and M. Poo, *Neuron*, **6**, 837 (1991).

[64] E.M. Silinsky, *Pharmacol. Rev.*, **37**, 81 (1985).

[65] R.C. De Lisle and J.A. Williams, *Annu. Rev. Physiol.*, **48**, 225 (1986).

[66] G.E. Palade, *Science*, **189**, 347 (1975).

[67] E. Neher and A. Marty, *Proc. Natl. Acad. Sci U.S.A.*, **79**, 6712 (1982).

[68] L.J. Breckenridge and W. Almers, *Proc. Natl. Acad. Sci. U.S.A.*, **84**, 1585 (1987).

[69] P. De Camilli and R. Jahn, *Annu. Rev. Physiol.*, **52**, 625 (1990).

[70] F. Valtorta, R. Jahn, R. Fesce, P. Greengard and B. Ceccarelli, *J. Cell Biol.*, **107**, 2717 (1988).

[71] J. Meldolesi and B. Ceccarelli, *Phil. Trans. R. Soc. Lond. B*, **296**, 55 (1981).

[72] F. Torri-Tarelli, C. Haimann and B. Ceccarelli, *J. Neurocytol.*, **16**, 205 (1987).

[73] F. Torri-Tarelli, A. Villa, F. Valtorta, P. De Camilli, P. Greengard and B. Ceccarelli, *J. Cell Biol.*, **110**, 449 (1990).

LOW MOLECULAR WEIGHT NEUROTRANSMITTERS

MONIKA Z. WRONA and GLENN DRYHURST

Department of Chemistry and Biochemistry
University of Oklahoma, Norman, OK 73019
U.S.A.

Contents

Bioelectrochemistry IV
Edited by B.A. Melandri *et al.*, Plenum Press, New York, 1994

1.   Introduction.

Our lives are critically dependent on our ability to sense the environment and world around us and to use this sensory information to guide our actions or physical movements. Our sensory system is designed to detect physical phenomena such as light, sound, heat or pressure and chemical stimuli by means of taste or smell. The devices employed to detect such stimuli are called primary sensory receptors. These sensory receptors convert the environmental stimulus into an electrical signal or nerve impulse which travels along a sensory neuron to a primary processing center where information from the particular form of sensing is initially evaluated. The frequency of the impulses transmitted along a particular sensory neuron reflects the intensity of the stimulation. For example, if a sensory receptor in the skin of the hand senses a very hot surface a burst of high frequency electrical impulses pass along the sensory nerve fiber, to which the receptor is attached, to the spinal cord which rapidly then initiates a burst of nerve impulses along appropriate motoneurons to muscle systems that act to remove the hand from the heat source. This is a very rapid reflex action designed to protect the hand from damage. However, the brain and spinal cord, *i.e.*, the central nervous system, continuously receives sensory information which is ultimately processed in cortical regions of the brain. A combination of voluntary decisions based upon past experience, which is stored in the central nervous system, and information provided from all sensory inputs is employed to decide upon an appropriate motor action. This information is transferred to the peripheral musculature *via* motoneurons which often project from the spinal cord and, to a somewhat lesser extent, directly from encephalic structures in the brain-system. The motoneurons carry nerve impulses to the muscles and cause the muscle to contract. Thus, neurons carry sensory information to the central nervous system. After appropriate evaluation and processing, neurons then carry instructions from the central nervous system to muscles to effect appropriate movements. Clearly, therefore, neurons or nerve cells are the biological entities responsible for the transfer and processing of all sensory and motor information.

2.   The neuron.

Each neuron is separated from direct physical contact with other neurons but, nevertheless, it can and must communicate with other neurons. The language of this communication is part electrical and part chemical. A schematic representation of a neuron is shown in Fig. 1. Thus, a neuron clearly consists of several distinct and specialized segments. The *cell body* or *soma* contains a nucleus which in turn contains the chromosomes that code the cellular genetic information in DNA. The nucleus also contains a nucleolus packed with RNA. The nucleus is surrounded by the cytoplasm which contains

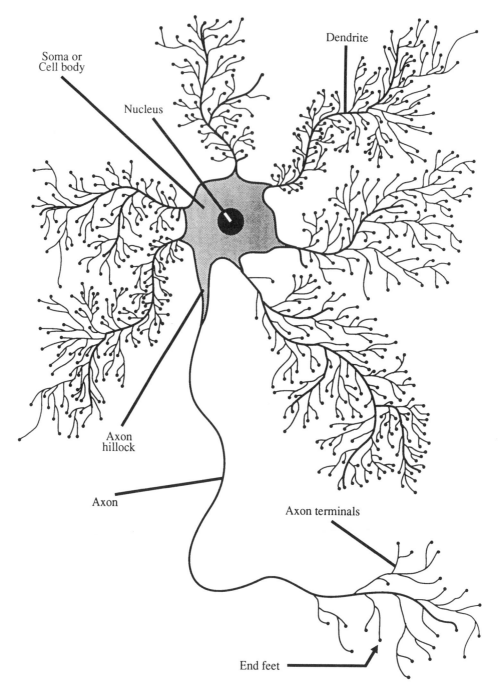

Soma or
Cell body

Nucleus

Dendrite

Axon
hillock

Axon

Axon terminals

End feet

Fig. 1. Schematic representation of a mammalian neuron or nerve cell.

the mitochondria, an endoplasmic reticulum, Golgi apparatus, ribosomes and lysosomes. The *dendrites* can be regarded as extensions of the soma. Some neurons have many dendrites while others have relatively few. Dendrites often branch profusely as shown in Figure 1. Projecting downward from the soma of the neuron shown in Fig.1 is a single axon that often ends in many small branches called *axon terminals* or, sometimes, *end feet*. In a general sense the dendrites and soma can be regarded as the input end of a neuron. The axon represents the output end of a neuron. The information received by the dendrites and soma is summed or integrated at the *axon hillock*. Under certain circumstances this summed information can trigger a signal or nerve impulse which is passed along the axon fiber to its terminals. This signal is then transmitted from these axon terminals to other neurons via connections generally with their dendrites or soma.

In the sections which follow several key aspects of neuronal signalling or transmission will be discussed:

(1) the mechanisms by which signals are collected by dendrites and soma;

(2) how this information is summed at the *axon hillock* and the conditions necessary for propagation of the signal along the axon fiber to its terminals; and;

(3) how information transfer or signalling occurs between the axon terminals of one neuron and the dendrites and soma of another neuron. It must be stressed that the intent of the information presented in these sections has been to provide the reader with only an introduction to the basic concepts of neuronal transmission. Many of the topics touched upon represent areas of intensive current research and are often only incompletely understood at this time. The interested reader is encouraged to consult more detailed and comprehensive monographs and the current research literature in order to gain a more thorough background.

2.1   The action potential or nerve impulse.

The membrane that ensheathes a nerve cell axon (and indeed the entire neuron) consists of a bilayer of phospholipid molecules. These phospholipids consist of a long hydrophobic hydrocarbon chain with a polar phosphate head. Thus, in the bilayer phospholipid membrane the polar heads face outward to the extracellular aqueous fluid and inward to the intracellular aqueous fluid (Fig. 2). The hydrophobic tails are directed to the inner regions of the membrane. The inner, hydrophobic region of the membrane is largely impermeable and hence provides a barrier to the free movement of ions. Proteins are present on the surfaces of the membrane, in part, to provide structural stability. However, some proteins completely penetrate the membrane and are involved with the transport of ions and other chemicals across the membrane. Other proteins act as receptors for substances that affect transport mechanisms and yet others are enzymes. A schematic representation of a neuronal membrane is presented in Fig. 2. The thickness of the membrane which ensheaths a neuron is about 7 nm [1–3].

The axon membrane separates the intracellular and extracellular fluids each of which

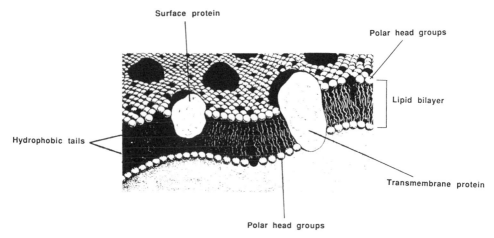

FIG. 2. Schematic representation of a neuronal membrane.

contains many charged ionic species. Of particular importance to an understanding of the *action potential* and electrical conduction along the axon fiber and to the effects of key receptor proteins are Cl^-, and large organic anions (*e.g.*, proteins), Na^+ and K^+ cations. The concentrations of these ions are normally quite different in the intracellular and extracellular fluids that bathe the membrane. For example, the concentration of organic anions in the intracellular fluid is approximately 300 times greater than in the extracellular fluid [4]. Similarly, the concentration of K^+ in the intracellular fluid is about 30 times greater than in the extracellular fluid. In contrast the intracellular concentrations of Na^+ and Cl^- are about 20 times and 10 times smaller, respectively, than their extracellular concentrations. There are several reasons which account for the unequal concentrations of these ions in the intraaxonal and extraaxonal fluids. Organic anions (*e.g.*, Protein⁻) are retained in the intracellular fluid simply because they are too bulky to pass through the membrane. The negative charge of these organic anions in the axoplasm must be balanced by reduced concentrations of Cl^- and/or by increased concentrations of Na^+ or K^+. In addition, the axonal membrane possesses a pumping system (active transport mechanism) called the sodium-potassium ion pump (Na^+-K^+ pump) that uses metabolic energy to pump Na^+ ions out of the neuron and K^+ ions into the neuron. The various ionic concentration gradients and the role of the Na^+-K^+ pump are illustrated in Fig.3. In the resting state the rate of transport of Na^+ out of the axon and K^+ into the axon against their concentration gradients by the ion pump is equal and opposite to the rate at which these ions diffuse across the semipermeable membrane down their concentration gradients. Because large organic anions, Cl^-, K^+ and Na^+ are distributed unequally across the axonal membrane there is a potential difference across the membrane which is often referred to as the *transmembrane potential*. The transmembrane potential can be measured for various large neurons by inserting one microelectrode through the membrane into the intracellular fluid and recording the potential difference between this electrode and a second identical electrode

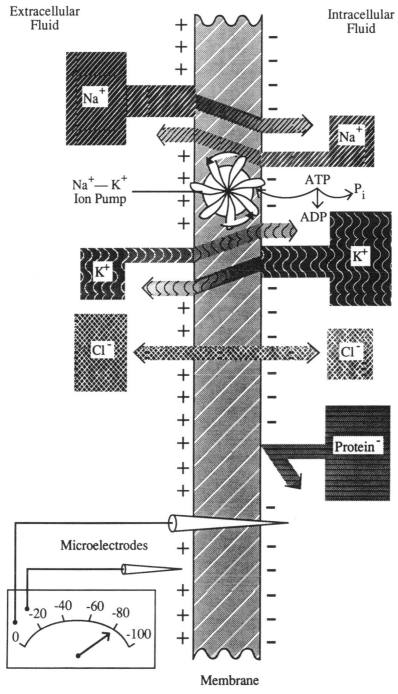

FIG. 3. Ion distrubutions across an axon membrane. The seizes of the boxes Na$^+$, K$^+$, Cl$^-$ and Protein$^-$ on either side of the membrane are indicative of the relative concentrations of these ions.

Stimulating electrode   Sensing electrode

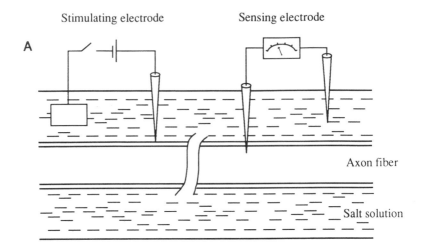

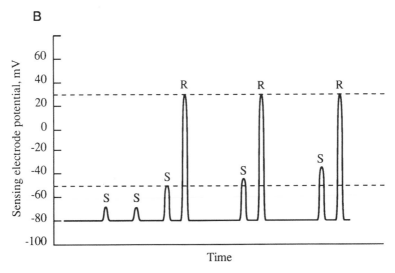

FIG. 4. A Experimental arrangements for electrical stimulation of the giant axon of the squid and recording, S representing the stimulating signal, R the response detected by the sensing electrodes.

immersed in the extracellular fluid. The measured potential difference is typically about 80 mV with the inside of the axon membrane being negative with respect to the outside. Thus, the transmembrane potential is designated – 80 mV. The *resting membrane potential* basically reflects the constant effect of the Na+–K+ pump, which maintains the unequal concentrations of these ions across the membrane.

The experimental arrangement illustrated in Fig. 4A represents an important approach to understanding the action potential or nerve impulse. In classic experiments of this type [5-6] the giant axon of the squid was employed which has a diameter of 0.5 – 1 mm and a membrane thickness of about 7 nm. The axon fiber in Fig 4A is immersed in a salt solution. To the left of the figure is a small stimulating electrode, which just touches the outer axon surface, and through which a small electrical current can be passed. To the right of the figure, and far away from the stimulating electrode, is a recording microelectrode which has penetrated into the intracellular fluid. The potential difference between the tip of this microelectrode and a second microelectrode in the extracellular solution is measured with a suitable high impedance potential–measuring device. Figure 4B shows that the initial potential difference recorded at the sensing electrode is – 80 mV, *i.e.*, the resting membrane potential. In order to understand the effects of electrical stimulation of the axon fiber it is necessary to introduce the concept of ion channels [6, 7]. Figure 5 is a representation of an axon membrane which shows the Na+ –K+ ion pump and the presence of a K+– channel and a Na+– channel. These channels are made of proteins which penetrate the membrane and, under certain conditions, are selectively permeable to either Na+ or K+ ions. In the resting state, *i.e.*, when the transmembrane potential is about – 80 mV the Na+- and K+- channels are closed.

Returning to Fig 4B, it can be seen that if the axon membrane is stimulated by a very small current with the stimulating electrode such that the transmembrane potential is decreased to about – 75 mV or – 60 mV there is no corresponding response detected by the sensing electrode system. However, when the stimulation is sufficiently large to de-

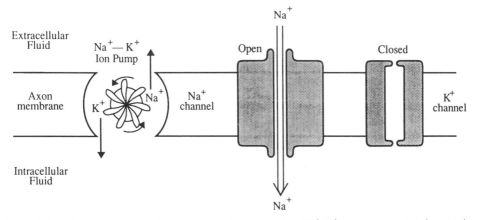

FIG. 5. Schematic representation of an axon membrane showing Na+–K+ ion pump and Na+–and K+ – channels or gates.

crease the transmembrane potential to about − 50 mV a very large response is noted at the sensing electrode (Fig. 4B). This response is caused by the fact that at − 50 mV the Na+– channels open and the axon membrane becomes completely permeable to Na+ ions at the site of stimulation. The K+–channels also open but much more slowly than the Na+– channels. The very rapid influx of Na+ ions causes the transmembrane potential to reach a value of about + 30 mV, *i.e.*, there is a reversal of potential difference The Na+ − channels rapidly inactivate but the K+– channels remain open so that the efflux of K+ ions continues. The combined effect of inactivation of the Na+– channels and the efflux of K+ ions through the open K+– channels causes the transmembrane potential to return towards its resting level. When the membrane is repolarized both the Na+– and K+– channels close, although the Na+–channel was already silent owing to inactivation. The Na+– K+ pump acts to restore the original ionic concentrations across the membrane. The potential difference which causes the Na+–channels to open and make the membrane completely permeable to Na+ ions is known as the *threshold potential*. The rapid reversal of potential difference (*i.e.*, shift from − 80 mV to + 30 mV) noted at the sensing electrode (due primarily to Na+ influx) and restoration of the resting potential (due primarily to K+ efflux) is called the *action potential*. Further increases in the stimulation applied to the axon by the stimulating electrode result in no additional increase in the measured response or action potential (Fig. 4B). This is a very important property of axons. Thus, every action potential or *nerve impulse* (see later discussion) fired by a particular axon is triggered by a fixed threshold potential. Once triggered the action potential always has an identical height or amplitude. However, the frequency of nerve impulses or action potentials can change. As noted previously, the frequency of nerve impulses increases with the intensity (or urgency) of the stimulus. This is an illustration of the *all-or-nothing* response of an axon to a stimulus. All that can be propagated along an axon is a full size response; the alternative is nothing. This all-or-nothing rule applies whether the axon is fired as a result of the interactions of receptors with neurotransmitters (see later discussion) or by a stimulating electrode. Finally, it should be noted that a decrease of the transmembrane potential (shift to more positive values) such as is caused by the stimulating electrode in Fig. 4 is known as *depolarization*. Thus, when the membrane is depolarized to a potential difference equal to (or more positive than) the threshold potential an action potential or nerve impulse is fired.

2.2   Conduction of the nerve impulse [5]

The electrical event that initially triggers an action potential normally originates in the soma (cell body). When the action potential fires at the junction between the axon hillock and the axon fiber (Fig. 1), with the transient fluxes of Na+ and K+ across the membrane, it causes adjacent segments of the membrane to become depolarized and so the action potential is propagated or glides along the axon. That this is so is clearly evident in Fig. 4A where the sensing electrode is located at a considerable distance from the stimulating electrode. This effect is illustrated in Fig. 6 which shows that the propa-

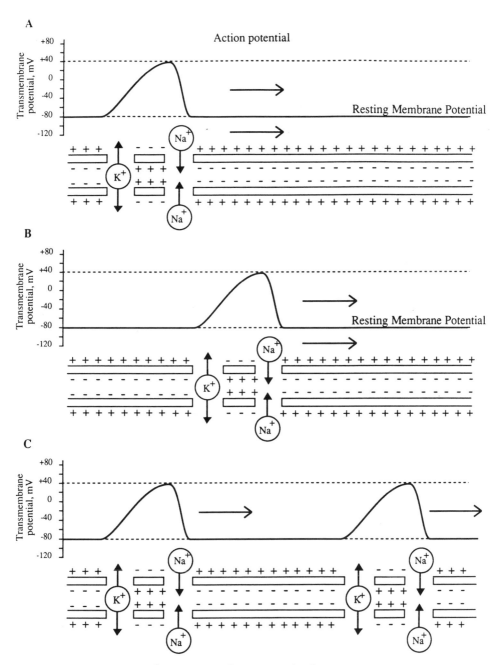

FIG. 6. Propagation of a nerve impulse along an axon.

gation of the nerve impulse coincides with the localized influx of Na^+ followed by of K^+ through their respective channels [8]. In Fig. 6C it can be observed that after a brief refractory period a second nerve impulse can follow the first down the axon fiber.

A nerve impulse does not travel at the same speed along all axons. The speed of the nerve impulse depends upon the resistance to current flow within the fiber and, obviously, impulse speed increases with decreasing resistance. The resistance to current flow decreases as the diameter of the axon fiber is increased. Thus, large axon fibers are faster nerve impulse conductors than small axon fibers. Clearly, much information must be passed along axons by nerve impulses often very rapidly. This, however, cannot generally be accomplished in practice simply by employing very large diameter fibers. Therefore, an alternative approach has evolved which employs narrow axon fibers which are wrapped in glial cells to speed nerve impulses. In the peripheral nervous system a type of glial cell known as Schwann cells are wrapped around some axons to form a compact, insulating sheath of myelin. In the central nervous system oligodendroglia are employed to form the myelin sheath. Between each glia cell sheath (*internode*, 1-2mm long) the membrane of the axon is exposed. Thus, along the length of the axon the membrane is periodically exposed by a gap in the myelin sheath. These gaps are known as the nodes of Ranvier (Fig. 7) . In such myelinated axon fibers the nerve impulses skip along the axon from one node to the next with the long internodal section being passive. This type of conduction is known as *saltatory conduction* [9, 10]. At each node the exposed nerve membrane has voltage–gated Na^+–and K^+–channels but at about a tenfold or greater density than they occur in an unmyelinated fiber [11]. As a general rule, a small myelinated axon can conduct a nerve impulse about as rapidly as an unmyelinated axon

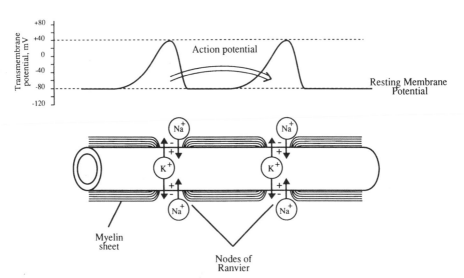

FIG. 7. Representation of a myelinated axon fiber and the process of saltatory conduction..

having a thirty times larger diameter. Mammalian peripheral nerves contain myelinated fibers with diameters of 0.5 to 20 μm which have conduction velocities of 3 to 120 m/s. By contrast, unmyelinated fibers with diameters of less than 2 μm have conduction velocities of 0.5 to 2 m/s. Clearly, therefore, saltatory conduction is an extremely effective method of transmitting nerve impulses along an axon. Nerve fibers responsible for carrying urgent information from peripheral sensory neurons to the central nervous system and from the central nervous system to peripheral muscles by motoneurons must necessarily be fast conductors of nerve impulses. Such neurons are therefore likely to be myelinated.

2.3. Dendrites

The preceding sections demonstrated that stimulation of an axon membrane by a small electrode can, provided it causes the transmembrane potential to reach the threshold potential, cause an action potential to discharge or fire and that this impulse can be conducted along the axon fiber. However, *in vivo*, electrical stimuli originating from quite different mechanisms are gathered by the dendrites and soma of the neuron. It is the stimuli gathered by these inputs to the neuron which ultimately decide if an action potential fires in the axon.

Dendrites and the soma have a membrane which in many respects is similar to that of an axon. Indeed, the resting potential across a dendritic membrane is essentially the same as that across an axon membrane. Similar to axon membranes, dendritic and somatic membranes also undergo changes in transmembrane potential when they are stimulated electrically or, *in vivo*, as a result of receptor activity (which will be discussed subsequently). When stimulated, dendrites do not generate an action potential, at least of the sort described above in connection with axons. (However, calcium spikes, *i.e.*, action potentials generated through sensitive calcium channels have been described). Rather, when a dendrite is appropriately stimulated the transmembrane potential changes from the resting potential in proportion to the intensity of the stimulation. In addition, the resulting potential decays as it moves away from the point of stimulation. The potential difference changes experienced by dendrites are known as *graded potentials* [4]. If the transmembrane potential difference decreases as a result of some stimulation (*i.e.*, becomes more positive than the resting membrane potential) it is referred to as *depolarization*. Conversely, when the transmembrane potential increases (*i.e.*, becomes more negative than the resting membrane potential) it is referred to as *hyperpolarization*. When a dendrite is stimulated such that it is depolarized at two points in very close proximity, the graded potentials generated at each point add. Similarly, if the two stimuli lead to hyperpolarization the two potentials would add. Thus, if the two stimuli produce identical depolarization (or hyperpolarization) the resulting graded potential will be twice as large as the two individual potentials. When the two similar stimuli occur at widely separated points on the dendrite the graded potentials dissipate before they reach each other and hence will not add. Stimuli given at intermediate distances will yield an additive graded potential but, because of the distance, the resultant graded potential will be less than the sum of the maximal individual potentials. When one stimulus leads to depolarization of the membrane at one point and a second results in hyperpolarization at another point the graded potentials subtract. The addition or subtraction of adjacent graded potentials is called spatial *summation.*

The graded potential resulting from stimulation of a dendrite always decays with time and ultimately disappears [4]. However, if a second similar stimulus occurs at the same site before the first potential has completely decayed away, the residual graded potential from the first stimulus will add to the second graded potential. Thus, the magnitude of the resulting graded potential will depend upon the intensity of the two individual stimuli and the time interval between them. This phenomenon is known as *temporal summation*. If the two stimuli applied at differerent times produce different graded potentials (*i.e.*, one depolarization, the other hyperpolarization) they will subtract to an extent dependent upon the time interval between them.

Each neuron normally receives many excitatory (depolarization) and inhibitory (hyperpolarization) stimuli at its dendrites and soma. When, as a result of both spatial and temporal summation, the resulting graded potential which arrives at the axon hillock is equal to the threshold potential (*e.g.*, *ca.* -50 mV), an action potential will fire in the axon fiber and be propagated along its length. The key site in the neuron, at which the decision to fire an action potential is arrived at, is the axon hillock (Fig. 1). If this area is depolarized to a potential at least equal to the threshold potential as a result of the spread of graded potentials the action potential will fire. Conversely, if *spatial* and *temporal* summation of individual graded potentials which reach the axon hillock are below the threshold potential an action potential will not fire.

3.   Chemical neurotransmission

The above discussion has demonstrated the ways in which an action potential fires and how the nerve impulse travels down an axon to its axon terminal. Furthermore, it is clear that the action potential is triggered as a result of spatial and temporal summation of electrical graded potentials resulting from stimuli at various points on the dendrites and soma of the neuron. It now, therefore, becomes necessary to understand the nature of the stimuli which lead to the generation of localized graded potentials. This, in turn, brings us to a discussion of the mechanisms by which neurons communicate with each other. Such communication is accomplished by a process called *chemical neurotransmission* which employs as its messengers special classes of chemicals known as *neurotransmitters*. As was noted earlier, a signal is carried along the axon fiber of a neuron by an electrical action potential or nerve impulse. The end of an axon fiber is swollen into a terminal or end foot. The membrane which encloses the axon terminal of the transmitting neuron is physically separated from the dendritic or somatic membrane of a second (receiving) neuron by a very narrow gap called a *synaptic cleft, synaptic gap* or *synapse*. A schematic representative of a synapse is shown in Fig . 8. The membrane of the axon terminal of neuron 1 is termed the *presynaptic membrane*. The membrane of the dendrite or soma (usually) of the second, receiving neuron is called the *postsynaptic membrane*. Within the body of the axon terminal are long fibers, or neurofibrils, structures which may be involved with the transport of chemicals, enzymes and other materials which are synthesized in the soma. This process is known as *axoplasmic flow* [12].The mitochondria are orga-

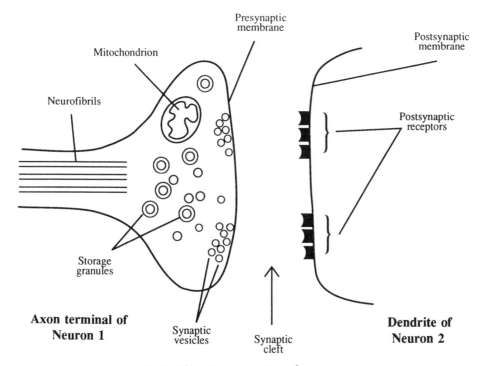

FIG. 8. Schematic representation of a synapse.

nelles responsible for providing the energy necessary for the metabolic processes which occur in the axon terminal. Granules and so-called large, dense-cored vesicles are thought to be used for the storage (and release) of peptide neurotransmitters or of protein-complexed low molecular weight neurotransmitters. Small, clear synaptic vesicles are believed to be employed for storage (and release) of small neurotransmitter molecules (*i.e.*, classical neurotransmitters). The postsynaptic membrane contains as an integral part of its structure complex proteins called *receptors* which possess the ability to bind weakly but very selectively with a particular neurotransmitter compound.

When the nerve impulse or action potential arrives at the axon terminal in neuron 1 (Fig.8) some of the synaptic vesicles release their neurotransmitters into the synaptic cleft. This rather complex sequence of events is accomplished by a process known as exocytosis [13]. Depolarization of the axon terminal membrane by a nerve impulse results in the opening of Ca^{2+}– channels in the presynaptic membrane with the result that Ca^{2+} in the vicinity of the synaptic cleft diffuses from the extracellular fluid into the nerve ending [14]. These Ca^{2+} ions combine with the protein *calmodulin* which is associated with the vesicle sheath [15, 16]. Calmodulin becomes activated by binding with four Ca^{2+} ions and in this form initiates a number of reactions [17, 18]. Some investigators have proposed that one of these reactions results in fusion of the synaptic vesicles to the presynaptic membrane to permit exocytosis of the neurotransmitters as conceptualized in Fig. 9. This is often referred to as the calmodulin hypothesis for exocytosis. However, calcium binding proteins specifically associated with synaptic vesicles have been characterized and may act as mediators of exocytosis. Nevertheless, the released neurotransmitter diffuses across the synaptic cleft and binds briefly to a receptor protein located in the postsynaptic membrane. Interaction of chemical neurotransmitters with postsynaptic receptors can result in the development of *excitatory postsynaptic potentials* (EPSPs) or *inhibitory postsynaptic potentials* (IPSPs) across the postsynaptic membrane [19]. (In some instances neurotransmitter-receptor interactions activate enzymatic cascades within the postsynaptic cell rather than ionic conductance changes on the postsynaptic membrane). In general the excitatory neurotransmitters evoke EPSPs by binding with their postsynaptic receptors which in turn results in Na^+ ion channels opening in the postsynaptic membrane. Thus, Na^+ enters the receiving neuron and hence reduces the transmembrane potential, *i.e.*, depolarizes the postsynaptic membrane. Depolarizing postsynaptic potentials increase the probability of the neuron firing an action potential as discussed earlier in connection with graded potentials. Inhibitory neurotransmitters interact with their postynaptic receptors and cause K^+– or Cl^- –channels to open. Thus, K^+ ions flow out of or Cl^- flow into the receiving neuron which has the effect of increasing the postsynaptic potential, *i.e.*, the transmembrane potential becomes more negative and the membrane is hyperpolarized. Hyperpolarizing potentials, or IPSPs, decrease the probability of a neuron ultimately firing an action potential. The channel-coupled receptors, of the type just described, are transmembrane proteins having a ligand-binding extracellular domain which constitutes a more or less selective ion channel. This channel is opened as a result of the conformational change that follows binding of the ligand (neurotransmitter) to the receptor site. The receptor site and channel constitute a protein complex known as a *ligand-operated channel*. These channels are distinct from voltage–sensitive channels described earlier.

It is clear from the above discussion that the graded potentials discussed with respect to dendrites and the soma can be traced to the release of neurotransmitters from the axon terminals of other neurons with which they synapse. Some neurotransmitter-receptor interactions result in EPSPs (depolarization) while others cause IPSPs (hyperpolariza-

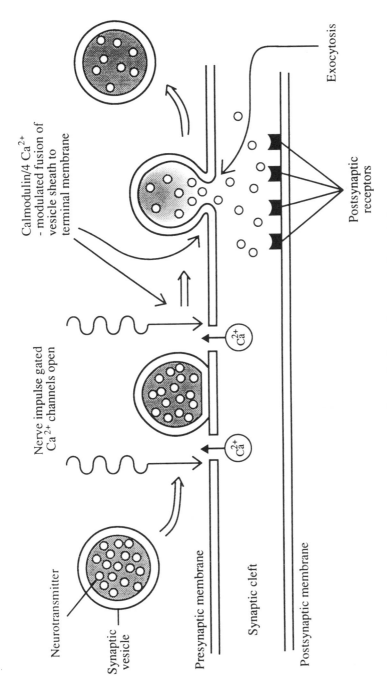

Fig. 9. Sequence of events leading to the process of exocytosis.

tion). There are often hundreds or even thousands of axon terminals synapsing with the dendrites (axo-dendritic), cell body (axosomatic) and sometimes even with the axons (axo-axonic) and synapses (axo-synaptic) of a single neuron. It has been estimated, for example, that the total number of synapses of all kinds on a single human cortical neuron can be as high as 40,000 [20]. Thus, it becomes clear that such a neuron can, potentially, receive an enormous number of EPSPs and IPSPs. These must be continuously integrated by the processes of spatial and temporal summation in order for the neuron to *decide* whether to fire an action potential along its own axon fiber. A very simplistic representation of this process is shown in Fig. 10. The transmembrane potential of the receiving neuron is measured with one microelectrode which is inserted into the axoplasm (intracellular fluid) with respect to a second microelectrode in the extracellular fluid. In Figure 10A none of the axon terminals which synapse with the neuron are firing and, hence the steady recorded potential corresponds to the resting membrane potential. In Figure 10B an impulse from excitatory fiber E_A results in a small EPSP which causes the transmembrane potential to decrease (become more positive). However, the magnitude of this EPSP is insufficient to reach the firing threshold potential and, hence, it simply decays away. When the excitatory impulse from fiber E_A is followed by a second excitatory impulse from excitatory fiber E_B (Fig 10C) the summed potential still fails to reach the threshold firing potential and the sum of the two potentials simply decays away. However, when impulses from excitatory fibers E_A, E_B and E_C follow sequentially, the summed potentials reaches the threshold potential (Fig. 10D) and, hence, an action potential fires and a nerve impulse glides along the axon fiber. In Figure 10E, the sequential excitatory impulses from fibers E_A, E_B and E_C are preceded by an inhibitory impulse from inhibitory fiber I_A. This inhibitory impulse generates an IPSP (*i.e.*, the transmembrane potential becomes more negative than the resting membrane potential) which, when substracted from the three excitatory impulses from fibers E_A, E_B and E_C prevents the summed potential from reaching the threshold value and an action potential does not fire.

It must be noted that neurons not only communicate with other neurons by the process of synaptic chemical neurotransmission but also with glands, other body organs and muscles. They accomplish this by releasing neurotransmitters from their axon terminals onto receptors present on the organs or muscles with which they synapse.

It has already been noted that the speed at which a nerve impulse travels along an axon fiber is dependent on the size (diameter) of the fiber and whether or not it is myelinated. In general nerve impulse speeds range from about 1 to 100 m/s [21]. The firing rate or frequency of nerve impulses ranges from less than 100 impulses per second to more than 1,000 impulses per second. Quite clearly, therefore, if the receptors on the postsynaptic membrane of the receiving neuron (or gland, organ or muscle) are to be capable of sensing each successive wave of neurotransmitter that is released as a result of such rapid firing rates, there must be mechanisms for very rapid removal or inactivation of the neurotransmitter substance. In other words once a neurotrans-

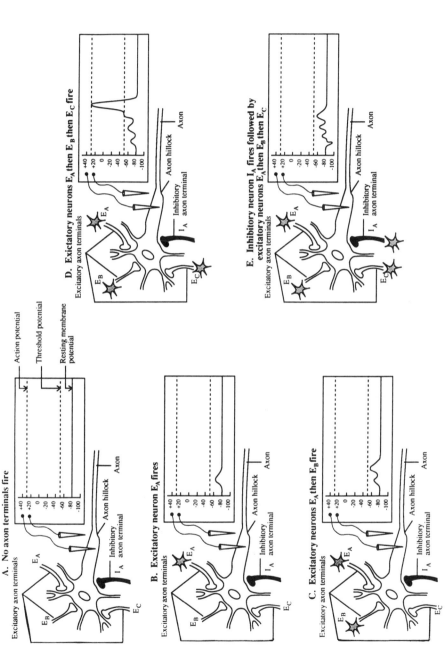

Fig.10. Nerve cell with the three excitatory nerve fibers (E$_A$, E$_B$, E$_C$, white) and one inhibitory fiber (I$_A$, black) forming synapses with its dendrites (E$_A$, E$_C$) and soma (E$_B$ and I$_B$). The symbol represents an axon terminal that fires and releases its neurotransmitter.

mitter has briefly interacted with its receptor protein to evoke its excitatory or inhibitory response on the postsynaptic membrane (*ca* 0. 5 ms) biochemical processes must exist to either remove or destroy the transmitter. Binding interactions between neurotransmitters and their receptors are quite weak and hence the transmitters are rapidly washed away by the extracellular fluid and are then either destroyed as a result of some enzyme-mediated reaction or taken back across the presynaptic membrane (*reuptake* process) into the axon terminal for reuse. These processes will be discussed in connection with individual neurotransmitters.

3.1   Chemical neurotransmitters.

In order for a chemical substance to be classified as a neurotransmitter in the nervous system it must meet the following requirements [22]:

1. The substance must be present in sufficient concentration in the nervous system.

2. The substance must be present in the neurons from which it is released and be concentrated in the nerve (axon) terminals.

3. Enzymes required for the biosynthesis of the substance must be present in the same neuron. Furthermore, the precursors and any other substance associated with the biosynthesis of the transmitter must be present in the neuron.

4. Ca^{2+}-dependent release of the substance from the nerve terminal to the extracellular fluid normally must occur when an action potential (nerve impulse) fires.

5. Mechanisms must be in place in the immediate vicinity of the nerve terminal for removal of the neurotransmitter once it has been released. These include a mechanism for selective uptake of the neurotransmitter back into the nerve ending, or the presence of enzymes which rapidly destroy the transmitter.

6. Receptors for the substance must be present in the region of the synapse associated with the axon terminals of a particular neuron.

In addition to the above criteria, particularly convincing evidence to support the contention that a suspected compound is indeed an endogenous neurotransmitter can be obtained experimentally by determining that an independently synthesized sample of the compound exactly mimics the postsynaptic action of the synaptically-released compound when added in the region of the synapse.

Pharmacological approaches can also provide evidence in support of a particular substance being a neurotransmitter. For example, with many known or putative transmitters pharmacological agents have been discovered which interfere with the biosynthesis, storage, release or physiological action of the transmitter on its receptor. Other pharmacological agents can mimic or even enhance the action of a transmitter or putative transmitter at its receptor or indirectly enhance its activity by other mechanisms. Thus, such compounds might act to stimulate the biosynthesis or release of the transmitter, inhibit its selective reuptake or inhibit its enzyme-mediated destruction.

When a neurotransmitter is released from an axon terminal into the synaptic cleft it diffuses to and binds briefly with a specific receptor on the post synaptic membrane of the receiving neuron. There it effects a brief action and communicates either an inhibitory or excitatory message. (It is, perhaps, worth noting here that receptors that activate enzymatic cascades or allow influx of Ca^{2+} may induce sustained or even long-term changes in the postsynaptic cell). This message is in the form of a transmembrane potential developed across the postsynaptic membrane which results from the opening of selective ion-channels across the membrane. There is more than one mechanism by which these ion channels are opened and, therefore, it is possible to classify neurotransmitters on the basis of these different mechanisms. However, a more straightforward classification derives from the chemical structure or chemical nature of the transmitter. Based upon the latter classification there are five groups or classes of chemical neurotransmitters: the amino acidergic (amino acids), cholinergic (acetylcholine), monoaminergic (aromatic amines), purinergic (purines), and peptidergic (peptides) systems. The names, commonly-used abbreviations and chemical structures of the transmitters which comprise these systems are presented in Table 1.

TABLE 1. STRUCTURES OF CATECHOLAMINES, INDOLEAMINES AND METABOLITES

Name	Abbreviation	Structure
Amino acidergic[1]		
L-Glutamic acid		$\begin{array}{c} NH_2 \\ | \\ HC.CH_2CH_2COOH \\ | \\ COOH \end{array}$
L-Aspartic acid		$\begin{array}{c} NH_2 \\ | \\ HC.CH_2COOH \\ | \\ COOH \end{array}$
γ-Aminobutyric acid	GABA	$H_2N.CH_2CH_2CH_2COOH$
Glycine		$H_2N.CH_2COOH$

continues TABLE 1. STRUCTURES OF CATECHOLAMINES, INDOLEAMINES AND METABOLITES

Name	Abbreviation	Structure

Cholinergic

| Acetylcholine | ACh | $CH_3\overset{O}{\overset{\|}{C}}\text{-O-}CH_2CH_2\overset{+}{N}(CH_3)_3$ |

Monoaminergic

| Dopamine | DA | |

| Norepinephrine (noradrenaline) | NE (NA) | |

| Epinephrine (adrenaline) | EPI | |

| 5-Hydroxytryptamine (serotonin) | 5-HT | |

| Histamine | | |

continues TABLE 1. STRUCTURES OF CATECHOLAMINES, INDOLEAMINES AND METABOLITES

Name	Abbreviation	Structure
Peptigdergic[2]		
Thyritropin-releasing hormone	TRH	**p–Glu–His–Pro–NH$_2$**
Methionine-enkephalin	met-Enk	**Tyr–Gly–Gly–Phe–Met**
Substance P		**Arg–Pro–Lys–Pro–Gln–Gln– Phe–Phe–Gly–Leu–Met–NH$_2$**
Somatostatin		**Ala–Gly–Cys–Lys–Asn–Phe–Phe–Trp** **|** **|** **Cys–Ser–Thr–Phe–Thr–Lys**
Neurotensin		**Glu–Leu–Tyr–Glu–Asn–Lys–Pro– Arg–Arg–Pro–Tyr–ILe–Leu**
ß - Endorphin		**Tyr-Gly-GLy-Phe-Met-Thr-Ser-Glu-Lys-Ser-Glu** **|** **Ala-Asn-Lys-Phe-Leu-Thr-Val-Leu-Pro-Thr** **|** **ILe-Val-Lys-Asn-Ala-His-Lys-Lys-Gly-Gln**

continues TABLE 1. STRUCTURES OF CATECHOLAMINES, INDOLEAMINES AND METABOLITES

Name	Abbreviation	Structure
Purinergic		
Adenosine		
Adenosine monophosphate	AMP	
Adenosine disphosphate	ADP	
Adenosine triphosphate	ATP	

1. Other aminoacids such as taurine, proline, serine, and ß-alanine have physiological activity and have been suggested to be neurotransmitters.
2. There have many other putative peptidergic neurotransmitters.

The amino acid neurotransmitters are employed by the overwhelming majority of all central neurons and, accordingly, they occur in much higher overall concentrations in the central nervous systems (μmoles/gram of tissue) than any other class of transmitter. The signal that is transmitted across the postsynaptic membrane as a result of the binding of an amino acid transmitter with its receptor is an electrical message in the form of a transmembrane potential change which develops as a result of the activation of selective ion channels. Such ion channels appear to exist in some form of very close physical association with the receptor protein or, in fact, are formed by the receptors themselves. Synapses which employ this mechanism in which ion channels are activated directly by the neurotransmitter-receptor interaction require only a few milliseconds for synaptic transmission to occur. This type of mechanism is known as *ionotropic transmission* [23] and evokes very rapid excitatory or inhibitory postsynaptic potentials. Some cholinergic neurons, which employ acetylcholine as their neurotransmitter, also effect rapid neuronal signalling as a result of ionotropic transmission. Such transmission occurs, for example, at the neuromuscular junction. It is worth noting, however, that many amino acidergic and cholinergic receptors are in fact not coupled to ionic pores but rather to metabolic pathways.

The monoaminergic neurotransmitters occur at much lower concentrations (nmoles/gram of tissue) in the central nervous system, *i.e.*, approximately one thousand times lower concentrations than the amino acid transmitters. When the monoaminergic transmitters bind to their postsynaptic receptors they evoke a signal that is much slower than is observed with ionotropic transmission and is generally known as *metabotropic transmission* [23]. In one form of metabotropic transmission the binding of the transmitter with its receptor causes a chemical message to be sent into the postsynaptic cell. This chemical message causes an enzyme system to be activated and the product of the resultant enzyme-mediated reaction is the so-called *second messenger* [24, 25]. A conceptualization of the second messenger mechanism is presented in Fig. 11. Thus, when the neurotransmitter binds to its postsynaptic receptor a closely associated adenyl cyclase enzyme is activated and catalyzes the conversion of ATP to cyclic 3',5' -AMP. The latter compound is the second messenger. Cyclic 3',5' -AMP then activates a specific protein kinase by causing its self-phosphorylation (autophosphorylation). This activated protein kinase then phosphorylates one (or more) protein(s) in the postsynaptic membrane. The structurally-modified protein(s) then alter the permeability of the postsynaptic membrane to specific inorganic ions, *i.e.*, an ion channel opens, and hence a change in transmembrane potential occurs. It is now thought possible that the structurally-modified (phosphorylated) proteins might also influence the rate of ion transport across the membrane *via* specific ion pumps hence causing changes in the local transmembrane potential. The actions initiated by the second messenger are terminated by rapid removal of the phosphate residue from the protein by a phosphoprotein phosphatase enzyme.

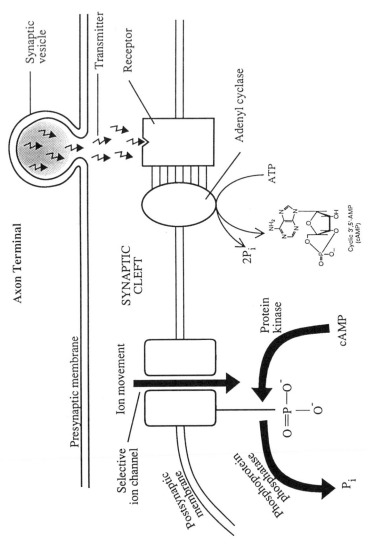

Fig. 11. Schematic representation of the second messenger concept in metabotropic neurotransmission.

The second messenger substance is sometimes cyclic 3',5' -guanosine monophosphate (c-GMP), which is activated by a guanylate cyclase enzyme, which itself is activated as a result of the chemical neurotransmitter binding to its postsynaptic receptor. The activation of metabotropic recepts does not necessarily elicit the production of cyclic nucleotides. For example, a widespread intervening step is the activation of GTP-binding proteins (G-proteins) directly coupled to the receptors.

Specific G-proteins may then either activate or inhibit, for example, adenylate cyclase. Other G-proteins directly activate ion channels or trigger the metabolic breakdown of membrane phospholipids (phosphoinositides) by activating phospholipase C, thus giving rise to diacylglycerol and phosphoinositol, two second messengers implied in intracellular calcium regulation and activation of protein-kinase C.

Metabotropic or second messenger signalling mechanisms are necessarily slower than the more direct ionotropic mechanisms and, therefore, a somewhat delayed transmembrane electrical signal is produced. Accordingly, synaptic transmission at synapses involving second messenger mechanisms takes up to 100 ms or longer.

The peptidergic neurotransmitters and putative transmitters are by far the largest group and include at least 25 and probably many more different peptides. However, this class of transmitters occurs at much lower concentrations (picomoles or sub-picomoles/gram of tissue) than any of the other neurotransmitters. Studies on peptidergic transmitters and their modes of action represent areas of intense current activity. It appears, however, that most neuropeptides evoke some form of metabotropic action. Several neuropeptides also seem to modulate the actions of the amino acidergic, monoaminergic and cholinergic transmitters.

That purely purinergic neurons actually exist is somewhat uncertain and, hence, it is not known whether purine derivatives such as adenosine, AMP, ADP and ATP are real neurotransmitters. There is much evidence, however, that adenosine, for example, can modulate certain aspects of synaptic function by its action on specific receptors in both the peripheral and central nervous systems [26].

The following discussion will concentrate on the low molecular weight cholinergic, monoaminergic and amino acidergic neurotransmitters.

3.2. Cholinergic neurotransmission.

Neurons and neuronal pathways which employ ACh as their chemical neurotransmitter comprise the cholinergic neuronal system. ACh was, in fact, the first neurotransmitter to be definitively identified and is also the compound that most completely meets the criteria demanded of a chemical neurotransmitter.

The biosynthesis of ACh involves a single-step reaction between acetyl-coenzyme A (acetyl-CoA) and choline (Ch) catalyzed by the enzyme choline acetyltransferease (ChAT) (Fig.12). The acetylation of Ch by acetyl CoA occurs predominantly in cholinergic nerve terminals. The key enzyme, ChAT, is synthesized in the soma and is transported to the terminal region by axoplasmic flow. The source of Ch is largely de-

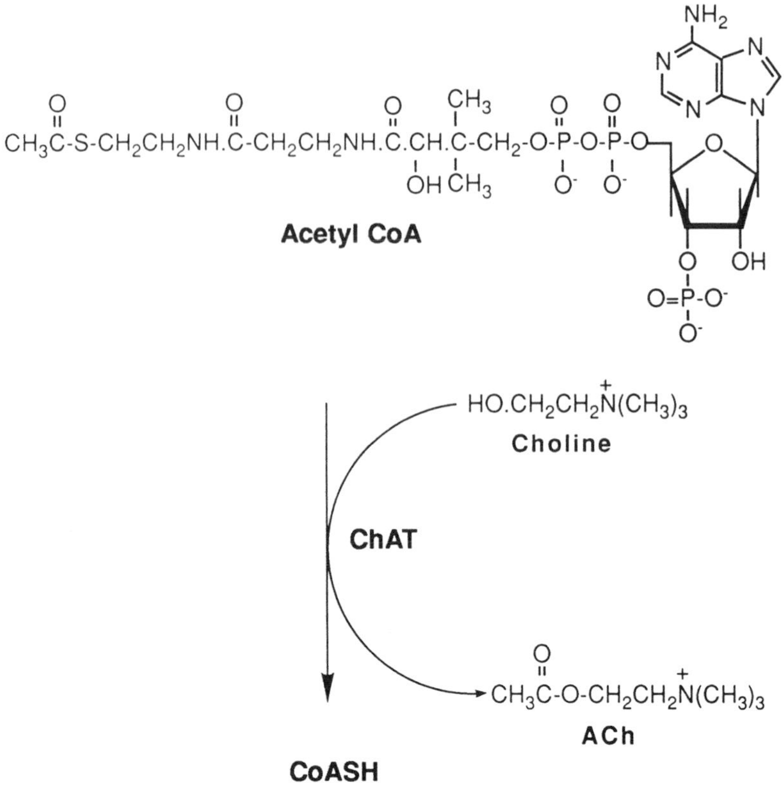

FIG.12. Biosynthesis of acetylcholine (ACh) from choline (Ch).

rived from the liver where it is produced as a result of the hydrolysis of phosphatidyl-choline (lecithin). Ch is transported to the brain and other organs and muscles in the blood. Free Ch is specifically taken up into cholinergic nerve terminals by a high-affinity membrane pump mechanism [27]. Acetyl-CoA is produced by a pyruvate dehydrogenase system in cholinergic mitochondria.

Following release of ACh from its presynaptic vesicles, as a result of the nerve impulse-induced process of exocytosis, biosynthesis of replacement transmitter becomes necessary. The rate of ACh biosynthesis is controlled by the availability of Ch and acetyl-CoA at the site of synthesis and the high affinity uptake of Ch. It is of particular interest to note that the high affinity Ch pump is inhibited by excess ACh and potentiated by low levels of ACh [28]. Thus, this biochemical pumping mechanism also acts as a control for the biosynthesis of ACh.

Similar to all neurotransmitters, once ACh is released into the synaptic cleft it binds only briefly with its postsynaptic receptors and must then be destroyed in order to prepare the receptors for the subsequent pulse of information. The destruction of ACh is performed extremely efficiently by the enzyme acetylcholinesterase (AChE) which is concentrated in the synaptic cleft. The reaction catalyzed by AChE is hydrolysis of ACh to Ch and acetic acid (Fig. 13). AChE is synthesized in the cholinergic

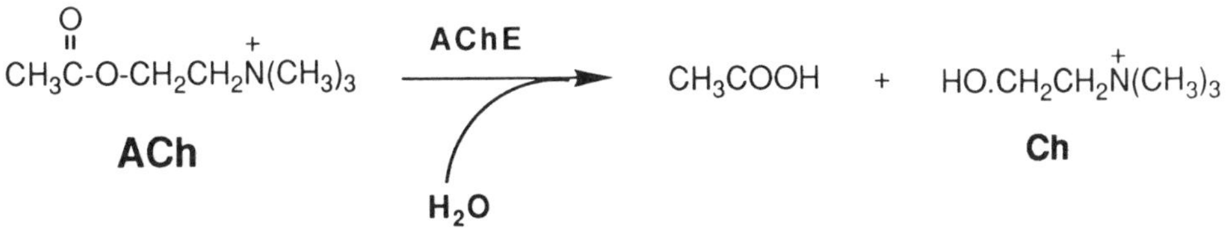

$$CH_3\overset{O}{\overset{\|}{C}}-O-CH_2CH_2\overset{+}{N}(CH_3)_3 \quad \xrightarrow{\text{AChE}} \quad CH_3COOH \quad + \quad HO.CH_2CH_2\overset{+}{N}(CH_3)_3$$

ACh

H_2O

Ch

FIG.13. Catabolism (hydrolysis) of acetylcholine (ACh) to choline (Ch) and acetic acid in the presence of acetylcholinesterase.

nerve cell body and is then transported to the nerve endings by axoplasmic flow. At least three forms of the enzyme are known. The Gl and G2 forms occur intracellularly while the G4 form is membrane bound such that its catalytic center points outward into the synaptic cleft [29]. AChE is very concentrated within the synaptic cleft of seve-

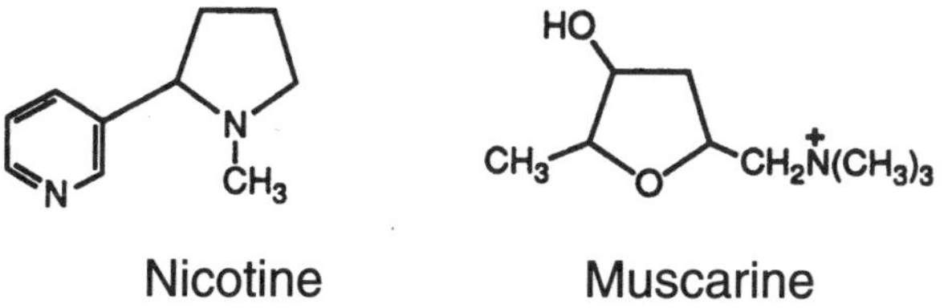

Nicotine   **Muscarine**

ral cholinergic synapses, including the neuromuscular junction. Here, its collagen-like tail anchors the enzyme to the basal lamina (membrane) in a strategic location for its rapid action.

There are two general types of cholinergic receptors: nicotinic receptors and muscarinic receptors. The naming of these two receptor types derives from the fact that the physiological effects of ACh at certain sites are mimicked by the alkaloid nicotine; at other sites the physiological effects of ACh are evoked by muscarine. The nicotinic receptors mediate fast ionotropic actions of ACh while the muscarinic receptors mediate much slower metabotropic (second messenger) actions. Nicotinic cholinergic synapses operate in vertebrate neuromuscular junctions and in the central nervous system. Muscarinic synapses are found in smooth muscle, cardiac muscle, glands and again in many brain regions. It has been estimated that there are 10 to 100 times more muscarinic cholinergic receptors in the brain than nicotinic receptors. It should be noted that there are multiple subtypes of both nicotinic and muscarinic receptors in the brain. One muscarinic receptor, M2, is also an autoreceptor.

Large motor neurons (motoneurons) which project from the ventral horn of the spinal cord are responsible for control of the voluntary muscles of vertebrates by means of the nicotinic cholinergic *neuromuscular junction* (or *motor end plate*). A schematic representation of a motoneuron is shown in Fig. 14. The soma and dendrites are located in the spinal cord and the axon (motor fiber) projects to the muscle fibers that it innervates. A motor axon typically branches to innervate hundreds of muscle fibers in a particular muscle. But each muscle fiber usually receives only one innervation from a motor axon fiber. A more detailed representation of the neuromuscular junction is provided in Fig. 15. Extensive folding of the sarcolemma (muscle fiber membrane) is employed to greatly increase the surface area of the postsynaptic (postjunctional) membrane in order to increase the depolarizing power of the synapse. An even more detailed view of the neuromuscular junction is shown in Fig. 16. Each presynaptic vesicle in the axon terminal is packed with about 10,000 molecules of the transmitter ACh. The nerve impulse when it arrives at the nerve ending results in the opening of Ca^{2+}–channels in the presynaptic membrane. Thus, Ca^{2+} ions diffuse into the nerve ending and activate calmodulin or some other calcium-binding protein(s) (see earlier discussion) with the result that the vesicle sheath binds to the nerve membrane followed by exocytosis which results in ACh being squirted into the synaptic cleft. Some of the released ACh molecules bind briefly to the nicotinic receptors on the muscle fiber. This binding activates and opens Na^+ and K^+ ion channels. These channels are actually constituted by the acetylcholine-receptor molecule itself and, in fact, are unselectively permeable to cations. Thus, Na^+, K^+ and Ca^{2+} flow through it. Because of the large amounts of ACh released from numerous axonal branches and the large area of the postsynaptic membrane, the flow of Na^+ and K^+

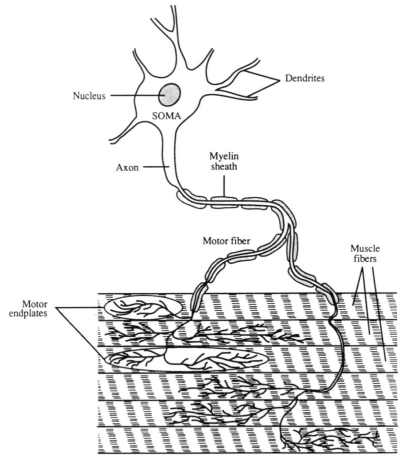

FIG.14. Schematic representation of a motoneuron showing its innervation of muscle fibers at the neuromuscular junction or motor end plate.

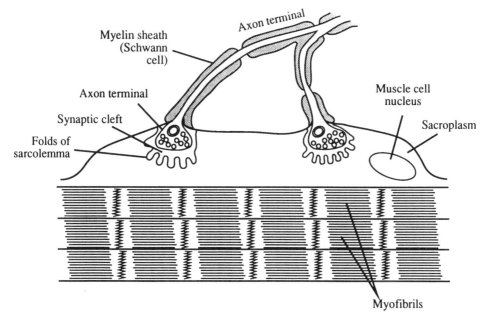

FIG.15. Neuromuscular junction or motor end plate.

drives the transmembrane potential of the postsynaptic membrane towards the Na$^+$–K$^+$ equilibrium potential-value. Thus, the muscle membrane is rapidly depolarized. Before this level of muscle membrane potential is reached, however, the depolarization triggers a muscle action potential that leads to contraction of the myofibrils. Thus, the muscle contracts. There are about 200,000 motoneurons issuing from the human spinal cord which are responsible for the contraction of all of the muscles of the limbs, body and neck, *i.e.*, all muscular performance except for that of the head.

3.3.   The monoaminergic neurotransmitters

The monoamine neurotransmitters include the catecholamines dopamine (DA), norepinephrine (NE; or noradrenaline, NA), and epinephrine (EPI; adrenaline). In general neuronal systems which employ the catecholamine neurotransmitters are known as the catecholaminergic system and can be comprised of dopaminergic (DA), noradrenergic (NE) and adrenergic (EPI) neurons. The indole 5-hydroxytryptamine (5-HT; serotonin) is the transmitter employed by serotonergic neurons. Histamine, an imidazoleamine, is employed by histaminergic neurons.

3.3.1 Catecholamine

The biosynthetic pathways leading to the catecholomines DA, NE and EPI are summarized in Fig. 17. The initial substrate in this pathway is the aromatic amino acid *L*-tyrosine which is available in tissue pools and can be transported to the brain from

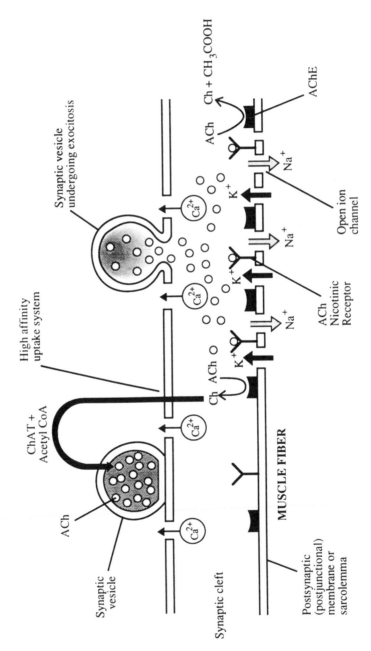

Fig.16. Detailed schematic representation of the neuromuscular junction or motor end plate.

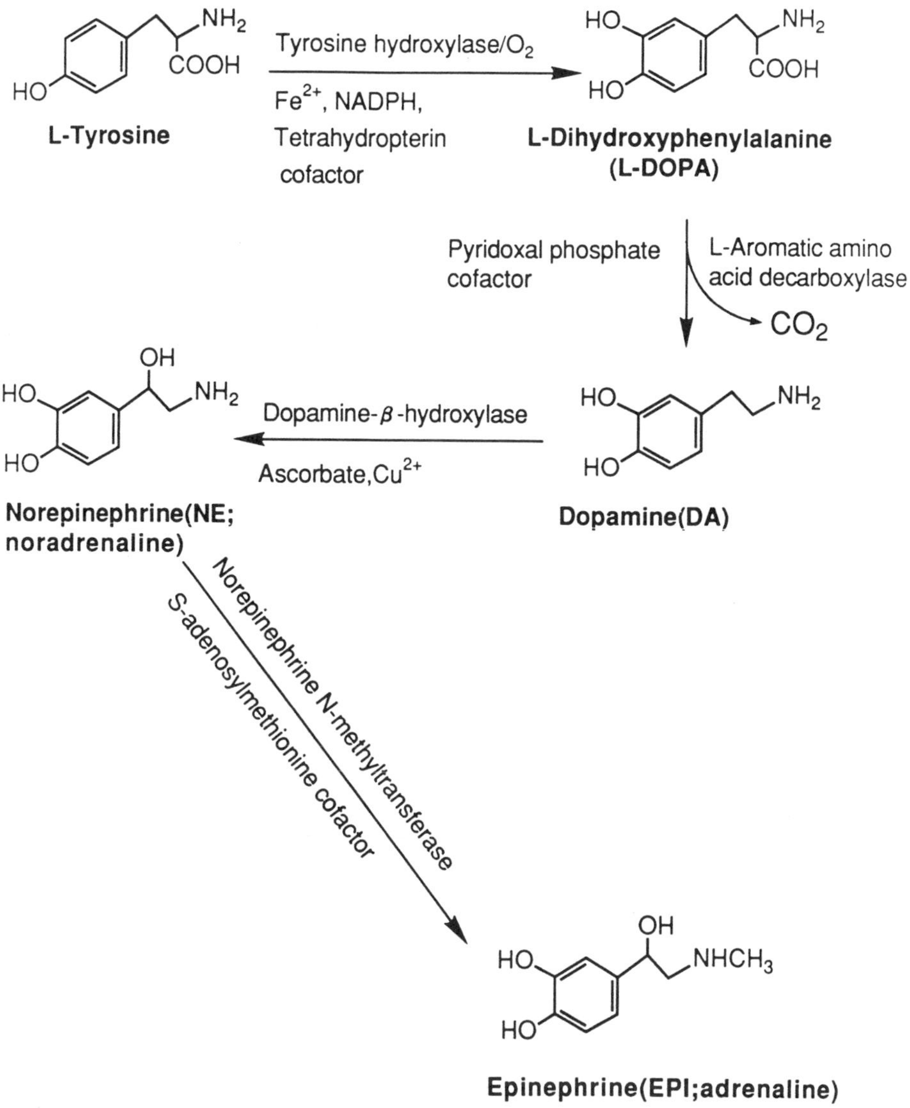

FIG.17. Biosynthesis of catecholamine neurotransmitters dopamine, norepinephrine and epinephrine.

the blood across the *blood-brain barrier.* Tyrosine can be taken up by catecholaminergic neurons in which it is converted into *L*-3,4-dihydroxyphenylalanine (*L*-DOPA) in a reaction catalyzed by tyrosine hydroxylase (tyrosine-3-monooxygenase) which also requires Fe^{2+}, tetrahydrobiopterin and molecular oxygen [30]. The enzyme *L*-aromatic amino acid decarboxylase (DOPA decarboxylase) then decarboxylates *L*-DOPA to give DA.This reaction requires pyridoxal phosphate as a cofactor. In dopaminergic neurons this biosynthetic route, of course, terminates at DA. However, noradrenergic and adrenergic neurons are characterized by the presence of the copper-containing enzyme dopamine-β-hydroxylase (dopamine-β-monooxygenase). This enzyme catalyzes a reaction which introduces a hydroxyl residue at the β-site of the side chain of DA to give NE. It is interesting to note that dopamine-β-hydroxylase is found in the synaptic vesicles of noradrenergic neurons. Thus, in such neurons, the final step in the biosynthesis of NE must occur in or on the surface of storage vesicles. Norepinephrine-N-methyltransferase (phenylethanolamine-N-methyltransferase) is an enzyme which occurs only in adrenergic neurons although it is in fact not associated with vesicles. This enzyme is responsible for N-methylation of NE to give EPI in a reaction which requires the methyl-group-donor S-adenosylmethionine.

Following the release of the catecholamine neurotransmitters into the synaptic cleft the principal mechanism for their inactivation is reuptake into the parent nerve endings [31, 32]. The uptake mechanism, often referred to as a pump, transports the catecholamine against a significant concentration gradient from outside the nerve terminal to the inside. A second method of inactivation derives from the degradation or catabolism of the catecholamines. Two principal enzymes are involved in this biological degradation: monoamine oxidase (MAO) and catechol-O-methyltransferase

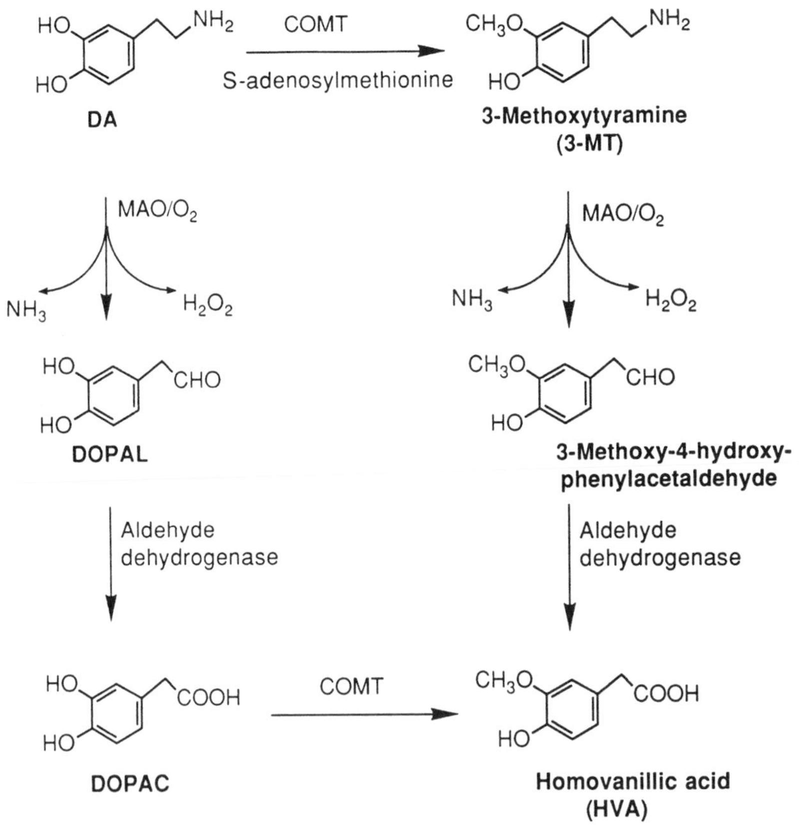

FIG. 18. Catabolism of dopamine (DA).

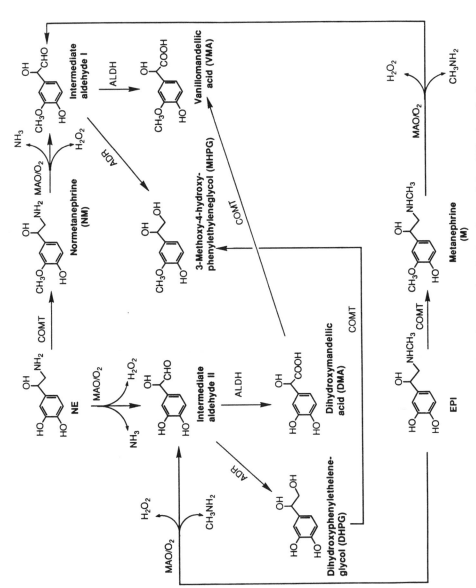

FIG. 19. Catabolism of norepinephrine (NE; noradrenaline) and epinephrine (EPI; adrenaline).

(COMT). MAO occurs both as a presynaptic enzyme on the surface of mitochondria and postsynaptically. The presynaptic enzyme probably plays a role in controlling the levels of the transmitter catecholamines within the neuron particularly as a result of leakage from vesicles. COMT, however, is attached to external cell membranes and acts only postsynaptically. A summary of the reactions mediated by MAO, COMT and related catabolic enzymes is shown in Fig 18. One route mediated by MAO oxidatively deaminates DA to 3,4-dihydroxyphenylacetaldehyde (DOPAL) which is rapidly further oxidized by aldehyde dehydrogenase (ALDH) to the corresponding carboxylic acid, 3,4-dihydroxyphenylacetic acid (DOPAC). Alternatively, COMT mediates the methylation of DA to 3-methoxytyramine (3-MT) which is then oxidatively deaminated by MAO to yield 3-methoxy-4-hydroxyphenylacetaldehyde. This reactive aldehyde is oxidized by ALDH to give homovanillic acid (HVA). An alternative route to HVA involves the COMT-mediated methylation of DOPAC (Fig. 18).

The catabolism of NE and EPI is summarized in Fig. 19. The key catabolic enzymes once again are MAO and COMT. The principal difference between the catabolism of NE and EPI on the one hand and DA on the other is that the aldehyde intermediates I and II, formed as a result of MAO-mediated oxidative deaminations, can be either oxidized to the corresponding carboxylic acids (VMA or DMA, respectively) by ALDH, or reduced by an aldehyde reductase (ADR) to the phenylethylene glycol derivatives MHPG or DHPG, respectively.

3.3.2 Dopamine receptors

There are several subcategories of DA receptors in the central nervous system which have been classified D_1, D_2, D_3, D_4 and, perhaps, D_5. Such classifications are based upon binding studies with various pharmacological agents and their physiological responses to various dopaminergic drugs [33, 34]. The most widely studied are the D_1 and D_2 receptors. Stimulation of D_1 receptors potentiates the formation of cAMP by adenylate cyclase activation and second messenger formation as discussed previously (see Fig. 11 and associated discussion). The D_2 receptors, however, couple to adenylate cyclase in a way to inhibit this enzyme or, perhaps, function independently of this second messenger system. Dopaminergic neurons also possess receptors located on the nerve ending membrane (i.e., presynaptic terminal) which are activated by the transmitter released by the parent terminal. These receptors are termed autoreceptors and act to reduce or control synaptic activation by decreasing the release of the transmitter.

3.3.3 Norepinephrine receptors

Based upon binding studies with pharmacological agents it appears that there are two subclassifications of NE receptors: α_1 and α_2 and β_1 and β_2. Binding of NE to the α receptors appears to result in the influx of Ca^{2+} ions followed by other trans-membrane ion fluxes (e.g., K^+ efflux). The β receptors, however, are coupled directly to adenylate cyclase through a G-protein and activation of this receptor type leads to the synthesis of cAMP which in turn often results in the generation of postsynaptic transmembrane potentials by the metabotropic mechanism. However, in many instances no significant change occurs in the electrical properties of the postsynaptic cell and the main effect of cAMP is the modulation of enzymes and metabolic reactions by means of activation of protein kinase A.

Considerably less is known about central EPI neurons simply because their occurrence is much lower than other catecholaminergic neurons. Adrenergic neurons also overlap considerably with noradrenergic terminals and there is currently an appreciable lack of distinction between noradrenergic and adrenergic receptors.

3.3.4 5-Hydroxytryptamine

The serotonergic transmitter 5-HT is biosynthesized from dietary L-tryptophan (L-TPP). The initial step in the biosynthetic pathway, once L-TPP is transported into the neuron, is catalyzed by L-tryptophan hydroxylase (L-tryptophan 5-monooxygenase) which produces L-5-hydroxytryptophan (L-5-HTPP) (Fig.20). The latter compound is then decarboxylated in a reaction catalyzed by L-5-HTPP decarboxylase (L-aromatic amino acid decarboxylase) to form 5-HT. The major catabolic route for 5-HT, initiated by a MAO-mediated oxidative deamination, forms 5-hydroxyindole-3-acetaldehyde (5-HIAD). Under normal conditions it is widely believed that 5-HIAD is further oxidized in a reaction catalyzed by an aldehyde dehydrogenase (ALDH) enzyme to give 5-hydroxyindole-3 -acetic acid (5-HIAA). A small amount of 5-HIAD, however, is reduced in the presence of an aldehyde reductase (ADR) to 5-hydroxtryptophol (5HTOL). Recent investigations, however, suggest that the catabolism of 5-HT might be considerably more complex than depicted in Fig 20 [35]. The primary method of inactivation of 5-HT, once released into the synaptic cleft, however, is reuptake [31, 32] across the membranes of neural cells by a high affinity Na+-dependent transport system.

3.3.5 5-HT receptors

The effects of 5-HT on target neurons are somewhat variable although in general the transmitter appears to be largely inhibitory in action as a result of metabotropic mechanisms. There are at least three major receptor types and several subtypes of each. A 5-HT autoreceptor is also probably present on serotonergic axon terminals which, upon interaction with the transmitter, inhibits additional release into the synaptic cleft.

3.4 Histamine

Only quite recently has histamine been seriously considered to be a

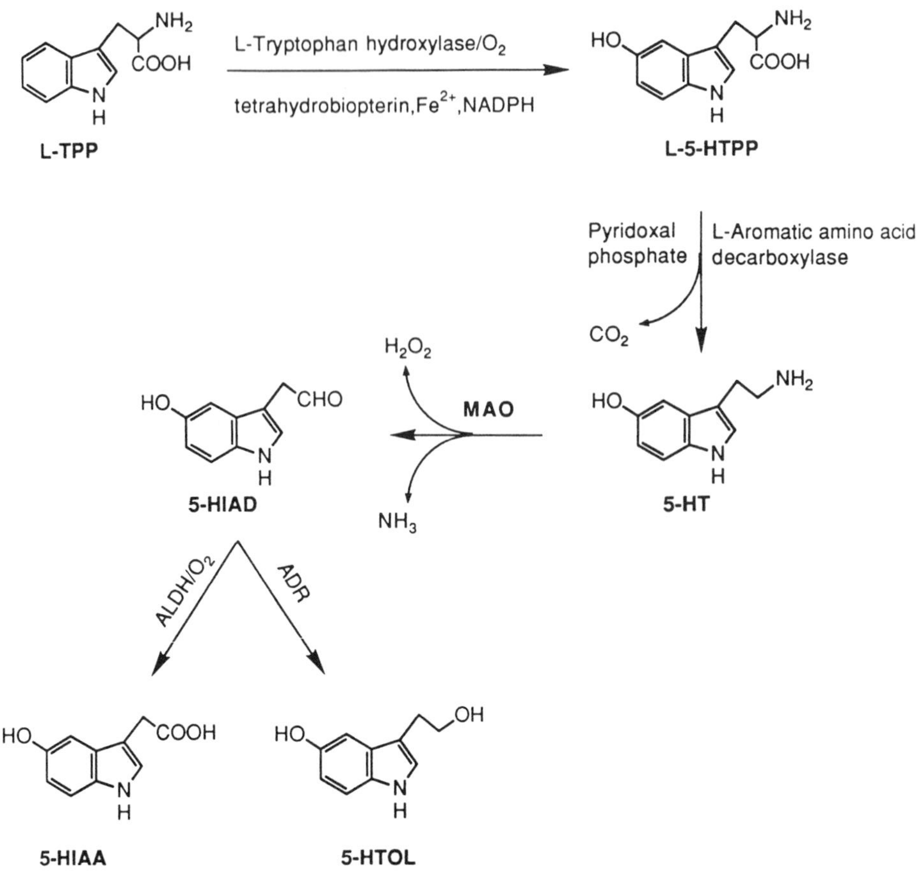

FIG.20. Biosynthesis and catabolism of 5-hydroxytryptamine (serotonin).

neurotransmitter. This compound is probably most widely known because of its release from certain cells as a result of allergic reactions and tissue damage.

Histamine does not readily cross the blood-brain barrier and hence histaminergic neurons biosynthesize this putative neurotransmitter. The immediate precursor of
histamine is L-histidine which is decarboxylated by the pyridoxal phosphate-depen-
dent neuronal enzyme L-histidine decarboxylase [36] (Fig. 21).

Since there is no known high affinity transport system to return histamine to histaminergic nerve terminals, it appears that enzyme mediated catabolic transforma-
tions are employed to inactivate the transmitter. The most likely catabolic route involves N-methylation of histamine by histamine N-methyltransferase, in a reaction
which requires S-adenosylmethionine as the methyl group source, to give 3-
methylhistamine (Fig. 22). The latter compound is then oxidatively deaminated by

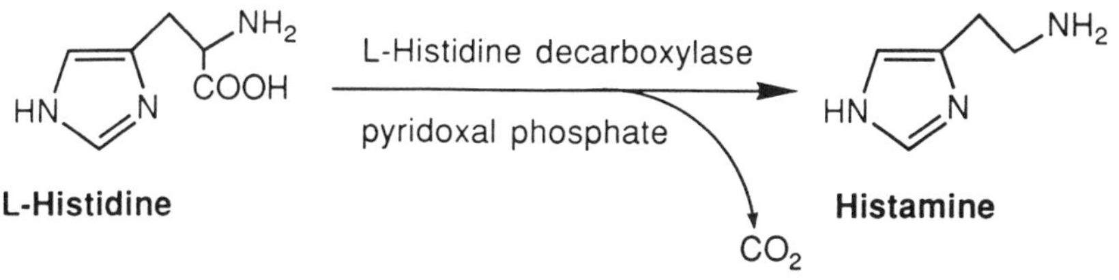

FIG.21. Biosynthesis of histamine.

the B-form of MAO to give 3-methylimidazole-5-acetic acid.

There currently seem to be two distinct types of histamine receptors: H_1 and H_2. The allergic reaction associated with peripheral histamine release is mediated by the binding of histamine to its H_1 receptors. In the central nervous sytem H_1 receptors are generally excitatory whereas H_2 receptors are inhibitory although there are apparently exceptions to this general rule. Histamine appears to evoke a metabotropic effect when it binds to its receptors.

3.5   The aminoacidergic neurotransmitter

3.5.1   Glutamic and aspartic acid

When either L-glutamate or L-aspartate are applied to brain or spinal cord neurons by iontophoresis (application by passing a current through a solution containing the substance in a micropipet) powerful excitatory actions are evoked. Glutamate in particular is a very potent excitatory transmitter and evokes a very rapid excitatory re-

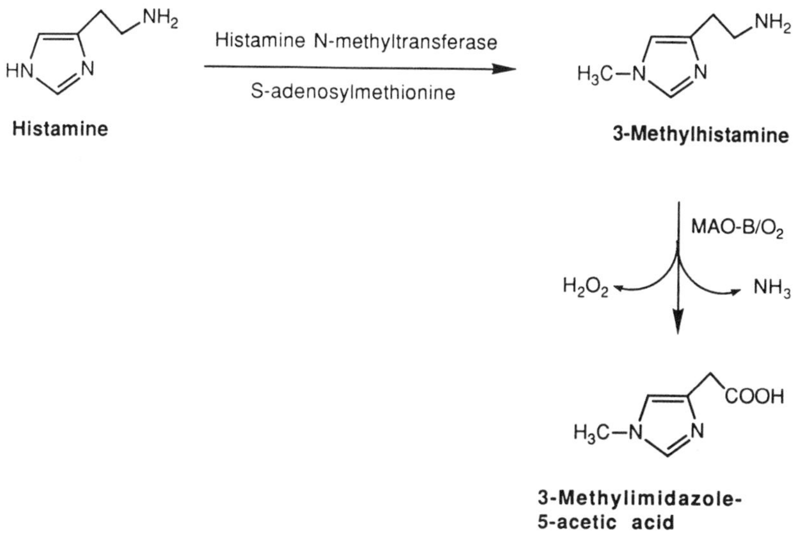

FIG.22. Catabolism of histamine.

sponse followed by rapid termination of action. Both glutamate and aspartate cause an increase in the permeability of the postsynaptic membrane to Na$^+$ and other ions and, therefore, are excitatory ionotropic neurotransmitters.

Neither glutamate nor aspartate can readily cross the blood-brain barrier and both transmitters are therefore biosynthesized in the central nervous system. There are several biosynthetic routes that are probably involved in the formation of glutamate and aspartate in the central nervous system. One involves the Krebs cycle which occurs in all mitochondria to permit, among other things, oxidative phosphorylation. In this cycle two key dicarboxylic acids are formed: α-ketoglutaric acid and oxaloacetic acid. The subsequent conversion of α-ketoglutarate to glutamate and oxaloacetate to aspartate is achieved with the help of the enzyme aspartate-aminotransferase as conceptualized in Fig. 23. There are many other potential routes which might be responsible for biosynthesis of the amino acid neurotransmitters glutamate and aspartate but no specific sources for these compounds have yet been definitively identified. Indeed, it is likely that many routes contribute to the biosynthesis of these transmitters.

The inactivation of glutamate and aspartate released into the synaptic cleft is primarily by reuptake across the membrane of neural cells by high-affinity, Na$^+$-dependent transport systems [37].

3.5.2. Glutamate and aspartate receptors

It is currently believed that glutamate is the main excitatory neurotransmitter in the

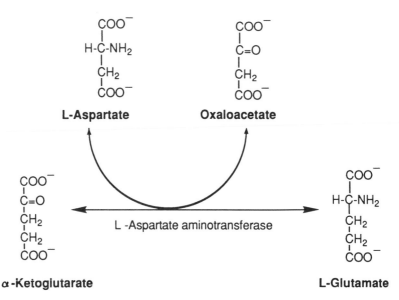

FIG.23. Biosynthetic routes to *L*-aspartate and *L*-glutamate.

CNS. Glutamate binds to at least five different types of receptors. Three ion-channel-coupled receptors can be distinguished pharmacologically by their relative selectivity for the glutamate analogs: quisqualic acid, kainic acid, and N-methyl-D-aspartate (NMDA) [38, 39]. At least two types of G-protein coupled receptors are also activated by glutamate.

Glutamate in particular, but to a lesser extent aspartate, excite central nervous system neurons so powerfully that if they are administered in sufficient excess they cause destruction of the neuron. This process is known as *excitotoxicity*. The excitotoxicity hypothesis [40] proposes that during excitation ion channels are wide open permitting excessive exchange of intracellular and extracellular ions. Currently available evidence suggests that exocytotoxicity is specifically related to the rise in intracellular Ca^{2+} concentration that follows activation of NMDA receptors in depolarized neurons. The NMDA receptor channel is permeable to Na^+ and Ca^{2+} but is blocked by Mg^{2+} from the outside at resting values of the membrane potential. In depolarized cells the electrochemical potential for Mg^{2+} drives it outwards and the pore becomes unblocked so that Ca^{2+} can flow into the cell. High concentrations of glutamate activate kainate and quisqualate receptors (coupled to Na^+–permeable channels) that induce membrane depolarization. Concurrent activation of the NMDA receptor under these conditions induces a massive inflow of Ca^{2+}. Sustained elevations of intracellular Ca^{2+} concentrations (μM range) are lethal to cells.

3.6. γ-Aminobutyric acid and glycine

Both GABA and glycine are inhibitory ionotropic amino acid neurotransmitters although they are not closely related to each other either structurally or from a metabolic viewpoint.

The immediate precursor of GABA is L-glutamic acid which, as noted in Fig 23, is derived from α-ketoglutarate whose remote progenitor is glucose *via* the Krebs (citric acid) cycle. An abbreviated view of the biosynthetic pathway leading to GABA is shown in Fig. 24. The first step in the reaction sequence is the transamination of α-ketoglutarate by L-aspartate in a reaction catalyzed by L-aspartate aminotransferase to give L-glutamate along with oxaloacetate as a byproduct. Decarboxylation of L-glutamate by the pyridoxal phosphate–dependent enzyme L-glutamate decarboxylase then yields GABA. Released GABA can be taken up by adjacent glial cells where it is transaminated in a reaction catalyzed by GABA-α-ketoglutarate transaminase whereby α-ketoglutarate is converted into L-glutamic acid and GABA is converted into succinic semialdehyde (Fig. 25). The latter compound is oxidized to succinic acid by succinic semialdehyde dehydrogenase. Succinic acid is a normal constituent of the Krebs cycle. The L-glutamic acid formed in the transamination reaction cannot be converted into GABA within glial cells because they lack the enzyme L-glutamate-decarboxylase. Rather, L-glutamate is converted into glutamine by the enzyme glutamine synthetase. Glutamine is then transported into the GABAergic nerve endings where it is con-

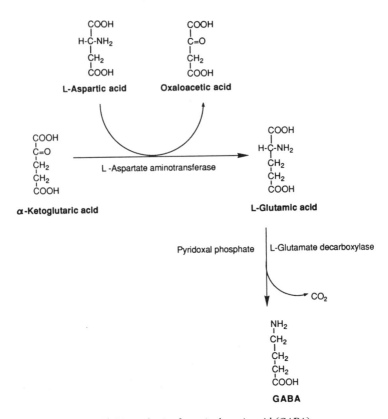

FIG.24. Biosynthesis of γ-aminoburyric acid (GABA).

verted to *L*-glutamate by the enzyme glutaminase. The *L*-glutamate can then be converted into GABA as demonstrated in Fig. 24. GABA can also be inactivated by reuptake mechanisms located in the nerve endings which release this neurotransmitter.

Binding of GABA to its postsynaptic receptor proteins results in the opening of Cl^--channels and Cl^- ions rapidly enter the postsynaptic cell. Thus, generally, GABA release leads to hyperpolarization and, therefore, postsynaptic inhibition. However, this is not the only action of this transmitter. For example, recent studies suggest that there are two types of GABA receptors: $GABA_A$ and $GABA_B$. Binding of GABA at GABA$_A$ sites causes hyperpolarization as a result of influx of Cl^- through selective ion channels. However, at the $GABA_B$ receptor, which is coupled to a G-protein, GABA acts to reduce the efflux of other neurotransmitters such as NE, DA, 5-HT or glutamate. Autoreceptors on GABA neurons are $GABA_A$ receptors and act to inhibit the release of GABA.

The second ionotropic inhibitory aminoacidergic neurotransmitter is the simplest of all amino acids, glycine. Glycine readily crosses the blood-brain barrier and

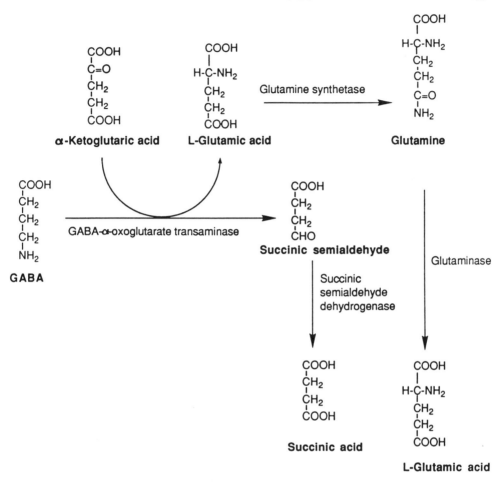

Fig.25. Catabolism of GABA.

hence can be transported to the central nervous sytem by the blood. Using radiolabelled tracers, however, it has been possible to demonstrate that some brain glycine is produced by *de novo* synthesis from glucose *via* serine. Two likely biosynthetic pathways from glucose are outlined in Fig. 26.

The principal mode of inactivation of glycine is by a high affinity transport system located on the neurons releasing this transmitter. Most aspects of the catabolism of glycine related to its activity as a neurotransmitter in the central nervous system are not well understood. The amino acid, however, is widely involved in the biosynthesis of very many cellular components and hence many pathways for utilization of this compound are available. Binding of glycine to its receptors evokes an inhibitory ionotropic response as a result of the opening of Cl^--channels and resultant hyperpolarization of the postsynaptic membrane.

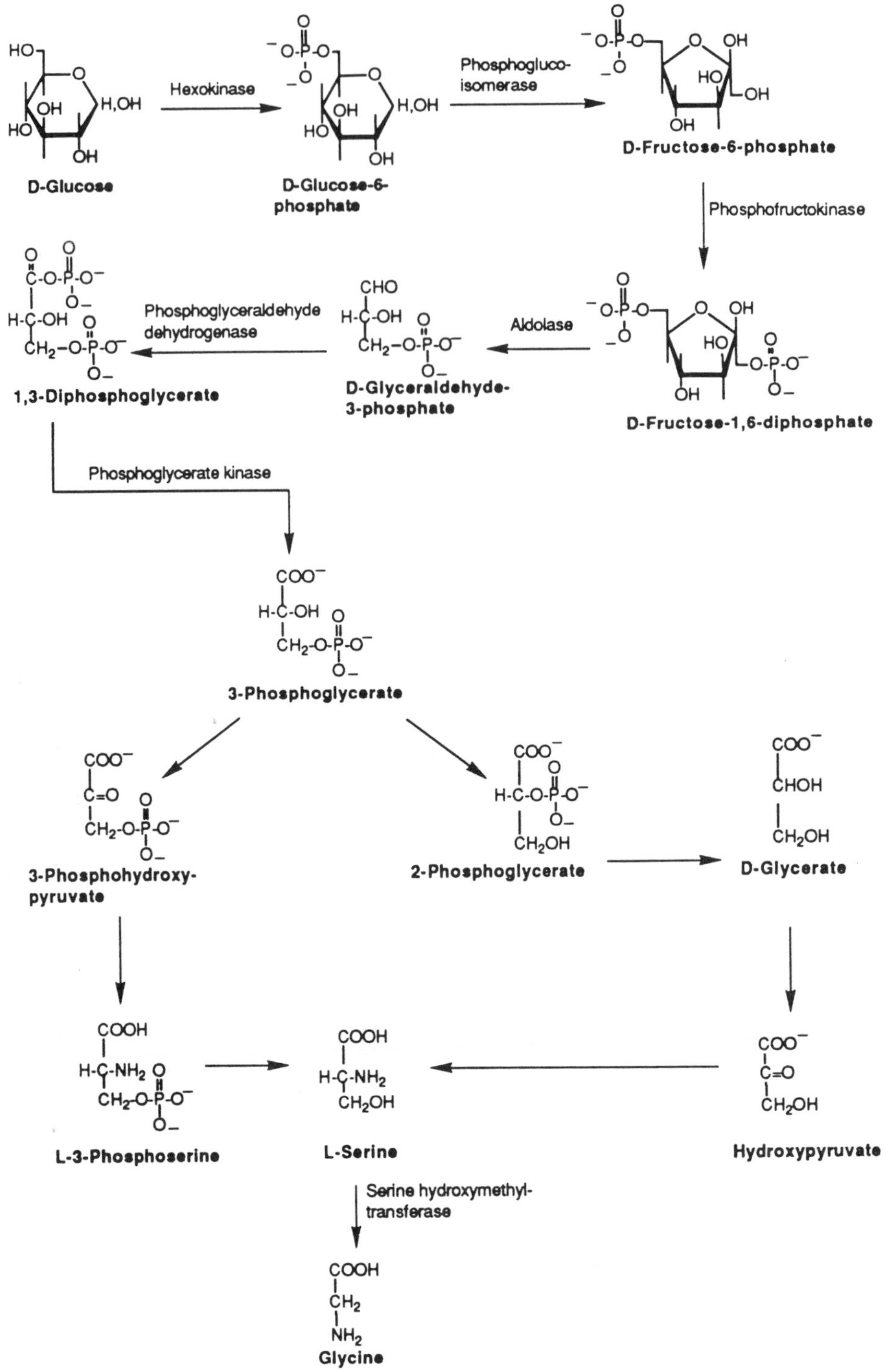

FIG.26. Biosynthetic pathways to glycine.

4. Nerve-muscle function

In the preceding sections the key components of neuronal communication have been described. It has been demonstrated that a neuron can receive signals from other neurons and that these signals are transmitted from the nerve ending (axon terminal) of one neuron by the release of a chemical neurotransmitter. This transmitter crosses the synaptic cleft and briefly binds with receptor proteins located on the membranes of the dendrites or soma (usually) of the receiving neuron. Interaction of the transmitter and its receptor results either in the development of a local, transient change in the transmembrane potential (or in some cases the activation of a cascade of enzyme reactions). Depolarization of the membrane (*i.e.*, shift of the transmembrane potential to more positive values) represents an excitatory signal. Hyperpolarization of the membrane (*i.e.*, a shift towards more negative values) represents an inhibitory signal. The receiving neuron performs a spatial and temporal integration of the incoming signals. When this summation at the axon hillock reaches a critical threshold potential an action potential fires and an electrical nerve impulse glides down the axon fiber to the nerve terminals which, in turn, results in the release of neurotransmitter which carries information to the synapsed neuron(s).

It is now necessary to consider how such neuronal signalling occurs in order to effect movement. One of the simplest neuromuscular processes is a spinal reflex (Fig. 27) which could represent withdrawal of one's hand from a very hot surface. Receptors on the skin surface sense the hot surface and emit a volley of nerve impulses, *i.e.*, action potentials. The frequency of these sensory nerve impulses increases with the hotness of the surface and provide an indication of the urgency of the transmitted signal. The *afferent neuron* (*i.e.*, axon which conveys impulses towards the spinal cord) enters the spinal cord through the dorsal root and synapses with a short *interneuron*. Such afferent fibers are always excitatory at their synapses. Thus, the afferent neuron synapses with an interneuron to produce an excitatory postsynaptic potential. A nerve impulse then travels along the axon of the interneuron which synapses with a motoneuron. The axon terminal of the interneuron synapses with the motoneuron in an excitatory fashion resulting in a nerve impulse firing along the axon of the motoneuron. The motoneuron exits the spinal cord through the ventral root. The nerve impulse travels along the motoneuron to its nerve endings which synapse at the muscle motor endplate. At this synapse ACh is released and interacts with fast-acting nicotinic receptors causing the muscle in the hand to contract and hence withdraw the hand from the heat source. Naturally, many skin receptors send similar signals and, in addition, many other neuronal connections are made so that the hand and arm move appropriately away from the hot surface. Again, it should be noted that the higher the frequency of the nerve impulses passing along the motoneuron to the neuromuscular junction, the stronger or more powerful is the resulting muscle contraction.

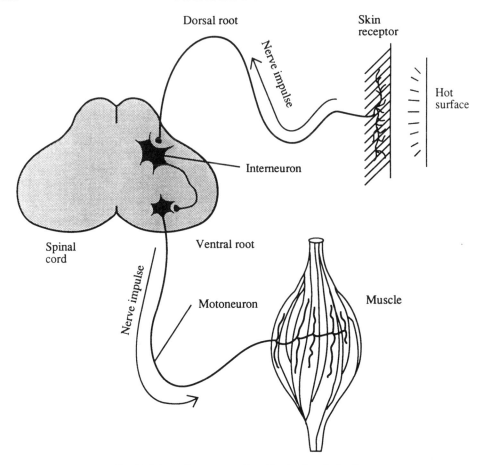

FIG.27. Schematic representation of a simple spinal reflex.

This kind of movement is under control of the spinal cord and is categorized as a spinal reflex. Thus, the movement decision – rapid hand withdrawal – is made at the level of the spinal cord.

The above example illustrates that the dorsal roots of the spinal cord (posterior tracts) are, with few exceptions, sensory in function. The ventral roots (anterior tracts) of the spinal cord are motor in function. Thus, the posterior tracts contain projections from the sensory receptors in the body and the anterior tracts consist of axons from cells that project to muscles or organs in the body. As a result, damage to the posterior regions of the spinal cord cause effects which are sensory in nature, whereas damage to the anterior portion of the cord causes changes that are selectively motor in nature.

A more complex example of neuromuscular control is illustrated in Fig. 28 which represents a slightly bent knee joint. Under such circumstances the extensor muscle holds the weight of the individual and, of course, is being stretched. Associated with the extensor muscle fibers are muscle spindles around which are *annulospiral endings* or stretch receptors. These stretch receptors fire nerve impulses to the spinal cord *via* a sensory neuron fiber and synapse directly or indirectly with and excite the knee extensor muscle motoneurons. These motoneurons, in turn, fire nerve impulses down their axon fibers to the extensor muscle so that it contracts sufficiently to hold the weight and maintains the desired knee bend position. However, if the muscle contraction is inadequate the knee bends more and stretches the extensor muscle

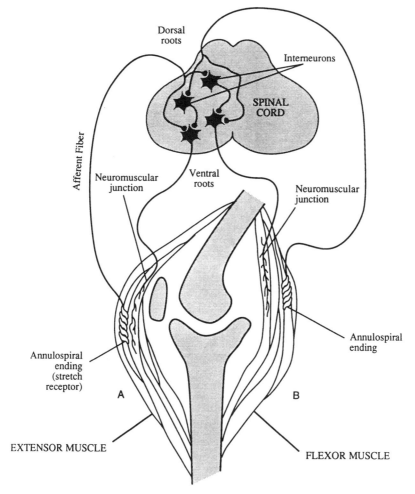

FIG.28. Representation of a slightly bent knee joint and sensory and motoneurons which control the position of this joint *via* the extensor and flexor muscles.

with resulting more rapid firing of nerve impulses from the extensor muscle stretch receptors. This evokes an increased reflex nerve impulse *via* the motoneuron to the extensor muscle which contracts the latter muscle to maintain a steady posture. The antagonist flexor muscle also has stretch receptors and motoneurons. Naturally, in order to hold the desired position the flexor muscle must not contract. Accordingly, the sensory fiber from the extensor muscle stretch receptor branches in the spinal cord and forms an excitatory synapse with a short interneuron which, in turn, synapses with the flexor muscle motoneuron. The latter synapse is inhibitory to the flexor muscle motoneuron and, hence, prevents this motoneuron from firing nerve impulses which would cause contraction of the flexor muscle. It must be emphasized that the above examples of neuromuscular control are extremely simplistic. In reality, posture is maintained as a result of concurrent modulated contractions of antagonistic muscle groups. Furthermore, it is very important to realize that the neuronal circuits described are normally under the control of higher nervous system centers in the brain.

Normally, each activated sensory receptor feeds sensory information into a chain of relays specific to that sense. These relays carry the signal higher and higher into the nervous system. At each level, however, the signal receives additional processing in order to evaluate its importance. The sensory system is, therefore, organized into an ascending hierarchy and information initially gathered at sensory receptors on the periphery moves up through a hierarchial system until it ultimately arrives at the cortex in the brain. Thus, as an example, a sensory receptor on the skin sends nerve impulses into the spinal cord through the dorsal root ganglia (collection of nerve cells). These sensory neurons then pass information (*i.e*, they synapse directly or indirectly) with neurons located in the dorsal horn of the spinal cord which represent the primary relay. These in turn, send nerve impulses along fibers in the spinothalamic tract, *i.e.*, to neurons in the thalamus, the secondary relay. Thalamic neurons project to the somatosensory cortex, which is the tertiary and highest processing center. At each level of this hierarchial system the signal, which originated at the periphery, receives additional processing. Furthermore, at the higher centers in this system sensory information from other sources is added as well as stored information about past experiences. The nature and importance of what has been sensed is ultimately determined by the process known as perception. At this stage any desired or imperative motor action is initiated. In each cerebral hemisphere there is a strip of cortex, adjacent to the somatosensory cortex, which is devoted to motor function called the *motor cortex*. Excitation of the motor cortex originates, in part, from the somatosensory cortex following assessment of all forms of sensory information. The motor cortex is responsible for originating and directing requirement movements. In order to do so the motor cortex must have access to motoneurons which control movements of the body, limbs and fingers. Cortical neurons which communicate directly with motoneurons of the spinal cord are known as Betz cells. The axons of Betz cells project from deep within the motor cortex and converge in a large bundle of fibers known as the *corticospinal* or *pyramidal tract.* It is interesting to note that as the Betz cell fibers descend to the spinal cord they cross over from the side of the cortex from which they originate to the opposite side of the spinal cord. From the spinal cord motoneurons carry instructions to the muscles. This arrangement is called the *primary pyramidal motor system* and, in a similar fashion to the sensory system, it is organized as a hierarchy but in a descending sense.

Motor cortex outputs are driven not only by input from the somatosensory cortex but also from two subcortical structures: the basal ganglia and cerebellum. Both basal ganglia and cerebellum send signals to the motor cortex *via* the thalamus, another subcortical structure. It has been demonstrated that neurons in the cerebellum fire before any muscle activity occurs in a subject trained to respond to a visual cue [8]. The exact role of the cerebellum in motor system function is not understood. However, experiments on subjects with an injured or stimulated cerebellum indicate its importance in holding a proper posture and in the performance of fast, consecutive, simultaneous movements. It has been suggested [41] that the cerebellum possesses a copy of the pattern of movement driven by the motor cortex and that during fine movements the cerebellum controls and adjusts the execution of these movements according to this stored program. The thalamus also relays signals from the basal ganglia to the motor cortex. Experiments indicate that activity in the basal ganglia begins just before initiation of certain types of volitional movements, *i.e.,* slow, directed movements from one region of space to another [41]. This is the kind of movement which is impaired in Parkinson's disease. A good example of the role of the basal ganglia and cerebellum in controlling movement is when one closes one's eyes and tries to touch the tip of one's nose. The large movement of bringing one's hand from its current position to the vicinity of one's nose is controlled by neural activity in the basal ganglia. But, it is the cerebellum which controls the final approach of the hand and finger to the tip of the nose. Figure 29 shows in a very simplified way some of the connections between the cerebellum, basal ganglia and thalamus and thence to the cerebral cortex and from the cortex *via* Betz cells and motoneurons to peripheral muscles.

The basal ganglia, cerebellum and thalamus together form the so–called extrapyramidal motor system. Lesions to the primary pyramidal motor system cause total paralysis of muscles whereas lesions to the extrapyramidal motor system cause uncoordinated, often involuntary, jerky movements. One of the major neurotransmitters in the basal ganglia is DA. Parkinson's disease patients have severaly reduced levels of DA in certain parts of the basal ganglia and exhibit several characteristic motor deficits. Other neurotransmitters associated with the basal ganglia include 5-HT, ACh, GABA and glutamate. The thalamus and cerebellum are innervated more uniformly by almost all of the low molecular weight neurotransmitters discussed earlier.

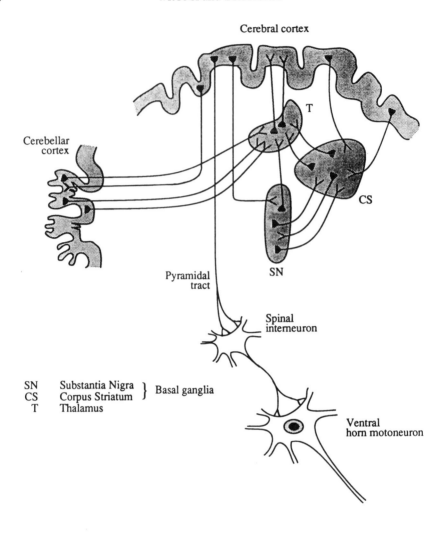

FIG29. Schematic representation of the interactions between the cerebral cortex, basal ganglia, thalamus and cerebellum and the motor system.

Acknowledgements

The authors would like to thank C. LeRoy Blank for help provided in the preparation of this paper and, particularly, anonymous editors who made many valuable suggestions. The illustrations were prepared by Kim Springer whose excellent work is greatly appreciated.

References

[1] L. STRYER, *Biochemistry*, 2nd Edition, Freeman, New York (1981), pp. 205–231.

[2] J. B. FINEAN, R. COLEMAN and R. H. MITCHELL, *Membranes and Their Cellular Functions*, 3rd Edition, Blackwell, Oxford (1984).

[3] M. S. BRETSCHER, *Science*, **181**, 622 (1973).

[4] B. KOLB and I. Q. WHISHAW, *Fundamentals of Human Neuropsychology*, W.H. Freeman and Company, San Francisco (1980), chapter 2, pp. 31–53.

[5] A. L. HODGKIN, *The Conduction of the Nervous Impulse*, Liverpool University Press, Liverpool U.K. (1964).

[6] B. KATZ, *Nerve, Muscle and Synapse*, McGraw–Hill, New York (1966).

[7] R.D. KEYNES, *Proc. Royal Soc. London Ser. B*, **220**, 1 (1983).

[8] E. V. EVARTS, *Scientific American*, **241**, 164 (1979).

[9] I. TASAKI, *Am. J. Physiol.*, **127**, 211 (1939).

[10] A.F. HUXLEY and R. STAMPFLI, *J. Physiol. (London)*, **108**, 315 (1949).

[11] P. McLcGEER, J. C. ECCLES and E. G. McGEER, *Molecular Neurobiology of the Mammalian Brain*, 2nd Edition Plenum Press, New York (1987), pp. 58–60.

[12] H.F. BRADFORD, *Chemical Neurobiology*, W. H. Freeman and Company, New York (1966), pp. 20–22.

[13] R.B. KELLY, J. W. DEUTSCH, S. S. CARLSON AND J. WAGNER, *Annu. Rev. Neurosci.*, **2**, 399 (1979).

[14] P. F. BAKER, A. L. HODGKIN, E. B. RIDGWAY, *J. Physiol. (London)*, **218**, 709 (1971).

[15] R. J. DELORENZO, *Cell Calcium*, **2**, 365 (1981).

[16] R. J. DELORENZO, in: *Neurotransmitter Interaction and Compartmentation*, H. F. BRADFORD (Editor), Plenum Press, New York (1982), pp. 101–120.

[17] W. Y. CHEUNG, *Science*, **207**, 19 (1980).

[18] A.R. MEANS, J. S. TASH AND J. G. CHAFOULEAS, *Physiol. Rev.*, **62**, 1 (1982).

[19] J. C. ECCLES, *The Physiology of Synapses*, Springer – Verlag, Heidelberg (1964).

[20] B. G. CRAGG, *Brain*, **98**, 81 (1975).

[21] H. BOSTOCK AND T. A. SEARS, *J. Physiol. (London)*, **280**, 273 (1978).

[22] P. L. McGEER, J. C. ECCLES, E. G. McGEER, *Molecular Neurobiology of the Mammalian Brain*, 2nd Edition, Plenum Press, New York (1987), pp. 151–154.

[23] J. C. ECCLES and P. L. McGEER, *Trends in Neurological Science*, **2**, 39 (1979).

[24] M. SCHRAMM and Z. SELINGER, *Science*, **225**, 1350 (1984).

[25] M. J. BERRIDGE, *Scientific American*, **253**, 142 (1985).

[26] T. W. STONE, *Neuroscience*, **6**, 523(1981).

[27] R. S. JOPE, *Brain Res.*, **180**, 313 (1979).

[28] D. S. DE BELLEROCHE and I. M. GARDINER, *Br. J. Pharmacol.*, **75**, 359 (1982).

[29] S. BRIMIJOIN, *Progr. Neurobiol.*, **21**, 291 (1983).

[30] R. H. ROTH, in: *The Neurobiology of Dopamine*, A. S. HORN, J. KORF and B. H. C.

WESTERINK, (Editors); Academic Press, London (1979), pp. 101–190.

[31] R. J. BALDESSARINI, in : *Handbook of Psychopharmacology*, L. L. IVERSEN, S. D. IVERSEN and S. H. SNYDER (Editors) Plenum Press, New York (1975), Vol. **3**, pp. 37–137.

[32] L. L. IVERSEN, in: *Handbook of Psychopharmacology*, L. L. IVERSEN, S. D. IVERSEN and S. H. SNYDER (Editors) Plenum Press, New York (1978), Vol. **3**, pp. 381–442.

[33] I. CREESE, *Trends in Neuroscience*, **5**, 40 (1982).

[34] P. SEEMAN, *Pharmacol. Rev.* , **32**, 229 (1980).

[35] R. SUSILO, H. ROMMELSPACHER and G. HÖEFLE, *J. Neurochem.*, **52**, 1793 (1989).

[36] J. C. SCHAWARTZ, H. POLLARD, and T. T. QUACH, *J. Neurochem.*, **35**, 26(1980).

[37] R. P. SHANK AND G. LE M. CAMPBELL, in: *Handbook of Neurochemistry*, 2nd Edition, A. LAJTHA (Editor), Plenum Press, New York (1983), Vol. **3**, pp. 381–404.

[38] G. E. FAGG, *Trends in Neuroscience*, **8**, 207 (1985).

[39] A. C. FOSTER and G. E. FAGG, *Brain Res. Rev.*, **7**, 103 (1984).

[40] J. W. OLLNEY, in: *Kainic Acid as a Tool in Neurobiology*, E. G. MCGEER, J. W. OLNEY and P. L. MCGEER (Editors), Raven Press, New York (1978), pp. 95–122.

[41] F. E. BLOOM and A. LAZERSON, *Brain , Mind and Behavior*, 2nd Edition, W. H. Freeman and Co., New York (1988), Chap.4.

General texts and reviews

P. L. MCGEER, J. C. ECCLES and E. G. MCGEER, *Molecular Neurobiology of the Mammalian Brain*, 2nd Edition, Plenum Press, New York (1987).

Z. L. KRUK and C. J. PYCOCK, *Neurotransmitters and Drugs*, Croom Helm, London (1979).

A. S. HORN, J. KORF and B. H. C. WESTERINK, *The Neurobiology of Dopamine*, Academic Press, New York (1979).

L. HERTZ, E. KVAMME, E. G. MCGEER and A. SCHOUSBOE, *Glutamine, Glutamate, and GABA in the Central Nervous System*, Alan R. Liss, New York (1985).

G. DICHIARA and G. L. GESSA (Editors), *Glutamate as a Neurotransmitter, Advances in Biochemical Psychopharmacology*, Raven Press, New York (1981), Vol. **27**.

F. FONNUM (Editor), *Amino Acids as Chemical Transmitters*, NATO ASI Series, Plenum Pess, New York (1978), vol. **48.**

H. F. BRADFORD, *Chemical Neurobiology*, W. H. Freeman AND Co., New York (1986).

S. D. ERULKAR, *Chemically Mediated Synaptic Transmission: An Overview*, in: *Basic Neurochemistry Molecular, Cellular and Medical Aspects*, G. J. B. SIEGEL, B. W. AGRANOFF, R. W. ALBERS and P. B. MOLINOFF, (Editors), 4th Edition, Raven Press, New York (1989), pp. 151–182.

F. E. BLOOM, *Neurotransmitters: Past, Present and Future Directions*, in *FASEB J.*, **2**, 32 (1988).

E. V. EVARTS, *Brain Mechanisms of Movement,* in *Scientific American,* September 1979, pp. 164–173.

B. KOLB and I. Q. WHISHAW, *Fundamentals of Human Neuropsychology,* W. H. Freeman and Co., San Francisco (1980).

F. E. BLOOM AND A. LAZERSON, *Brain Mind and Behavior,* 2nd Edition, W. H. Freeman and Co., New York, 1988.

F. H. NETTER, *Nervous System,* Parts I and II, Ciba Collections of Medical Illustrations (1983).

EXPERIMENTAL TECHNIQUES TO MEASURE LOW MOLECULAR WEIGHT NEUROTRANSMITTERS AND THEIR PRECURSORS AND METABOLITES

GLENN DRYHURST

Department of Chemistry and Biochemistry
University of Oklahoma, Norman, OK 73019
U.S.A.

Contents

Bioelectrochemistry IV
Edited by B.A. Melandri *et al.*, Plenum Press, New York, 1994

Acronyms

Ach	acetylcholine
AchE	acetylcholine esterase
ALA	alanine
ASP	aspartic acid
t-BT	$tert$-butyl thiol
C_{18}	octadecyl
Ch	choline
CNS	central nervous system
CSF	cerebrospinal fluid
DA	dopamine
DHBA	dihydroxybenzylamine
DOPA	L-dihydroxyphenylalanine
DOPAC	3, 4-dihydroxyphenylacetic acid
DOPEG	3, 4-dihydroxyphenylethylene glycol
EHC	ethylhomocholine = N, N-dimethyl-N-ethyl-3-amino-1-propanol
EPI	epinephrine
EPIN	epinine
GABA	γ-aminobutyric acid
GC	gas chromatography
GC-MS	gas chromatography-mass spectrometry
GLN	glutamine
GLU	glutamic acid
GLY	glycine
5-HIAA	5-hydroxyindole-3-acetic acid
HPLC	high performance liquid chromatography
5-HT	5-hydroxytryptamine
5-HTOL	5-hydroxytryptophol
5-HTP	5-hydroxytryptophan
HVA	homovanillic acid
LCEC	liquid chromatography (with) electrochemical detection
3-MT	3-methoxytyramine
N-Ac-5-HT	N-acetyl-5-hydroxytryptamine
NE	norepinephrine
NM	normetanephrine
N-MET	N-methyl-5-hydroxytryptamine
OPA	o-phthalaldehyde
TAU	taurine
THF	tetrahydrofuran
VMA	vanillomandelic acid

1.   Introduction

Most biochemical and pharmacological studies of the brain and nervous system depend to a considerable extent on analyses designed to measure levels of neurotrasmitters and, often, their chemical precursors and metabolites. Many analytical approaches have been employed to measure the levels of such substances in nervous system tissues and fluids. These include fluorescence methods, gas chromatography (GC) and gas chromatography-mass spectrometry (GC-MS) of suitably derivatized compounds, radioenzymatic methods, immunohistochemistry, radioimmunoassay, autoradiographic methods and immunocytochemical techniques. The application of such techniques to the analysis of endogenous neurochemicals has recently been reviewed [1]. Such methods continue to find applications in neurochemical investigations but, in fact, generally suffer from one or more serious limitations. For example, procedures based upon these analytical approaches are often tedious, slow, subject to interferences and require expensive and rather complex equipment. In the early 1970s methods based on a combination of the powerful separation technique of high performance liquid chromatography (HPLC) and highly sensitive and selective thin-layer electrochemical detectors began to be adapted for the analysis of endogenous neurochemicals in brain tissue, cerebrospinal fluid (CSF), blood, and urine [2]. Indeed, high performance liquid chromatography with electrochemical detection (LCEC) has now emerged as the method of choice for such analyses because it is very sensitive, selective and relatively inexpensive. The selectivity of the method derives not only from the separation power of modern HPLC columns but also from the fact that only compounds which can be oxidized or reduced at the potential applied to the working electrode of the detector give rise to an electrochemical response or signal.

In the sections which follow, a brief introduction to LCEC will be provided. This will include a discussion of the necessary instrumentation and the applications of the technique to analyses for endogenous neurotransmitters and related compounds in neurochemical investigations.

2.   Instrumentation for LCEC

A schematic representation of the equipment required to perform LCEC analyses is presented in Fig. 1. This experimental arrangement consists of two key components: a high resolution separation system, the heart of which is the analytical chromatographic column, and a very high sensitivity thin-layer electrochemical detector. The mobile phase solvent is pumped through the system at a known and constant flow rate by a high pressure reciprocating piston pump. In LCEC applications the mobile phase flow must be as smooth and pulseless as possible simply because at the high detector sensitivities em-

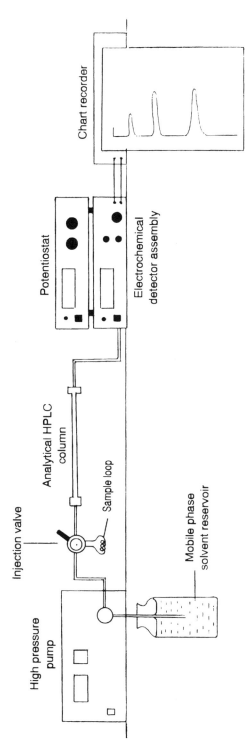

FIG. 1. Schematic representation of equipment for high performance liquid chromatography with electrochemical detection (LCEC).

ployed any pulsations cause large baseline oscillations on the chromatograms obtained. Pumps having dual pistons or the use of pulse dampening devices generally alleviate most mobile phase pulsation effects. Typical flow rates employed in analytical LCEC with conventional packed columns range from about 0.5 to 2.0 cm^3/min and require pump pressures ranging from about 500 to 2000 p.s.i., (3.5 – 14.0 MPa). Electrochemical detectors require mobile phases having relatively high electrochemical conductivities. Accordingly, in LCEC it is usual to employ aqueous buffers containing small amounts of organic solvents (known as organic modifiers) such as methanol, acetonitrile or tetrahydrofuran as the mobile phase solvent. Such highly polar mobile phases, in turn, dictate that the analytical chromatographic column employed must be of the reversed phase, ion exchange, or polar bonded type. In fact, it will become obvious in due course that the vast majority of LCEC applications in neurochemistry have employed reversed phase columns, $i.e.$, particles of silica (3–5 μm) covalently bonded to an octadecyl (C$_{18}$) outer layer.

The mobile phase is initially pumped through an injection valve assembly which is a device by means of which it is possible to introduce a known volume of sample into the mobile phase as a very concentrated plug. Typical injection volumes range from 5 to 100 μl. The sample is then carried by the mobile phase onto the chromatographic column where the individual components are separated. As each compo-

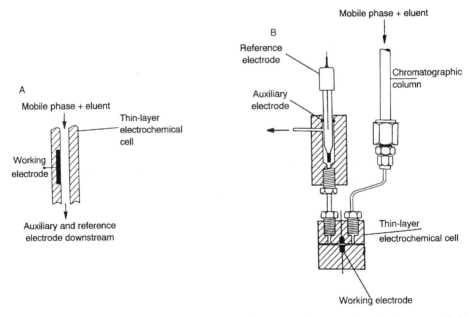

Fig. 2 (A) Geometry for a thin-layer electrochemical cell for use as a detector in LCEC. (B) Commercial thin-layer LCEC detector (Reprinted from P.T. Kissinger and W.R. Heineman, (Eds), *Laboratory Techniques in Electroanalytical Chemistry*, Marcel Dekker, New York, (1984) with permission of the publisher).

nent sequentially emerges from the column it passes into the thin-layer electrochem-
cial detector. Many designs and configurations of these electrochemical detectors have been employed [3]. However, a typical design is shown in Fig. 2. This detector
employs a planar working electrode (*i.e.*, the electrode at which the analyte of in-

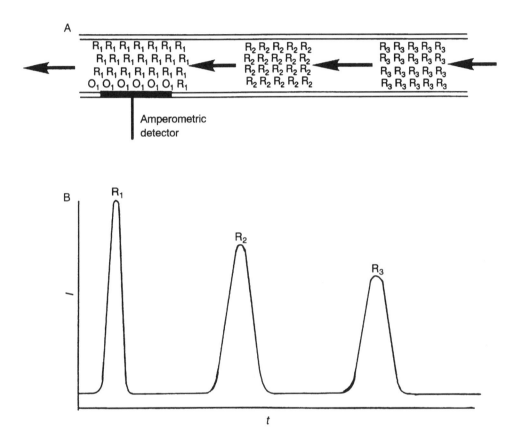

FIG. 3 (A) Representation of chromatographically-separated compounds R_1 R_2 and R_3 being detected at an
amperometric electrochemical detector in a thin-layer cell. (B) Resultant chromatogram.

terest undergoes an oxidation or reduction reaction) and its response is based on
the technique of controlled potential amperometry. The basic function of this detector is illustrated in Fig. 3. Thus, three electroactive substances R_1, R_2 and R_3
emerge sequentially from the analytical chromatographic column and pass into
the thin-layer electrochemical detector unit. The working electrode is held at a
constant potential with reference to a suitable reference electrode by means of a
potentiostat. In order to apply this potential difference a third electrode, known
as an auxiliary or counter electrode is necessary. The placement of the working

reference and auxiliary electrodes in a typical electrochemical detector is shown in Fig. 2. In the example shown in Fig. 3 the working electrode is held at a constant potential at which R_1, R_2 and R_3 can be electrochemically oxidized. Hence, as component R_1 passes over the electrode, the layer of molecules immediately adjacent to the electrode surface is oxidized to O_1, *i.e.*, electrons are transferred from R_1 to the electrode to form O_1 and a current is produced which is proportional to the concentration of R_1 which enters the detector cell. It is important to note that only the layer of R_1 molecules adjacent to the working electrode are electrochemically transformed. Typically <5 % of the total R_1 molecules are oxidized in the detector which is, therefore, generally referred to as an amperometric detector. Detector electrodes can, however, be designed which have very much greater surface areas such that essentially all analyte molecules are electrochemically transformed. These are known as coulometric detectors. Such detectors are intrinsically more sensitive than amperometric detectors, although for a variety of practical reasons the latter detectors are very much more widely employed [3].

When component R_1 has passed over the working electrode in the thin-layer detector the current returns to the baseline level (Fig 3). Subsequently, component R_2 passes over the working electrode and the surface layer is electro-oxidized to O_2 and a current is detected which is proportional to the concentration of R_2. Finally, component R_3 passes over the working electrode and a current signal proportional to its concentration is recorded (Fig. 3). A variety of materials have been employed as electrodes in thin-layer electrochemical detectors in LCEC. The most commonly employed materials are glassy carbon, carbon paste (*i.e.,* a paste of graphite particles in a viscous water-insoluble liquid such as Nujol or bromoform), platinum, and mercury although several other electrode materials have been used [4].

3.   LCEC analysis of the biogenic amines

The initial development of thin-layer LCEC in the form employed today resulted from a need for rapid, highly sensitive and selective analyses for the catecholamine neurotransmitters in brain tissue, CSF and other biological fluids [5]. The selectivity of the LCEC methods for the biogenic catecholamines and indoleamines and their precursors and metabolites derives in part from the powerful separation properties of modern analytical HPLC columns and in part from the ease of electrochemical oxidation of these compounds. The sensitivity of the methods similarly derive from the facile oxidation electrochemistry of the biogenic amines. The electrochemical oxidations of the biogenic catecholamines and metabolites have been studied extensively by the group of R.N. ADAMS at the University of Kansas. These studies have been recently reviewed [6]. The primary step in the electrochemical oxidation of the catecholamines such as dopamine,

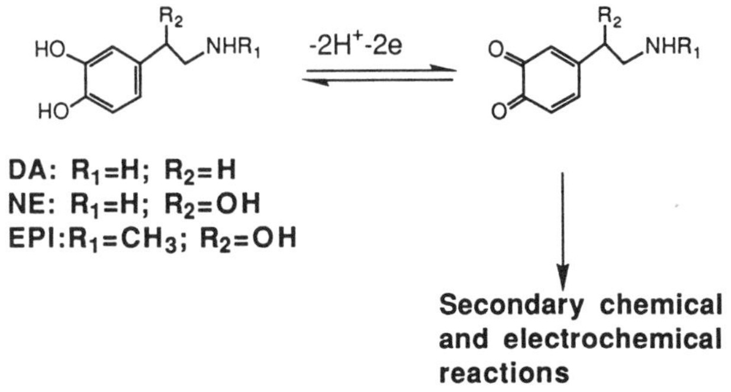

DA: R_1=H; R_2=H
NE: R_1=H; R_2=OH
EPI:R_1=CH$_3$; R_2=OH

epinephrine and norepinephrine is a $2e^-$, $2H^+$ reaction giving the corresponding o-quinone [equation (1)].

The reaction shown in equation (1) probably represents the process that occurs at the thin-layer working electrode in LCEC because of the very short contact times involved. However, during longer time-scale conditions complex follow-up chemical and electrochemical reactions follow the initial formation of o-quinone intermediates.

The 3-methoxy substituted metabolites of the biogenic catecholamines, formed as a result of catechol-o-methyltransferase activity, are initially electro-oxidized to oxonium ion intermediates which demethylate to give o-quinones [equation (2)], [6].

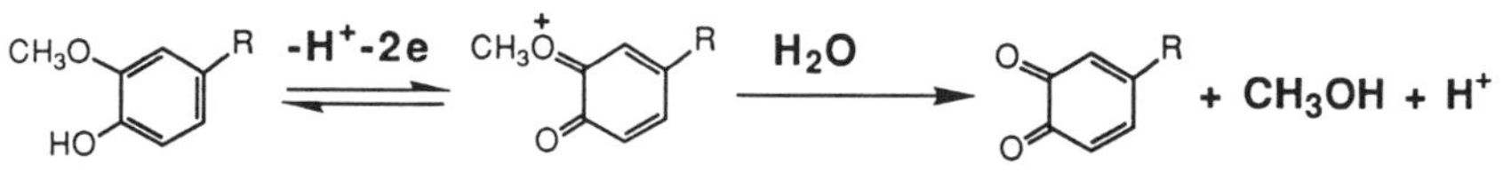

Electrochemical oxidations of the indolic neurotransmitter 5-hydroxytryptamine (5-HT; serotonin) [7–9] and related endogenous indoles [10, 11] in aqueous solution are very complex processes. However the initial step in the electro-oxidation reaction involves a 1 e^-, 1 H^+ reaction to a radical intermediate which is rapidly further oxidized (1 e^-) to a carbocation [equation (3)]. While many secondary chemical and electrochemical reactions can occur, the oxidation conceptualized in equation 3 is largely responsible for the current response monitored in LCEC detectors.

Following the initial report of an LCEC method for the analysis of the biogenic catecholamines and some of their metabolites in nervous system tissue and fluids [5], similar methods were described to determine 5-HT [12] and catecholamine [13–18]

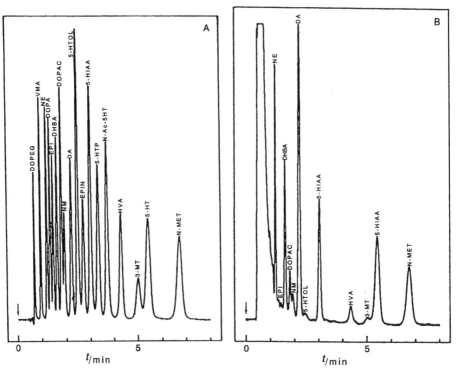

$$\text{HO}\underset{\text{N}}{\overset{R_1}{\underset{R_2}{\bigotimes}}} \xrightarrow{\text{-e -H}^+} \underset{\text{H}}{\overset{R_1}{\underset{R_2}{O=}}} \xrightarrow{\text{-e}} \underset{\text{H}}{\overset{R_1}{\underset{R_2}{O}}}^+ \tag{3}$$

5-HT: R_1=NH$_2$; R_2=H
5-HTP: R_1=NH$_2$; R_2=COOH
5-HTOL: R_1=OH; R_2=H

Secondary chemical and electrochemical reactions

and indoleamine [19–21] metabolites. Rather than discuss the individual methods which have been employed to determine these different classes of compounds, the power and utility of LCEC will be illustrated by, perhaps, the most comprehensive method presently available. This LCEC method can be employed for the simultaneous

FIG. 4 (A) LCEC chromatogram of a mixture of 18 catecholamines, indoleamines and related compounds. 5 mm^3 injection contained each compound at a concentration of *ca.* 1 μ*M*. (B) 5 mm^3 of whole rat brain homogenate. Conditions: Perkin-Elmer HC-18 reversed phase column (100 x 4.6 mm; 3 μm particle size). Mobile phase: 0.1 *M* citric acid, 0.05 m*M* ethylenediaminetetraacetic acid, 7.5 % acetonitrile (vol:vol), 0.255 m*M* sodium octyl sulfate, 0.06 % diethylamine (wt/vol), pH 2.45. Flow rate: 1.85 cm^3/min. Carbon paste thin-layer electrode set at + 0.85 V *vs.* Ag/AgCl electrode. (Reprinted from LIN *et al. J. Liq. Chromatogr.,* by courtesy of Marcell Dekker, Inc).

determination of all of the catecholamine and indoleamine neurotransmitters and many of their important precursors and metabolites [22]. The selectivity and sensitivity of this method, which employs a short reversed phase column packed with 3 μm particles is illustrated in Fig.4A. This shows an LCEC chromatogram of a mixture of 16 different catecholamine and indoleamine and related species along with two compounds employed as internal standards, *i.e.*, 3,4-dihydroxybenzylamine (DHBA) and N_ω-5-methyl hydroxytryptamine (N-MET). Clearly, all 18 compounds can be resolved in less than 7 minutes. The structures of the components separated and analyzed in Fig. 4 are shown in Table I. The chromatogram shown in Fig. 4B was obtained on a sample of rat brain homogenate. This analysis is particularly elegant since samples could be analyzed after only homogenization of the brain, centrifugation, and clarification by filtration [22]. No further purification steps were required. Thus, in a very short period of time it is possible to determine at least 10 endogenous catecholamines, indoleamines and metabolites.

In order to carry out quantitative determinations of the compounds of interest it is necessary to prepare standard solutions. These are prepared by replacing the tissue (*e.g.*, brain) sample with a roughly equivalent volume of a stock external standard solution containing the species to be quantitated at concentration levels appropriate to the unknown samples. The internal standard DHBA is employed for catechol analyses (*i.e.*, for the early eluting components) while N-MET is used as an internal standard for analyses of indolic components (*i.e.*, for the late eluting components). The level of a specific endogenous compound in a tissue sample, for example, is calculated using the expression:

$$c\,(\text{nmole/g}) = \frac{R_{sample}}{R_{external\ standard}} \quad X \quad \frac{\text{nmol compound in external standard}}{\text{weight sample (g)}} \quad (4)$$

where R_{sample} = ratio of the peak height of the compound of interest to that of the internal standard in a single tissue sample; $R_{external\ standard}$ = average ratio of the peak height of the compound of interest to that of the internal standard in all of the standard samples chromatographed; nmol compound in external standard = the number of nmol of the compound of interest in a single external standard sample; weight of sample, g = the weight of the tissue sample in grams.

A careful inspection of Fig. 4A and 4B reveals that in the brain sample analysis the chromatographic peaks of DOPEG and VMA are masked by a large background/solvent front peak ratio. In order to determine such metabolites further purification of the sample prior to injection is necessary [22].

The LCEC methodology outlined above for the analysis of catecholamines, indoleamines and related substances in nervous tissue can also be employed to determine the activities of key enzymes. This is accomplished by measuring the product of the enzyme reaction of interest under conditions where further metabolism of the product is blocked.

TABLE 1. STRUCTURES OF CATECHOLAMINES, INDOLEAMINES AND METABOLITES

Name	Abbreviation	Structure
Dopamine	DA	
Norepinephrine	NE	
Epinephrine	EPI	
L-3, 4-Dihydroxyphenylalanine	DOPA	
3, 4-Dihydroxyphenylthyleneglycol	DOPEG	
Vanillomandellic acid	VMA	
3, 4-Dihydroxyphenylacetic acid	DOPAC	

continues TABLE 1. STRUCTURES OF CATECHOLAMINES, INDOLEAMINES AND METABOLITES

Name	Abbreviation	Structure
Normetanephrine	NM	
Epinine	EPIN	
Homovanillic acid	HVA	
3-Methoxytyramine	3-MT	
5-Hydroxytryptamine	5-HT	
5-Hydroxytryptophol	5-HTOL	
5-Hydroxyindole-3-acetic acid	5-HIAA	

continues TABLE 1. STRUCTURES OF CATECHOLAMINES, INDOLEAMINES AND METABOLITES

Name	Abbreviation	Structure
5-Hydroxytryptophan	5-HTP	
N-Acetyl-5-hydroxy-tryptamine	N-Ac-5-HT	
N$_\omega$-Methyl-5-hydroxy- tryptamine[1]	N-MET[1]	
3, 4-Dihydroxybenzylamine[1]	DHBA[1]	

[1]Internal standards employed for quantitation purposes.

As an illustration, in order to determine the activities of *L*-tyrosine hydroxylase and *L*-tryptophan hydroxylase *in vivo,* the experimental animal is treated with the drug NSD 1015 which blocks the enzymes *L*-DOPA decarboxylase and *L*-5-hydroxytryptophan hydroxylase (Scheme I). The consequent build up of *L*-DOPA and *L*-5-hydroxytryptophan in brain tissue can be measured by LCEC and, accordingly, can be used to determine the *in vivo* activities of tyrosine hydroxylase and tryptophan hydroxylase. LIN *et al.* [22] have developed similar LCEC methodology to determine the activities of *L*-DOPA decarboxylase, *L*-5-hydroxytryptophan decarboxylase, monoamine oxidase and catechol-*O*-methyltransferase.

It should be noted that catecholamine transmitters and most of their metabolites are all very polar compounds and, hence, have very low retention times on reversed phase columns even using 100 % aqueous buffers as the mobile phase [23]. Thus, separation of mixtures of these compounds on reversed phase columns is normally impossible. In order to resolve this problem a long chain detergent such as sodium octyl

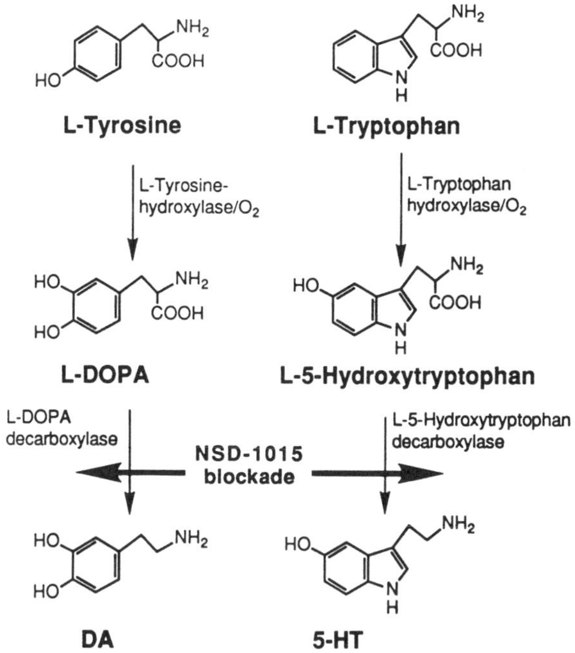

Scheme I

sulfate is added to the mobile phase (0.1-10 mM). After equilibration (*ca.* 48 h) the detergent is adsorbed on the reversed phase (*e.g.*, octadecyl, C_{18}) stationary phase and forms, in effect, a layer of negative charges on the surface. Thus, the reversed phase column is effectively converted into an ion exchange column and, as a result, the catecholamines, indoleamines and related endogenous compounds can be separated. The advantage of this approach over using a conventional ion-exchange analytical column is that the reversed phase column–detergent system provides superior peak symmetry and chromatographic efficiency [23].

4. LCEC analysis of acetylcholine and choline

Levels of acetylcholine (ACh) and choline (Ch) in nervous tissue and fluids can be determined by a variety of methods including GC-MS of the appropriately derivatized compounds, radioenzymatic assay and by bioassay techniques [24]. These methods, however, are time consuming and/or expensive to perform. However, recently developed LCEC techniques [25–28], which are relatively simple, rapid and inexpensive, have largely replaced these methods. The original method, developed by POTTER *et al.* [25], was based on the separation of ACh and Ch by reversed phase HPLC. Then, as the effluent emerged from the analytical column it was mixed with acetylcholinesterase (AChE) and choline-oxidase (Ch oxidase). Ch oxidase converts the Ch and Ch produced by hydrolysis of ACh (by AChE) to betaine and H_2O_2 (Scheme II). The H_2O_2 byproduct is then detected by its electrochemical

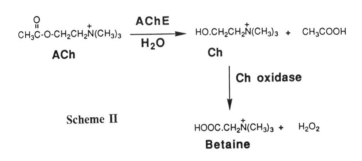

Scheme II

oxidation at a platinum working electrode housed in a thin-layer cell. The more recent methods which developed from this work employ either a reversed phase column [27] or a polystyrene column [27, 29] to initially separate ACh and Ch. The analytical column, however, is coupled directly to a reactor which contains AChE and Ch oxidase either adsorbed to a short anion exchange column [27] or covalently bound to a proprietory inert polymeric material packing [29]. The experimental arrangement, employed by EVA *et al.* [27] and depicted in Fig. 5, employed an analytical column packed with polystyrene particles. The initial compound to elute from this column is Ch which enters the reactor column where it is oxidized by the action of Ch oxidase to betaine and H_2O_2. The H_2O_2 so formed is detected at the thin-layer cell Pt working electrode where it is oxidized. The current response so measured is proportional to the concentration of H_2O_2 which, in turn, is proportional to the Ch concentration.

Ethylhomocholine (EHC; N,N-dimethyl-N-ethyl-3-amino-1-propanol [25]), used as an internal standard for ACh and Ch analyses, is the second compound to elute from the analytical column. EHC is oxidized by the catalytic action of Ch oxidase in the reactor column to generate H_2O_2 which is electrochemically detected. The final compound to elute is ACh which in the reactor column is first hydrolyzed to Ch by the action of AChE which is oxidized to betaine and H_2O_2 by Ch oxidase. Again, the H_2O_2 produced is detected electrochemically.

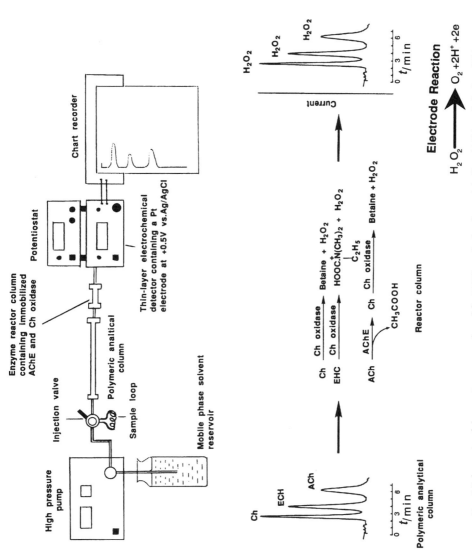

FIG. 5 Schematic of the equipment employed for LCEC analysis of acetylcholine (ACh) and choline (Ch).

A significant problem associated with analyses of ACh and Ch in brain derives from the fact that this tissue contains very high AChE activity. Accordingly, when conventional methods of animal sacrifice are employed, such as decapitation, virtually all endogenous ACh is hydrolyzed to Ch before analyses can be performed. Thus it becomes necessary to rapidly deactivate AChE. This is very conveniently accomplished by the use of the focused microwave irradiation technique for animal sacrifice [30, 31]. With this technique a beam of microwave radiation is focused on the head of the experimental animal. The localized, very rapid heating caused by the microwave radiation rapidly kills the experimental animal and denatures AChE and, indeed, other CNS enzymes *in situ*. Brain tissue can then be excised using conventional surgical techniques. Representative LCEC chromatograms obtained using the experimental arrangement shown in Fig. 5 are presented in Figure 6.

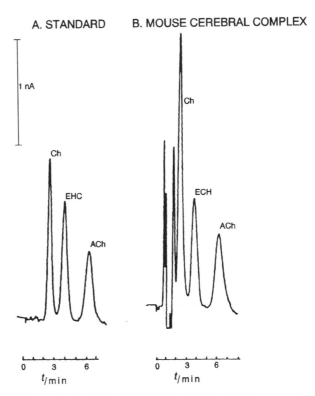

FIG. 6. Chromatograms of: (A) a standard solution of ACh, EHC and Ch and (B) mouse cerebral cortex extract using a HAMILTON PRP-1 polystyrene cartridge analytical column (100 x 4.6 mm, 10 μm particles) and an enzyme-loaded (AChE, Ch-oxidase) postcolumn reactor (Brownlee Aquapore A x 300, 30 x 2.1 mm, 10 μm particles). Mobile phase: 20 mM tris acetate, 200 μM octane sulfonate, 1 mM tetramethylammonium chloride, and 2.5 % acetonitrile, pH 7.0. Standards: 20 mm^3 of a solution containing Ch (125 pmol), EHC (250 pmol) and ACh (250 pmol) was injected. Tissue extract: 50 mm^3 containing EHC as internal standard was injected. (Reprinted from EVA *et al.*, *Anal. Biochem.*, **143**, 320 (1984) by courtesy of Academic Press, Inc.).

5. LCEC analysis of amino neurotransmitters

One of the earliest methods devised to separate and analyze mixtures of amino acids was described in 1958 by Sparkman *et al.* [32]. These workers employed ion-exchange chromatography to separate the amino acids which were then reacted with ninhydrin to give a colored complex which could be detected. This technique lacks the speed and sensitivity required for the analysis of amino acid neurotransmitters and putative transmitters in nervous tissue and fluids. HPLC, particularly using reversed phase columns and gradient mobile phase systems has recently been developed to rapidly separate mixtures of amino acids. However, because of the difficulties associated with the detection of most amino acids even by spectroscopic methods, it is still necessary to employ strategies to convert these compounds to derivatives which can be easily detected. Derivation strategies have largely been based on the reaction of amino acids with *o*-phthalaldehyde (OPA) [33–36]. Perhaps the most important of these methods involves the reaction of OPA with amino acids in the presence of a thiol to form S-substituted isoindole derivatives which are fluorescent (Scheme III).

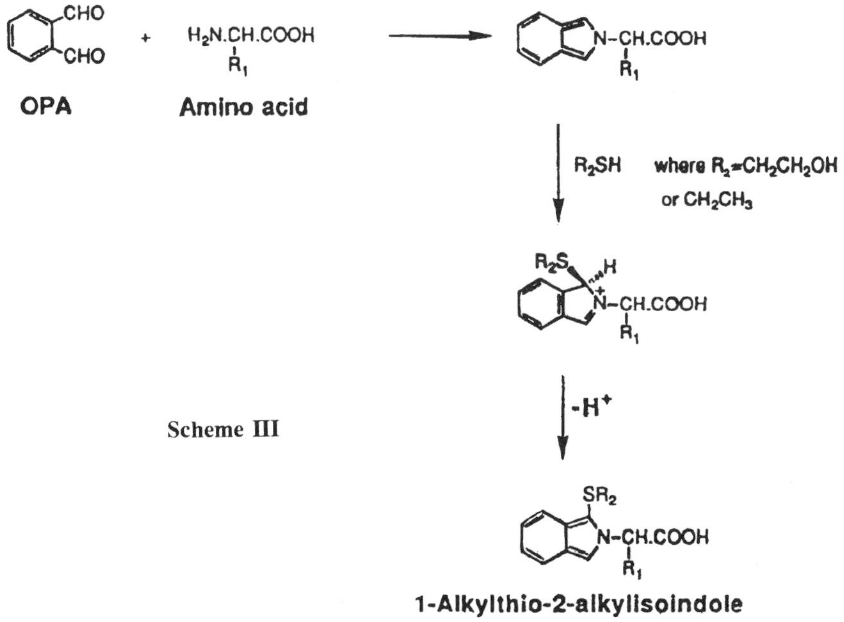

Scheme III

1-Alkylthio-2-alkylisoindole

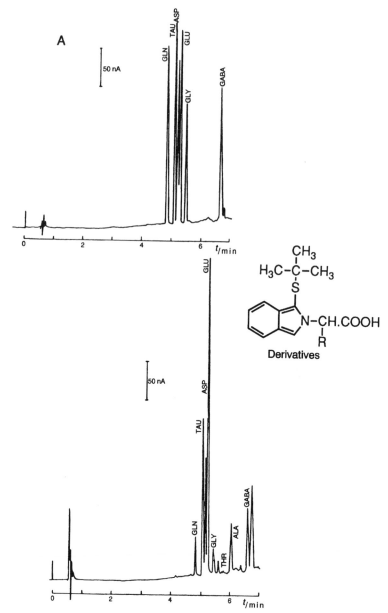

FIG. 7 . (A) Chromatogram of OPA–*t*-butylthiol derivatives of 167 pmol each of glutamine (GLN), taurine (TAU), aspartic acid (ASP), glutamic acid (GLU), glycine (GLY) and γ-aminobutyric acid (GABA). (B) Chromatogram of neurotransmitter amino acids present in rat brain homogenate. Concentrations in homogenate were: GLN, 0.216 m*M*; TAU, 0.808 m*M*, ASP, 0.726 m*M*; GLU, 1.63 m*M*; GLY, 0.241 m*M*; GABA 0.50 m*M*. Chromatographic conditions: Perkin-Elmer HS-3 reversed phase (C₁₈) column (100 x 4.6 mm, 3 μm particle size); initial mobile phase: 67.2 % perchlorate–citrate buffer (0.05 *M* NaClO₄, 0.005 *M* trisodium citrate, pH 5.00), 30.8 % methanol and 2 % tetrahydrofuran THF. The gradient progressed in linear segments to 33.6 % buffer, 64.4 % MEO and 2 % THF.at 2 min from injection and then to 20 % buffer, 45 % MeOH and 35 % THF at 5 min. The latter conditions were maintained until 8 min and then the initial conditions were resumed. Flow rate: 1.5 cm³ min⁻¹. Detector: glassy carbon thin-layer electrode with an applied potential difference of + 0.80 V *vs*. Ag/AgCl electrode. (Reprinted from ALLISON *et al.*, *Anal. Chem.*, **56**, 1089 (1985), with permission of the American Chemical Society).

These S-substituted isoindoles have most widely been formed using β-thioethanol or ethanethiol. Generally, the derivatives are formed prior to reversed phase HPLC (*i.e.*, precolumn derivation) [35–37] but because of their instability analytical procedures require very careful control. Recently, however, ALLISON *et al.* [38] have demonstrated that the stability of the S-substituted isoindole derivatives of amino acids are greatly improved when *t*-butylthiol (*t* BT) is used in the derivatization reaction. The OPA–*t* BT derivatives can be readily separated using a gradient mobile phase system and a reversed phase column and detected at a glassy carbon,working electrode housed in a thin-layer electrochemical cell. The OPA–*t* BT derivatives are electrochemically oxidized at quite low potentials permitting their easy detection and quantitation by LCEC. Stable derivatives of glutamate, aspartate, γ-aminobutyric acid, glycine and many other putative amino acid neurotransmitters can be formed. Figure 7A shows an LCEC chromatogram of the OPA–*t* BT derivatives of glutamine (GLN), taurine (TAU), asparatic acid (ASP), glutamic acid (GLU), glycine (GLY) and γ–aminobutyric acid (GABA). A chromatogram of amino acids present in a rat brain homogenate is presented in Fig. 7B. Quite clearly this gradient mobile phase LCEC method permits the very rapid analysis of brain tissue for endogenous amino acids.

Using the OPA–β-thioethanol or OPA–*t* BT derivatives and gradient HPLC separations, detection limits of less than 500 femtomoles can be easily achieved [38]. However, subsequent modifications of these methods have permitted the determinations of ASP, GLU, GLY, TAU and GABA by means of LCEC of their OPA–*t* BT derivatives with detection limit of 50–100 femtomols [39]. A rapid gradient LCEC method to determine ASP and GLU in, for example, very small volume microdialysis samples has also been described [40]. An isocratic LCEC determination of GABA by means of its OPA–*t* BT derivative having a detection limit of 100–200 femtomoles has recently appeared [41].

6.   Conclusions

The recently developed technique of LCEC provides a very rapid, relatively simple and highly sensitive analytical method to determine the levels of the biogenic amines and their metabolites, ACh and Ch and many of the amino acid neurotransmitters.

References

[1] A.A. BOULTON, G.B. BAKER and J.M. BAKER, *Neuromethods,* Volumes **2** and **3**, Humana Press, Cifton, N.J., (1985).

[2] R.E. SHOUP, *High Performance Liquid Chromatography,* **4**, 91 (1986).

[3] P.T. KISSINGER in: *Laboratory Techniques in Electroanalytical Chemistry,* P.J. KISSINGER and W.R. HEINEMAN, (Editors), MARCEL DEKKER, New York, (1984), Chapter 22.

[4] R.E. SHOUP in: *Recent Reports on Liquid Chromatography with Electrochemical Detection,* BAS Press, West Lafeyette, IN, (1981).

[5] P.T. KISSINGER, C.J. REFSHAUGE, R. DREILING and R.N. ADAMS, *Anal. Lett.,* **6** 465 (1973).

[6] G. DRYHURST, K.M. KADISH, F. SCHELLER and R. RENNEBERG, *Biological Electrochemistry,* Academic Press, New York, (1982), Chapter 2.

[7] M.Z. WRONA and G. DRYHURST, *J. Electroanal. Chem. Interfacial Electrochem.,* **278**, 249 (1990).

[8] M.Z. WRONA and G. DRYHURST, *Bioorg. Chem.,* **18**, 291 (1990).

[9] G. DRYHURST, *Chem. Rev.,* **90**, 795 (1990).

[10] K. HUMPHRIES and G. DRYHURST, *J. Pharm. Sci.,* **76,** 839 (1987).

[11] F-C CHENG, M.Z. WRONA and G. DRYHURST, *J. Electroanal. Chem. Interfacial Electrochem.,* **310,** 187 (1991).

[12] S. SASA and C.L. BLANK, *Anal. Chem.,* **49,** 354 (1977).

[13] P.T. KISSINGER, L.S. BRUNTLETT, G.C. DAVIS, L.J. FELICE, R.M. RIGGIN and R.E. SHOUP, *Clin. Chem.,* **23,** 1449 (1977).

[14] C.R. FREED and P.A. ASMUS, *J. Neurochem.,* **32**, 163 (1979).

[15] I.N. MEFFORD, *J. Neurosci. Methods,* **3,** 207 (1981).

[16] L.J. FELICE and P.T. KISSINGER, *Anal. Chem.,* **48**, 794 (1976).

[17] R.E. SHOUP and P.T. KISSINGER, *Clin. Chem.,* **23**, 1268 (1977).

[18] R.M. WIGHTMAN, P.M. PLOTSKY, E. STROPE, R.J. DELCORE and R.N. ADAMS, *Brain Res.,* **131**, 345 (1977).

[19] D.D. KOCH and P.T. KISSINGER, *J. Chromatogr. Biomed. Appl.,* **164,** 441 (1979).

[20] I.N. MEFFORD and J.D. BARCHAS, *J. Chromatogr. Biomed. Appl.,* **181**, 187 (1980).

[21] F. PONZIO and G. JONSSON, *Dev. Neurosci.,* **1,** 80, (1979).

[22] P.Y.T. LIN, M.C. BULAWA, P. WONG, L. LIN, J. SCOTT and C.L. BLANK, *J. Liq. Chromatogr.,* **7,** 509 (1984).

[23] R.E. SHOUP, *High Performance Liquid Chromatography,* **4**, 91 (1986).

[24] I. HANIN in: *Modern Methods in Pharmacology,* S. SPECTOR and N. BACK, (Editors), ALAN R. LISS, New York, (1982), pp. 29-38.

[25] P.E. POTTER, J.L. MEEK and N.H. NEFF, *J. Neurochem.,* **41,** 188 (1983).

[26] J.L. MEEK and C.EVA, *J. Chromatogr.,* **317**, 343 (1984).

[27] C. EVA, M. HADJICONSTANTINOU, N.H. NEFF and J.L. MEEK, *Anal. Biochem.,* **143,** 320 (1984).

[28] G. DAMSMA, B.H.C. WESTERINK and A.S. HORN, *J. Neurochem.,* **45**, 1649 (1985).

[29] R.E. SHOUP, *Current Separations,* **9**, 61 (1989).

[30] Y. MARUYAMA, Y. IKARASHI and C.L. BLANK, *Biogenic Amines,* **4,** 55 (1987).

[31] W.B. STAVINOHA in: *Study of Brain Neurochemistry Utilizing Rapid Inactivation of Brain Enzymes by Heating with Microwave Irradiation,* C.L. BLANK, W. STAVINOHA and Y. MARUYAMA, (Editors), Pergamon Press, New York, (1983), pp. 1-12.

[32] D.H. SPACKMAN, W.H. STEIN and S. MOORE, *Anal. Chem.,* **30**, 1190 (1958).

[33] D.W. HILL, F.H. WALTERS, T.D. WILSON and J.D. STUART, *Anal. Chem.,* **51**, 1338 (1979).

[34] T.P. DAVIS, C.W. GEHRKE, C.W. GEHRKE, T.D. CUNNINGHAM, K.C. KUO, K.O. GERHARDT, H.D. JOHNSON and C.H. WILLIAMS, *J. Chromatogr.,* **162**, 293 (1979).

[35] D.L. HOGAN, K.L. KRAEMER and J.I. ISENBERG, *Anal. Biochem.,* **127,** 17 (1982).

[36] P. LINDROTH and K. MOPPER, *Anal. Chem.,* **51**, 1667 (1979).

[37] D.C. TURNELL and J.D.H. COOPER, *Clin. Chem.,* **28,** 527 (1982).

[38] L.A. ALLISON, G.S. MAYER and R.E. SHOUP, *Anal. Chem.,* **56,** 1089 (1984).

[39] P.A. SHEA and W.A. JACOBS, *Current Separations,* **9,** 53 (1989).

[40] W.A. JACOBS and P.A. SHEA, *Current Separations,* **9,** 59 (1989).

[41] P.A. SHEA and W.A. JACOBS, *Current Separations,* **9,** 57 (1989).

THE ELECTROCHEMICAL SIGNAL TRANSMISSION BY THE ACETYLCHOLINE RECEPTOR: SINGLE CHANNEL CONDUCTANCE EVENTS AND OLIGOCHANNELS

EBERHARD NEUMANN and THEO SCHÜRHOLZ

Faculty of Chemistry, University of Bielefeld,
P.O. Box 100131, D-4800 Bielefeld 1, Germany

Contents

Bioelectrochemistry IV
Edited by B.A. Melandri *et al.*, Plenum Press, New York, 1994

1.   Introduction

The rapid signal transmission between the nerves or a nerve and the target cells of a muscle or of a fish electric organ is mediated electrochemically. In cholinergic synapses (Fig.1) the electrochemical signal transfer, resulting in muscle contraction or in electric discharge, is initiated by nerve impulses triggering the release of the neurotransmitter acetylcholine [1,2].

The agonist aceylcholine (ACh) is cationic; it binds to the postsynaptic (nicotinic) acetycholine receptor (AChR) causing a conformational change in the AChR protein [2]. The ACh-induced structural change implies the transient opening of a pathway for the flow of Na^+ and K^+ ions within the allosteric AChR [3]. The ion flow

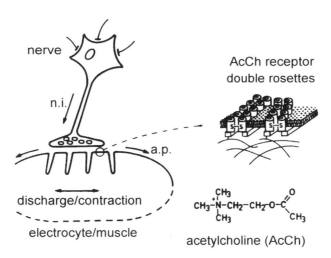

FIG. 1. Scheme for the rapid signal transmission through the synapse of a cholinergic nerve and a muscle cell or an electrocyte of an electric organ. A nerve impulse (n.i.) triggers the release of acetylcholine (ACh⁺), stored in synaptic vesicles. ACh–binding to the ACh receptors (AChR), organized in double rows in the postsynaptic membrane of the target cell (muscle or electrocyte), finally causes action potentials (a.p.) leading to muscle contraction or electrocyte discharge.

through the receptor channel effects the electric depolarization of the postsynaptic membrane. The depolarization, in turn, triggers the action potential which is necessary to cause the coherent contraction of the muscle cells or the concerted electric discharge of the electrocytes in fish electric organs.

2.   Oligomeric AChR organization and action potential

The triggering of the action potential not only requires that the postsynaptic membrane is depolarized by at least 20 mV (vertebrate muscle cells), but also that this potential difference change proceeds within a short time interval of about 0.3 ms [1]. It is presumed that in cholinergic synapses this *steep* membrane depolarisation is related to the particular organization of the nicotinic acetylcholine receptor. In *Torpedo* fish electrocytes the AChR is organized in linear rows of dimer rosettes [4,5]. The functional significance of these dimers and of the linear dimer clusters is not known.

2.1.   Isolation of receptor proteins

The AChR of *Torpedo* and *Electrophorus electricus* can be detergent-solubilized and isolated in two macromolecular forms [6–11]. When sulfhydryl-alkylating agents are present in the first tissue homogenization step, the dimer ($M_r \approx 290,000$) consisting of two disulfide-bridged monomers is the predominant species. In contrast, if reducing agents such as dithioerythriol (DTE) are used, the monomer ($M_r \approx 290,000$) of subunit composition ($\alpha_2\beta\gamma\delta$) will prevail.

It is known that the presence of higher concentrated detergents such as 2 % Triton X-100 during gel electrophoresis and ultracentrifugation causes cleavage of the δ - δ subunit disulfide-bridge [10], probably by SH/SS-rebridging [12]. After termination of the centrifugation run, the protein of the monomer peak however contains appreciable amounts of covalent dimers (V. VOLZ, unpublished results). There is apparently a partial reformation of covalent dimers in the sample fractions collected from the monomer peak position.

On the other hand, at low detergent and lipid concentrations, the formation of receptor multimers or even of large aggregates was observed in detergent solution [10,13) as well as during AChR reconstitution into vesicles [14]. The large channel events or oligochannels of reconstituted AChR are suggestive of receptor oligomers which open and close simultaneously on the time scale of channel current resolution [15,16]. It is remarked that oligochannels caused by cooperatively operating clusters of densely packed receptors would lead to a steeper depolarisation than randomly distributed, uncoupled receptor channels.

So far, isolation and purification of intact AChR species from mammalian muscle membranes are impaired by proteolytic nicking. The dimer disulfide linkage appears to be especially sensitive to proteolytic attack [17]. The dimer species of the *Electrophorus* AChR could only be preserved when the preparation was performed in the presence of several protease inhibitors [10]. The failure to isolate AChR dimers from mammals might therefore be due to the high lability of the intermonomeric disulfide bond. On the other hand, most of the *in vivo* patch-clamp data on the nicotinic AChR were gained at myoballs from rat or frog as well as from human cells lines. Therefore, only the *Torpedo* AChR expressed from cloned cDNA in *Xenopus laevis* oocytes [18] can be taken as the reference for the purified *Torpedo* AChR reconstituted in vesicles.

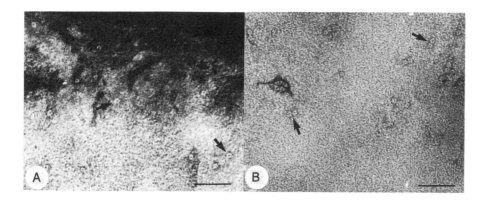

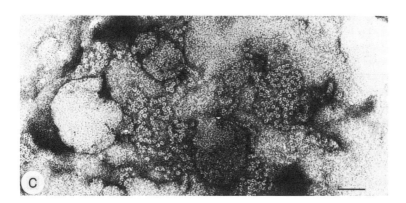

Fig. 2. Electron micrographs of lipid vesicle-reconstituted AChR species of *Torpedo californica* electric organ. The arrows point at monomers (A) and dimers (B), reconstituted at 0.1 g/dm^3 receptor protein, respectively. A loose association of receptor proteins is obtained at a higher protein concentration (0.5 g/dm^3). The bars refer to 50 nm.

2.2. Functional differences of the Torpedo AChR species

There is now conclusive evidence that there are also functional differences between the isolated AChR monomer and dimer species. In flux studies with vesicle-reconstituted AChR the monomer shows an almost ten–fold lower apparent agonist binding constant, as well as a less pronounced desensitization behavior [19,20]. In addition, the AChR monomer reconstituted in planar lipid membranes has an almost two–fold lower single–channel conductance than the reconstituted dimer species [15].

Evidence is accumulating that it is the dimer species, which is the functional unit that is quantitatively related to the Ca^{2+}– dependent conductance levels of the cloned *Torpedo* AChR channels expressed in oocytes. The data suggest that the single channel events of the dimer reflect the highly cooperative, and thus simultaneous, switching of the two constituent monomeric parts of the *Torpedo* AChR in both the reconstituted vesicl membrane and the *Xenopus* oocytes. The larger macrochannel events of the reconstituted AChR have conductance values which are multiples of the dimer or of the monomer conductances. Such oligochannels caused by cooperatively functioning AChR oligomers are suggestive to guarantee the steep depolarisation required for the initiation of the postsynaptic action potentials causing muscle contraction and discharge of electric organs.

3.   Reconstitution of *Torpedo* AChR into lipid membranes

Negative stain electron micrographs (Fig. 2) may serve to characterize the protein species of the AChR reconstituted in soybean lipid vesicles. It is seen that both the AChR monomer and dimer species, respectively, are incorporated into the vesicles *as single units*. In vesicles containing relatively high protein concentrations apparently loosely-associated AChR species are seen (Fig. 2C). The concentration of monomers in the vesicles is considerably lower compared to reconstituted dimers because a major part of the monomeric AChR molecules is not incorporated into vesicles, but forms large ribbon-like aggregates [21].For patch-clamp measurements the vesicle size has to be increased by 2-3 freeze and thaw cycles; addition of pure lipid vesicles to the protein-lipid vesicles (formed by detergent dialysis) strongly increased the vesicle patch stability.

4.   Vesicle patch-clamp of reconstituted *Torpedo* AChR

In vesicles of AChR–dimer preparations the dominant single channel event has a conductance value of 84 ± 6 pS at 100mM $CaCl_2$ and 20 °C (Fig. 3A). In addition, higher conductance values occur, frequently 170 ± 9 pS and 255 ± 9 pS. These macrochannel events or oligochannels are multiples of the 84 pS dimer channels event. The conductance

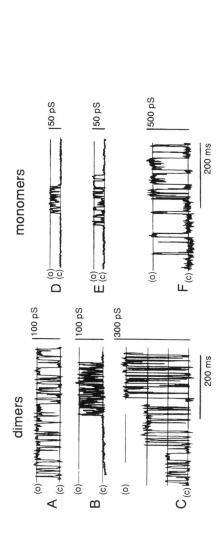

FIG. 3. Single channel events (K$^+$-conductances) of vesicle-reconstituted *Torpedo californica* AChR, induced by suberoyldicholine (SCh) in the patch-pipette (diameter Ø ≈ 0,5µm). Pipette and external solution: 0.1 M KCl, 0.1 mM CaCl$_2$, 10 mM HEPES, pH 7.0; 20 °C. A, B, C: dimer species, C: oligochannels (multiples of the events in A). D, E, F: monomer species, F: oligochannels (multiples of events in D, E) [SCh] = 0.5 µM (A, D); [SCh] = 5 µM (B, E, C, F). The monomers were prepared with DTT (D, E) or separated in a sucrose density gradient by centrifugation (F). Patch membrane potential difference: 100 mV (A, B, C) 200 mV (C, D) and 25 mV (F). The patch currents were measured with an EPC amplifier usually 2-3 min after patch formation, digitized with a modified Sony PCM 701 at a sampling rate of 40 kHz, stored on a video tape recorder and analysed with a computer oscilloscope (E. Bablock, Augsburg, Germany) or with a CED 1401 patch-clamp system (open–time distribution). (c) closed channel, (o) open channel.

values of all the channels events observed in one patch, normally were stable during the whole record of on average 20 min. The AChR channel activity has been induced by the agonist suberoyldicholine (SCh), present in the patch pipette in concentrations ranging from 0.2 to 5 μM.

At [SCh] < 1 μM, frequent switching of the channels between the open and closed state has been observed (Fig.3). Two main channls patterns are apparent. In the first two minutes after patch formation the channels are predominantly in the open state. Open-time and closed-time analysis of up to 10,000 channels yields a mean open-time τ_o = 4.0 ± 1.6 ms and a mean closed-time τ_c = 0.7 ± 0.2 ms, fitted to exponential distributions, respectively; hence $r = \tau_o/\tau_c$ > 1. At times ≥ 3 min after patch formation, τ_o = 1.9 ± 0.6 ms and τ_c = 4.5 ± 1.3 ms (Fig. 4); r thus reverts to r < 1. Transitions between the two configurations, r > 1 and r < 1, either develop continuously with recording time or happen suddenly.

At higher agonist concentrations, [SCh] ≥ 1 μM, the AChR channels preferentially open in burst of flickers with short closed-time, followed by long lasting quiet periods (Fig. 3B). The average open-time of a flicker burst is 0.8 ± 0.2 ms and the closed-time is 2.3 ± 0.3 ms.

Single channel events of AChR-monomers, prepared by reductive splitting of the S-S bond of the dimer, have a mean conductance value of 42 ± 3 pS at 100 μM KCl, 0.1 mM CaCl$_2$ and 20 °C (Fig. 3D). This low conductance value is rather seldom seen with the dimer preparations (which contain ≥ 10 % monomers as judged from SDS-PAGE analysis). Multiple conductance steps, as observed with the dimer preparations, are rare in the case of the monomer, which is consistent with the low monomer concentration in the vesicles. The channel openings of the monomers only appears as bursts; the quiet periods between two bursts range from several seconds to minutes. Both, monomers and dimers show a linear I-U-dependence over the whole U-range analyzed (Fig.5).

4.1   Channel histograms and oligochannels

Whereas single channel traces of preferably one conductance value are instrumental for the straightforward analysis of open-times and closed-times, multiple channels records more obviously demonstrate the different conductance value of the monomer and dimer preparations under the same conditions. A channel histogram obtained from traces of up to 10 minutes is shown in Fig. 6. It can be clearly seen that the respective conductance levels of the monomer channels are only half of the conductance levels of the dimers. It appears that occasional subconductance events are statistically not important.

Monomers, enriched by separation of a mixture of monomers and dimers in sucrose gradient centrifugation (i.e., without reductive splitting) show dominantly

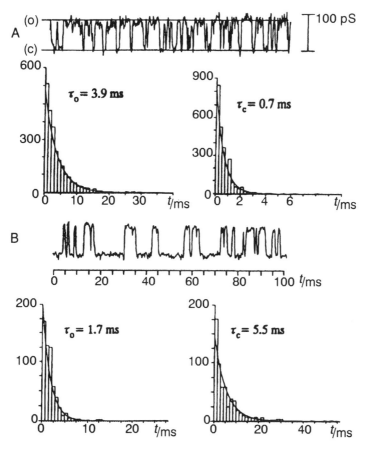

FIG. 4. The two typical opening-closing pattern of AChR dimers with different ratios of the mean open-time τ_o and the mean closed-time τ_c from single exponential fits. A, $r > 1$, current recording 2 r min after patch formation; B, $r < 1$, current recording 5 min after patch formation. The frequency-of-occurrence histograms refer to 10-20 min current recordings. Membrane potential difference 200 mV, 20 °C, see legend to Fig. 3.

higher conductance values, *e.g.* 400 pS (Fig. 3F). In addition, single channel events of 80, 120, 160, and 240 ($\pm$ 7) pS are measured. The numerical differences between the conductance values confirm that, also with sucrose gradient proteins, the lowest unit of the monomer conductance is 42 $\pm$ 3 pS at 0.1 M K⁺, at 0.1 mM Ca²⁺ and 20 °C.

Applications of the agonist as well as of the channel inhibitor α-bungarotoxin are only effective when applied on the inside of the patch pipette.

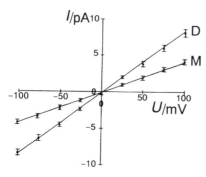

FIG. 5. Single channel I-U diagram of the AChR dimer (D) and monomer (M) species, respectively. Experimental conditions as in Fig. 3. The slope conductances are 84 pS (for D) and 42 pS (for M) at 0.1 M KCl, 0.1 mM Ca^{2+}, 20 °C.

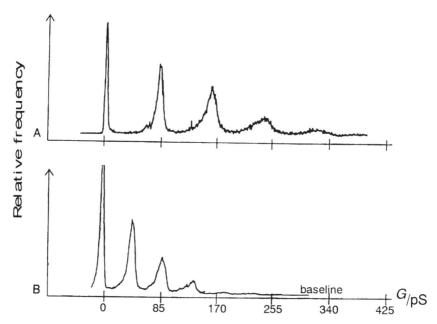

FIG. 6. Histograms of single channel events of AChR dimer (A) and monomer (B) preparations, respectively. The monomers were prepared by reductive splitting (DTT) of dimers. Conditions as in Fig. 3.

4.2 Dependence of the K⁺–conductances on Ca²⁺–concentration

The influence of Ca^{2+} on the channel conductance of the reconstituted *Torpedo* AChR can be studied experimentally only between 0.1 mM and 2 mM Ca^{2+}. At higher Ca^{2+}– concentrations, the lipid vesicles aggregate in the bath; below 0.1 mM Ca^{2+} the life time of the membrane patch is too low. At 2 mM Ca^{2+} the K⁺–conductance of the reconstituted AChR dimer channel (84 pS) is decreased to about half its initial value (42 pS) at no added Ca^{2+} (Fig. 7). The mean open – and closed – times, as measured for 0.1 mM Ca^{2+}, do not change, when the Ca^{2+}–concentration is increased. If the Ca^{2+}–concentration is increased only on the outside of the patch–pipette (bath) the channel conductance remains constant.

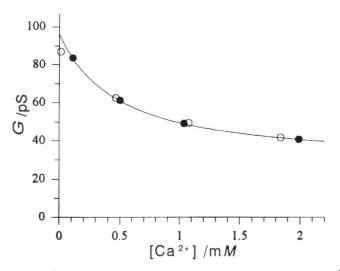

Fig. 7. Dependence of the K⁺– conductances G of *Torpedo californica* AChR on the external Ca^{2+} concentration $[Ca^{2+}]$ at 0.1 M KCL, 20 °C. ●, vesicle–reconstitued dimer; ■, vesicle–reconstituted monomers. Dimer conductance fit with equation (4) of the text yields G_0 = 98 pS, G_∞ = 27 pS and K_{Ca} = 0.48 mM. The conductance data (○) of *Xenopus* oocyte–expressed *Torpedo californica* AChR at 0.1 M KCl, 12 °C, by Iмото *et.al.* (1986) fit to those of the vesicle–reconstituted *Torpedo* AChR dimer.

5. Comparison of the AChR channel events

If one compares the single channel events of the reconstituted AChR monomer and dimer species, the different reconstitution efficiency of the two receptor species has to be taken into account. Since the monomer concentration in the vesicle is low, the probability of receptor aggregation is also low in accordance with the mass action law. Indeed, at variance with the AChR dimers, multiple channel events of DTE-treated monomers, indicating receptor aggregation are generally absent. The increase of AChR channel events after repeated freeze and thaw cycles of the reconstituted receptor vesicles can be rationalized by the redistribution of receptor molecules concentrated in a part of the vesicles during the reconstitution by detergent dialysis, leaving many vesicles void of protein [22-24].

5.1   Channel activity and receptor desensitization

Since the agonist has to be applied from the inside of the patch pipette, the AChR is exposed to the agonist about one minute before channel recording is technically perfmable. Therefore most of the channel events probably reflect the channel activity of the desensitized receptor states, even at the beginning of the U-clamp measurements. Often, the single channel events were rather rare. The decrease in channel open-times and the concomitant increase of the closed–times at low agonist concentration develops with time and can be interpreted as one of the functional features of the desensitization phenomenon. Another desensitization feature is that, similar to observations at intact muscle cells [25], at agonist concentrations > 1 μM the openings of the channels occur in bursts. Interestingly, after chemically splitting of the δ–δ disulfide bridge only burst–like channel openings were observed with reconstituted AChR.

Work with chimaeras between the *Torpedo californica* and bovine AChR δ–subunits expressed in *Xenopus laevis* oocytes has shown that the δ–subunit is primarily responsible for the different gating kinetics and conductance values of bovine and *Torpedo* receptors [26,18].

5.2   Channel–pair cooperativity of the AChR dimer

The Ca^{2+} dependence of the K^+– conductance indicates that the channel events of the various AChR species must be compared at the same ionic conditions.

The single channel events of the *Torpedo* AChR dimer (84 pS at 0.1 mM Ca^{2+}) is twice as large as that of the monomer species (42 pS at 0.1 mM Ca^{2+}). On the same line of interpretation of the different conductance values found for *Torpedo* AChR monomers and dimers in planar lipid bilayers [15], the dimer single channels event reflects the highly cooperative opening and closing of the two monomer channels in the dimer species. The *planar* bilayer AChR conductance values are however about twofold lower (20 pS and 40 pS) than those of the *vesicle*–reconstituted AChR (Fig.3). The lower values of the AChR species in planar bilayers are probably due to irreversible structural changes resulting from the exposure of AChR protein to the air/water interface during the formation of the planar bilayer, or to the differences in the previous protein preparation and reconstitution. In addition the higher ion concentration (0.5 M NaCl) of the planar bilayer system increases the cohesive pressure of the lipid bilayer, which can cause a reduction of the channel conductance. On the other hand, AChR reconstitution into vesicles does not expose the proteins to a water/air interface.

5.3. AChR cluster and oligochannels

The multiple channels of the dimer preparation (170 pS and 255 pS) very often appear as single channel events within the time resolution of the channel recording. The apparent cooperative interaction between channel molecules resulting in macro– or oligochannels does not require the formation of a covalent bond between the species [15,16]. In solution the AChR monomers appear to have a higher tendency to form aggregates than the dimers. Large aggregates of non–incorporated protein were formed during the reconstitution process of DTE–treated monomers. Surprisingly, when monomers were separated from dimers by centrifugation in a sucrose gradient (as a less invasive alternative to the DTE treatment) massive oligomeric channels were observed. However, because there is partial S–S bond reformation after termination of the centrifugation field, only 35 % – 50 % monomers are obtained by this method from the monomer peak as proved by SDS–PAGE analysis (V. VOLZ, unpublished results). The observed oligochannels suggest that the sucrose density centrifugation conditions obviously foster the formation of protein oligomers leading to multiple channel events.

5.4. HILL cooperativy and AChR gating scheme

The HILL coefficient of the carbamoylcholine (CCh) induced Rb^+–uptake into *Torpedo* AChR–reconstituted vesicles is lowered from $n_H = 1.8 \pm 0.2$ to $n_H = 1.2$ after reduction of the AChR with dithiothreitol (DTT); in addition, the agonist binding constant is reduced six-fold [19,20]. These data indicate that structural as well as functional changes are connected with the reductive dissociation of the dimers to the monomers.

Corresponding to the two α–subunits per monomer species also two agonist binding sites have been inferred [27]. At the usually applied toxin concentrations of ≤ 10 μM, freshly prepared receptor-rich *Torpedo* membrane fragments have been found to bind only *one* α–bungarotoxin (α–BTX) molecule per AChR monomer [28]. If detergent is added the second site is gradually occupied [28,29]. The presence of detergent however does not change the number of ACh molecules bound per monomer species. In the range [ACh] ≤ 1 mM, ACh always occupies only one of the two α–sites. Therefore, in membrane fragments without detergents the ACh:α–BTX binding ratio is 1:1. In the presence of detergents the toxin binding reference for membrane fragments as well as for the isolated receptor species is different: the ratio ACh:α–BTX is 1:2 per monomeric AChR species. Interestingly, the investigators reporting the stoichiometry ratio in membrane fragments to be 1:1 have assayed for toxin binding in the absence of detergent [30-33] while the ratio 1:2 appears to refer to the presence of 0.1 % Triton X-100 [34].

The apparent discrepancy between the binding of only one ACh per ACh monomer and the HILL coefficient $n_H = 2$ for channel activity and Rb^+–ion flux can be readily rationalized in terms of the cooperatively synchronized channel–pair of the dimer species.

According to the pair–cooperativity concept, channel activation results from the successive association of two ACh molecules (A), with the two monomeric low affinity conformers R_l of the dimer species $R_l R_l$ according to

$$2 A + R_l \cdot R_l \leftrightarrow AR_l \cdot R_l + A \leftrightarrow AR_l \cdot AR_l$$
$$\updownarrow$$
$$AR_l^* \cdot AR_l^*$$

The functionally dominant step is the concerted transition of the dimer complex $AR_l \cdot AR_l$ to the ion conducting conformer complex $AR_l^* \cdot AR_l^*$ yielding $n_H = 2$ for channel activation. The presence of DTE or DDT appears to split the dimer into monomers, consistent with the observed $n_H = 1$ for the Rb^+–flux [20].

The occupation of only one of the α–subunits within the monomeric parts of the dimer compares well with the channel activation by irreversibly binding agonits. It is sufficient for channel activation when only one *irreversible agonist* binds, after reduction of disulfide bridges in the vicinity of a binding site for ACh, on one of the two α–subunits [35–38]. The covalent agonist labeling of a second α–subunit requires much higher concentrations of irreversible agonists [39].

There are still other arguments supporting an activation scheme in terms of the dimer. For instance, flux amplitude analysis of both the CCh–induced Li^+–efflux [40,41] and the $^{23}Na^+$–efflux [42] from sealed *Torpedo* membranes (microsacs) shows that the *functional* unit of ion transport gating comprises two α–BTX sites. Since in the *absence* of detergents only one α–BTX molecule binds with high affinity per monomeric AChR [28,29] the functional unit of ion transport gating must be the dimer (with $n_H = 2$).

Receptor inactivation proceeds in at least two phases (see, *e.g.* Ref. 3): a fast cooperative one [43,44] and a slow non cooperative one [42,44]. Dialysis data show that at low ACh concentrations (≤ 50 mM) the slow transition (time constant $\tau = 3$ min) to the final equilibrium state of very high affinity, denoted by R_{vh} ($K_{vh} \approx 5$ nM), is induced by agonist binding [29]. The preexisting high affinity receptor state Rh to which the agonist directly binds has a smaller affinity ($K_h \approx 0{,}1$ μM for ACh).

Channel activation from desensitized ACbR. – Channel activity is also observed under conditions where the AChR protein is usually assumed to be inactivated or desensitized [25,15]. Whereas the responses immediately after agonist addition are frequently occurring short single channel events, longer exposure to agonists gives rise to occasional bursts of repetitive opening and closing events [25,45,46,15]: *Nachschlag* or flickering [25]. The occasional dimer–channel activation under desensitizing condizions requires an extension of the receptor activation scheme to include the reversible cooperative coupling ($n_H = 2$) of the open conformation $AR_l^* \cdot AR_l^*$ to inactivated states:

$$AR_l^* \cdot AR_l^* \leftrightarrow AR_h \cdot AR_h$$

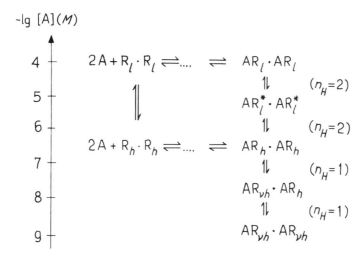

$-lg\ [A]\,(M)$

FIG. 8. Gating scheme for the neurotransmitter–induced channel activation of the *Torpedo* AChR in terms of the channel-pair cooperativity of the dimer species. Two main patterns of channel appearances after addition of agonists (A). Rapid activation form the low-affinity receptor conformers R_1 ($K_1 = 10^{-4} - 10^{-5}$ M for ACh, $[R_1]$ / $[R_h] \approx$ 1:5 to 1: 10). The opening-closing behaviour is decribed by the transitions $AR_l\ AR_l \leftrightarrow AR^*_l\ AR^*_l$ coupled to binding of A with $n_H = 2$ and $r = \tau_o/\tau_c > 1$. (The rare transitions of unliganded conformers $R_1\ R_1 \leftrightarrow R^*_l$ R^*_l are not included. The channel activations form desensitized conformers R_h (preexisting, $K_h \approx 10^{-7}$ M for ACh) and R_{vh} ($K_{vh} \approx 10^{-9}$ M for ACh) are modelled by the transition $AR_h\ AR_h \leftrightarrow AR^*_l\ AR^*_l$ which is coupled to two reaction steps. The $AR_h\ AR_h$ conformer is coupled to direct activator binding yielding the rapid desensitization mode ($n_H = 2$ and $r < 1$) and to intramolecular non-cooperative transitions $AR_h \leftrightarrow AR_{vh}$, rate limiting for the slow desensibilization phase ($n_H = 1$) and the flicker bursts.

Gating scheme – Based on the dimer cooperativity concept, the reaction scheme depicted in Fig. 8 is an abbreviated form of the possibly more complicated reaction network of AChR activation and inactivation processes. In detail, the inactive (intermediate) state AR_h may be populated either *via* the open receptor states or by direct binding of activator to the R_h conformers. The slow non cooperative phase of inactivation to the final state AR_{vh} of very high AcCh affinity involves structural transition of the type $AR_h \leftrightarrow AR_{vh}$, where each of the two AR_h conformers within the dimer $AR_h \cdot AR_h$ can independently convert to the AR_{vh} state ($n_H = 1$). The scheme does not include occasional, very rare openings ($R_l \cdot R_l \leftrightarrow R^*_l \cdot R^*_l$) in the absence of ACh [47]. The intermediate states of half activation $R^*_l \cdot R^*_l$ and $R^*_l \cdot R_l$ are the structural candidates for conductance substates.

Monomer. – The isolated AChR monomer in detergent solution as well as reconstituted in planar lipid bilayers appears to have a looser structure compared to the dimer [10,15]. The high affinity ACh binding to the monomer species is a factor of about ten ($K \approx 5 \times 10^{-8}$ M) less stable [6] than to the dimer ($K \approx 5 \times 10^{-9}$ M). The hydrodynamic properties of the monomer suggest a more extended, perhaps partially unfolded structure [10]. As seen in Fig. 3, the occurrence of substates of lower conductance are much more frequent for the monomers than for the dimers. The monomer channels are thus more *flexible* than the dimer channels. The looser structure of the monomer may also be the reason why membrane fragments in the presence of detergents (0.1 % Triton X–100) bind two α–BTX molecules per monomer compared to only one high affinity bound α–BTX per monomer in the absence of detergents.

6.   The Ca²⁺ effect on the *Torpedo* AChR channel protein

The sidedness of the Ca^{2+} effect on the K^+–conductance of vesicle–reconstituted *Torpedo* AChR matches that of the divalent ion effect on *Xenopus* oocyte–expressed *Torpedo* AChR [18], suggesting that the extracellular (synaptic) side of the receptor protein is particularly sensitive to the presence of divalent ions.

As demonstrated in Fig. 7, within the margin of experimental error the conductance values of the oocyte-expressed *Torpedo* AChR (recorded at 12 °C) [18], and the [Ca^{2+}] dependence are the same as those of the vesicle-reconstituted *Torpedo* AChR dimer (recorded at 20 °C).

Due to the great tendency of the AChR monomers to aggregate to dimers and higher oligomers, in particular at higher Ca^{2+} concentration (> 0.1 mM), the conductance value of the monomer could reliably be only measured at 0.1 mM Ca^{2+}.

In conclusion, to the extent to which the vesicle-reconstituted *Torpedo* AChR can be compared to that expressed in *Xenopus* oocytes, the conductances and their Ca^{2+}–dependences suggest that the AChR single channel events recorded in oocyte membranes are actually synchronous double–channel events caused by the dimer species.

6.1.   *Ca²⁺ binding model*

The detergent–solubilized *Torpedo* AChR binds large amounts of Ca^{2+} ions in the Ca^{2+} concentration range of the dissociation equilibrium constant $K_{Ca} \approx 0.33$ mM at 20 °C [48]. ACh addition (10^{-5} M) releases part of the bound Ca^{2+}. Because of the ACh–AChR interaction, the Ca^{2+} is most probably released from the protein part of the AChR–lipid-detergent micelles.

One explanation for the dependence of the AChR channel conductance on divalent cation concentrations is that it reflects changes in the specific Ca^{2+} association to negatively charged groups of the AChR. The partial *inhibitory* effect of Ca^{2+} (Fig. 7) suggests that the K^+–ion conductance G is proportional to free Ca^{2+}–binding sites R of the AChR and that

$$\frac{G - G_\infty}{G_0 - G_\infty} = \frac{[R]}{[R_T]} = 1 - \beta_{Ca} \tag{2}$$

where $[R_T]$ is the total site concentration, $\beta_{Ca} = [CaR] / [R_T]$ is the degree of Ca^{2+}–binding, G_0 refers to $[Ca] \rightarrow 0$ and G_∞ is the K^+–conductance at saturating Ca^{2+}–concentrations. If the several binding sites are considered independent, β_{Ca} is given by

$$\beta_{Ca} = \frac{[Ca]}{[Ca] + K_{Ca}} \tag{3}$$

where K_{Ca} is the apparent dissociation equilibrium constant. Combining the equations (2) and (3) yields

$$G = G_0 - (G_0 - G_\infty). \frac{[Ca]}{[Ca] + K_{Ca}} \tag{4}$$

The data in Fig. 7 are fitted with equation (4) and yield: $G_0 = 98$ pS, $G_\infty = 27$ pS, $K_{Ca} = 0.48$ mM. Due to the large excess of monovalent ions near the AChR ($[K^+]_{out} = 0.36$ M, see below), K_{Ca} refers to $[K^+]_{out} \approx 0.4$ M and is therefore practically independent of Ca^{2+} in the range $[Ca^{2+}] \leq 3$ mM. The K_{Ca} value is, as expected, slightly larger than that of 0.33 mM, measured at a lower monovalent ion content ($[Na^+] \approx 0.1$ M) [48].

6.2. Electrostatic screening model

It is known that the *Torpedo* AChR protein has an isoelectric point of pI = 4.9, at 20°C [49], suggesting that at pH 7 the protein is anionic. Site directed mutagenesis and patch–clamp data analysis have revealed that the *Torpedo* AChR has three anionic rings probably on the inner surface of the tubular vestibule [50]. At low divalent cation concentration permeant cations would be concentrated in the tubular vestibule of the channel protein as compared with the bulk solution. Negatively charged lipids in the vicinity of the channel protein will also contribute to the cation accumulation.

Divalent cations will effectively screen the anionic surface charge and thereby reduce the locally higher concentration of permeant ions. Since the K^+–conductance of the AChR may be assumed to be proportional to $[K^+]$, the conductance ratio G_0/G_∞ reflects the ratio of the locally higher K^+–concentration $[K^+]_0$ and the bulk solution value $[K^+]_b$. The *charged channel model* relates G with the electrostatic surface potential Ξ_0 caused by the anionic groups of the channel vestibule [51, 52]. For K^+ as the permeant cation we obtain with $z = z(K^+) = 1$:

$$\frac{G_o}{G_\infty} = \frac{[K^+]_o}{[K^+]_b} = \exp\frac{-\mathcal{F}\varphi_o}{RT} \tag{5}$$

where $\mathcal{F}$ is the FARADAY constant, R the gas constant and T the absolute temperature.

The data in Fig. 7 yield $[K^+]_o = 0.36\ M$ and $\varphi_o = -32.8$ mV. Thus the negative surface potential causes an increase of the K^+–concentration at the channel mouth and in the vestibule by a factor of 3.6 as compared with the bulk.

The surface charge density is given by

$$\sigma_o = \frac{\varepsilon_o\,\varepsilon\,q_o}{l_D} \tag{6}$$

where ε_o is the vacuum permittivity, ε the dielectric constant and l_D the DEBYE screening length. At $[K^+]_o = 0.36\ M$, $l_D \approx 0.5$ nm. Using $\varepsilon = 80$ (at 20 °C) in equation (6) we obtain $\sigma_o = -0.04$ C/m². If the anionic groups are monovalent ($z = -1$) the group density $N_o = -\sigma_o/e_o = 0.29$/nm², where e_o is the elementary charge. Thus the average distance $<x>$ between the anionic groups is $<x> = (N_o)^{-1/2} = 1.8$ nm, which is about four times the screening length.

If we approximate the tubular vestibule geometry by a hollow-cylinder of surface $A_{hc} = \pi\phi_i\,l_i$ with $\phi_i \approx 3$ nm as the average inner diameter and $l_i \approx 6$ nm as the average length of the tubular vestibule [53, 29], the effective number of negatively charged groups in the AChR monomer vestibule is $N = N_o\,A_{hc} = 16.4$, which appears plausible in the light of the rather low pI-value and of the anionic sugar residues at the channel mouth.

7.   Comparison with mammalian muscle AChR

The single–channel conductance events of the purified, vesicle–reconstituted AChR dimers of *Torpedo californica* (27 pS at 0.1 M K$^+$, [Ca^{2+}] > 5 nM, Fig. 7) are very similar to the conductance events of unextracted AChR of membrane microsacs from *Torpedo marmorata,* incorporated in planar bilayers (25 pS, at 0.25 M Na$^+$, [Ca^{2+}] = 4 mM, [Mg^{2+}] = 2 mM [54]). The close similarity suggests that the structural unit underlying the single–channel events in the biomembrane of electrocyte microsacs is also the synchronized double–channel of the AChR dimer.

In any case, a comparison of conductances is only meaningful at similar conditions for the permeant monovalent cations and for the divalent ions. At present it is not known whether the model of concerted dimer gating also applies to other cholinergic systems. Remarkably, the single–channel conductances of frog and rat muscle AChR [25,55] are closer correlated with the conductance of the *Torpedo* AChR dimer than with that of the monomer species. However, comparison is obscured by the different ionic conditions apart from possible species differences.

On the other hand the channel conductance of oocyte-expressed rat muscle AChR (38 pS at 0.115 M Na$^+$, 1.8 mM Mg^{2+}, 20 °C) [56]) compares well with that of the *Torpedo* AChR dimer (42 pS, 0.1 M KCl, 2 mM Ca^{2+}, 20 °C) at similar concentrations of the monovalent permeant ions and of the divalent ions. Therefore the rat muscle AChR single–channel events most likely also reflect the AChR dimer. Similar to the early channel data of *Torpedo* AChR [15], the single channel events of rat myotube AChR have also been interpreted as the simultaneous open–closed switching of two channels [57].

8.   Experimental details

Electric organ tissue from *Torpedo californica* was received on dry ice from Pacific Bio–Marine Laboratories, Inc., Venice, CA and was stored in liquid nitrogen. ^{125}J–α–bungarotoxin (^{125}J–BTX) was obtained from Du Pont, NEN Research Products. Benzamidine HCl, phenylmethylsulfonyl fluoride (PMSF), iodoacetamide (IAA) dithioerythriol (DTE), N-ethylmaleimide (NEM), carbamoylcholine and gallamine trithiodide (Flaxedil) were purchased from Sigma. The soybean phosphatide extract ($\approx$ 20 % (w/w) phosphatidylcholine) were from Avanti Polar Lipids. The affinity gel, methyl [N–6–aminocaproyl–6'aminocaproyl)–3–amino] pyridinium bromide (dicaproyl–MP) / Sepharose 4B was a generous gift from Prof. H.W. CHANG, New York. The detergents CHAPS (3–[(3–cholamidopropyl) dimethylammonio]–1–propanesulfonate) and sodium dodecylsulfate were from Sigma, sodium cholate was purchased from Serva.

8.1   Preparation of Torpedo AChR species

All procedures except the first homogenization step were carried out at 4 °C. The buffers were adjusted to pH 7.4, Millipore filtrated and deaerated. The purity of the AChR protein was analyzed by sodium dodecyl sulphate polyacrylamide gel electrophoresis (SDS-PAGE). Protein concentrations were estimated by absorbance at 280 nm ($A_{280}^{0.1\%}$ = 1.8) [6] and measured with the BCA-protein assay (Pierce).

Method A. – Frozen *Torpedo* electric organ (100 – 200 g; all buffer volumes refer to 100 g of tissue) was homogenized in 300 cm^3 of 10 mM EPES, 5 mM EDTA, 5 mM EGTA, 3 mM NaN$_3$, 5 mM IAA, 3 mM benzamidine HCl, and 0.2 mM PMSF and centrifuged 1 h at 25000 g. The pellet then was homogenized in 200 cm^3 10 mM HEPES, 1 M NaCl, 1 mM EDTA, 3 mM NaN$_3$, and was centrifuged 1h at 25000 g. The pellet

was resuspended in 150 cm³ 10 mM HEPES, 1 mM, EDTA and 3 mM Na N$_3$ followed
by centrifugation 1 h at 25 000 g. Membrane proteins were extracted by shaking the
resuspended pellet 1.5 h with 100 cm³ of 10 mM HEPES, 1 mM EDTA, 3 mM NaN$_3$, 1
% (w/w) CHAPS, and 5 g/dm³ soybean phospholipids (SBL). Then the non–solubilized material was separated by 1 h centrifugation at 45 000 g. The supernatant was immediately applied to the affinity column. The gel was washed successively with 100
cm³ buffer W1 (10 mM HEPES, 0.1 mM EDTA, 0.5 % CHAPS and 5 g/dm³ SBL), 50 cm³
buffer W2 (buffer W1 and 80 mM NaCl), and 50 cm³ buffer W3 where the [NaCl] was
reduced to 60 mM. The gel-bound receptor protein was eluted with 150 μM Flaxedil
in buffer W3 and stored at –80 °C.

   Monomer. – After the first centrifugation step of the dimers A the pellet was resuspended in 50 cm³ of 50 mM HEPES, 1 mM EDTA, 10 mM carbamoylcholine chloride and 3 mM NaN$_3$. DTE was added to yield 7.5 mM and the homogenate was shaken
for 30 min at 4 °C. After the reductive splitting of the S–S bond of the dimer, sulfhydryl
group alkylation was performed with 18 mM IAA, again for 30 min, to stabilize the
monomeric form. The suspension was diluted with 100 cm³ 50 mM HEPES, 1 mM
EDTA, 1,5 M NaCl, 3 mM NaN$_3$, and centrifuged for 1 h at 45 000 g. The pellet was homogenized in buffer W3, pelleted and extracted as described above.

   AChR monomers in a sucrose gradient. – Samples of 1 nmol AChR, prepared
without alkylating reagent, are successively layered on top of a 11 ml gradient of 5 %
to 20 % sucrose containing 1 % CHAPS, 10 mM HEPES, 5 mM EDTA, 3 mM NaN$_3$, pH
7.0. The gradients were centrifuged for 10–14 h at 4 °C in a Kontron TST 41 rotor at
200 000 g. Fractions of 0.8 cm³ were collected and analyzed for protein content. The
upper band, containing the enriched monomers, was dialyzed against the reconstitution buffer and the AChR was reconstituted into vesicles.

8.2   Reconstitution into vesicles

   For reconstitution of AChR in lipid vesicles, detergent solubilized receptor protein was suspended in a lipid/cholate mixture to yield a final concentration of 0.5 –1
g/dm³ protein, 1.2 % cholate and 25 g/dm³ lipid. The soybean phospholipids were
dissolved in chloroform together with 20 % (w/w) cholesterol, and the solvent was removed before storage. The homogeneous suspension was dialyzed for 2 days with 3
changes against 500 volumes of 10 mM HEPES, 100 mM NaCl, 3 mM NaN$_3$, and 0.1
mM CaCl$_2$, at 4 °C. Samples analyzed by electron microscopy had been frozen in liquid nitrogen and stored at –80 °C.

8.3   Assay of AChR function by Li$^+$-flux

   Functionally active AChR species were analyzed by agonist-induced uptake of
Li$^+$ into reconstituted vesicles. After detergent removal the vesicles were dialyzed against

2x500 volumes of 100 mM NaCl in buffer A, followed by dialysis for the same period against 2x500 volumes of 145 mM sucrose in buffer A at 4 °C. For the flux assay, samples of 115 mm^3 were mixed with 5 mm^3 of 2 M LiCl and 5 mm^3 of 5 mM carbamoylcholine chloride. The control values were determined by substituting distilled water for the agonist solution in the assay. Entrapped Li$^+$ was separated from external Li$^+$ by passage through a 1.5 cm^3 Dowex 50 WX-8-100 cation-exchange column preequilibrated with 3 cm^3 sucrose solution (170 nM, 3.3 mg/cm^3 BSA). The column was immediately eluted with 1.6 cm^3 sucrose solution (175 mM), and fractions of 200 mm^3 were analyzed by atomic absorption (AAS) at 670 nm. All assays were performed in duplicate.

8.4   Electron microscopy and patch-clamp recordin

For negative staining, vesicle samples were diluted to about 1g/dm^3 lipid. A carbon-shadowed formvar filmed grid was applied to a sample drop of 50 mm^3, blotted, and stained with 1 % aqueous uranyl acetate. The samples were viewed at 50 kV in a Zeiss TEM-109 elctron microscope.

Patch-clamp signals were measured with an EPC 7 amplifier. The digitized data (modified Sony PCM 701) were stored on a video tape recorder and were analyed with a computer oscilloscope (E. Bablock, Augsburg, FRG) or with a CED 1401 patch clamp system (open-time distribution).

Acknowledgement

We thank Frau M. Hofer for careful typing of the manuscript and we gratefully acknowledge financial support by the Deutsche Forschungsgemeinschaft (SFB 223/C01, C02).

References

[1]   B. Katz, *The release of neural transmitter substances,* Liverpool University Press, Liverpool (1969).

[2]   D. Nachmansohn, *Harvey Lect.,* **49**, 57 (1955).

[3]   J. P. Changeux, *Trends Pharmacol. Sci,* **11**, 485 (1990).

[4]   J. Cartaud, E. Benedetti, J.B. Cohen, J.C. Meunier asnd J.P. Changeux, *FEBS Lett.,* **33**, 109 (1973).

[5]   E. Nickel and L.T. Potter, *Brain Research,* **57**, 508 (1973).

[6]   H. W. Chang and E. Bock, *Biochemistry,* **18**, 172 (1979).

[7]   S.L. Hamilton, M. McLaughlin and A. Karlin, *Biochem. Biophys. Res. Commun.,* **79,** 692 (1977).

[8]   B.A. SUÀREZ-ISLA and F. HUCHO *FEBS Lett.*, **75**, 69 (1977).

[9]   J. CARTAUD, J.L. POPOT and J.P. CHANGEUX, *FEBS Lett.*, **121**, 327 (1980).

[10]   R. RÜCHEL, D. WATTERS and A. Maelicke, *Eur. J. Biochem.*, **119**, 215 (1981).

[11]   M. CRIADO and F.J. BARRANTES, *Biochim. Biophys. Acta*, **798**, 374 (1984).

[12]   A. AHARANOV, R. TARRAB-HAZDAI, I. SILMAN and S. FUCHS, *Immunochemistry*, **14**, 129 (1977).

[13]   F. SPILLECKE, *Thesis, University of Bielefeld (Germany)*, 395 (1983).

[14]   J.L. POPOT, J. CARTAUD and J.P. CHANGEUX, *Eur. J. Biochem.*, ***118***, 203 (1981).

[15]   H. SCHINDLER, F. SPILLECKE and E. NEUMANN, *Proc. Natl. Acad. Sci. Usa*, **81**, 6222 (1984).

[16]   T. SCHÜRHOLZ, J. WEBER and E. NEUMANN, *Bioelectrochem. Bionerg.*, **21**, 71 (1989).

[17]   J. LINDSTROM, J. COOPER and S. TZARTOS, *Biochemistry*, **19**, 1454 (1980).

[18]   K. IMOTO, C. METHFESSEL, B. SAKMANN, M. MISHINA, Y. MORI, T. KONNO, K. FUKUDA, M. KURASAKI, H. BUJO, Y. FUJITA and S. NUMA, *Nature (London)*, **324**, 670 (1986).

[19]   J. W. WALKER, R.J. LUKAS and M.G. MCNAMEE, *Biochemistry*, **20**, 2191 (1981).

[20]   J. K. WALKER, C.A. RICHARDSON and M.G. MCNAMEE, *Biochemistry*, **23**, 2329 (1984).

[21]   T. SCHÜRHOLZ, J. KEHNE, A. GIESELMANN and E. NEUMANN, *Biochemistry*, **31**, 5067 (1992).

[22]   R. ANHOLT, J. LINDSTROM and M. MONTAL, *J. Biol. Chem.*, **256**, 4377 (1981).

[23]   R. ANHOLT, D.R. FREDKIN, T. DEERINCK, M. ELLISMAN, M. MONTAL and J. LINDSTROM, *J. Biol. Chem.*, **257**, 7122 (1982).

[24]   B.A. SUÀREZ-ISLA, K. WAN, J. LINDSTROM and M. MONTAL, *Biochemistry*, **22**, 2319 (1983).

[25]   B. SAKMANN, J. PATLAK and E. NEHER, *Nature (London)*, **286**, 71 (1980).

[26]   B. SAKMANN, C. METHFESSEL, M. MISHINA, T. TAKAHASHI, T. TAKAI, M. KURASAKI, K. FUKUDA and S. NUMA, *Nature (London)*, **318**, 538 (1985).

[27]   J.-P. CHANGEUX, A. DEVILLERS-THIÉRY and P. CHEMOULLI, *Science*, **225**, 1335 (1984).

[28]   H.W. CHANG, E. BOCK and E. NEUMANN, *Biochemistry*, **23**, (1983).

[29]   E. NEUMANN, E. BOLDT, B. RAUER and H. WOLF, *Biocelectrochem. Bioenerg.* **20**, 45 (1988).

[30]   M. WEBER and J.-P. CHANGEUX, *Mol. Pharmacol.*, **10**, 14546 (1984).

[31]   H. SUGIYAMA and J.-P. CHANGEUX, *Eur. J. Biochemistry*, **55**, 505 (1975).

[32   G. WEILAND, B. GEORGIA, S. LGNAPPI, C.F. CHINGELL and P. TAYLOR, *J. Biol. Chem.*, **252**, 7648 (1977).

[33]   R.R. NEUBIG and J.B. COHEN, *Biochemistry*, **18**, 5464 (1979).

[34]   M. SCHIMERLIK, U. QUAST and M.A. RAFTERY, *Biochemistry*, **18**, 1884 (1979).

[35]   I. SILMAN and A. KARLIN, *Science,* **164,** 1420 (1969).

[36]   R.N. COX, M. Kawai, A. Karlin and P.W. BRANDT, *J. Membr. Biol.,* **51,** 145 (1989).

[37]   H.A. LESTER, M.E. KROUSE, M.M. NASS, N.H. WASSERMANN and B.F. ERLANGER, *J. Gen. Physiol.,* **75,** 193 (1984).

[38]   A. MAELICKE, *Angew. Chem.,* **96,** 193 (1980).

[39]   J.M. WOLOSIN, A. LYDDIATT, J.O. DOLLY and E.A. BARNARD, *Eur. J. Biochem.,* **109,** 495 (1980).

[40]   J. BERNHARDT and E. NEUMANN, in *Neuroreceptors,* F. HUCHO (Editor) W. DE GRUYTER, Berlin, (1982) pp. 221-232.

[41]   J. BERNHARDT and E. NEUMANN, *Biophys. Chem.,* **15,** 327 (1982).

[42]   R.R. NEUBIG, N.D. BOYD and J.B. COHEN, *Biochemistry,* **21,** 3460 (1982).

[43]   P. PENNEFATHER and D.M. QUASTEL, *Br. J. Pharmacol.,* **77,** (1982).

[44]   T. HEIDMANN, J. BERNHARDT, E. NEUMANN and J. P. CHANGEUX, *Biochemistry,* **22,** 5452 (1983).

[45]   G. BOHEIM. W. HANKE, F.J. BARRANTES, H. EIBL, B. SAKMANN, G. FELS and A. MAELICKE, *Proc. Natl. Acad. Sci. U.S.A.,* **78,** 3586 (1981).

[46]   M. MONTAL, P. LABARCA, D.F. FREDKIN, B.A. SUÀREZ-ISLA and J. LINDSTROM, *Biophys. J.,* **45,** 165 (1984).

[47]   M.B. JACKSON, *Proc. Natl. Acad. Sci, U.S.A.,* **81,** 3901 (1984).

[48]   H.W. CHANG and E. NEUMANN, *Proc. Natl. Acad. Sci, U.S.A.,* **73,** 3364 (1976).

[49]   M.A. RAFTERY, J. SCHMIDT and D.G. CLARK, *Arch. Biochem. Biophys.,* **152,** 882 (1972).

[51]   J.A. DANI, *Biophys. J.,* **49,** 607 (1986).

[50]   K. IMOTO, C. BUSCH, B. SAKMANN, M. MISHINA, T. KONNO, J. NAKAI, H. BUJO, Y. MORI, K. FU UKA and S. NUMA, *Nature (London),* **334,** 645 (1988).

[52]   J.A. DANI and G. EISENMAN, *J. Gen. Physiol.,* **89,** 959 (1987).

[53]   C. TOYOSHIMA and N. UNWIN, *Nature (London),* **336,** 247 (1988).

[54]   H. SCHINDLER and U. QUAST, *Proc. Natl. Acad. Sci. U.S.A.,* **70,** 3052 (1980).

[55]   O.P. HAMILL and B. SAKMANN, *Nature (London),* **294,** 462 (1981).

[56]   C. METHFESSEL, V. WITZEMANN, T. TAKAHASHI, M. MISHINA, S. NUMA and B. SAKMANN, *Pflügers Arch.,* **407,** 577 (1986).

[57]   E. YERAMIAN, A. TRAUTMANN and P. CLAVERIE, *Biophys. J.,* **50,** 253 (1986).

NERVOUS CONTROL OF CARDIAC FUNCTION: MODULATION OF PACEMAKER ACTIVITY

DARIO DI FRANCESCO

Università di Milano, Dipartimento di Fisiologia e Biochimica Generali,
Elettrofisiologia, via Celoria 26, 20133 Milano,
Italy.

Contents

Bioelectrochemistry IV
Edited by B.A. Melandri *et al.*, Plenum Press, New York, 1994

Introduction

Modulation of pacemaker activity in mammalian heart is mediated by the hyperpolarization-activated current (I_f), an inward, mixed Na^+ and K^+ current activated at potential membrane (U) more negative than about –40 mV (DI FRANCESCO, [1]). This current is involved in the generation of the slow diastolic depolarization, which underlies spontaneous action potential oscillations of cardiac pacemaker cells of the sino-atrial node (DI FRANCESCO *et al.*, [2]). The slope of the diastolic depolarization phase, and consequently the frequency of spontaneous oscillations, depend on the extent of I_f activation, which is in turn regulated (in opposite ways) by both adrenergic and cholinergic input. This mechanism is at the basis of neurotransmitter regulation of the heart rate (DI FRANCESCO and TROMBA, [3,4]; DI FRANCESCO, DUCOURET and ROBINSON, [5]).

Both adrenergic and cholinergic neurotransmitters modulate I_f *via* a modification of adenylyl-cyclase activity and of the intracellular cAMP level, which acts as a second messenger in the I_f control. Indeed, the level of intracellular cAMP determines the position of the I_f activation curve on the U-axis and hence the degree of current availability to U-dependent activation. As several other cardiac (typically, Ca^{2+}- and delayed K^+-) and noncardiac channels are known to be modulated *via* phosphorylation induced by cAMP-dependent protein kinase (PKA), a phosphorylation process could also be involved in the I_f channel control. However, recent data indicate that, surprisingly, cAMP does not act on I_f channels *via* phosphorylation of the channel protein, but activates I_f channels directly (DI FRANCESCO and TORTORA, [6]). This chapter briefly reviews the I_f modulation in cardiac pacemaker cells as studied in whole-cell conditions in excised inside-out patches.

2. Neurotransmitter-induced I_f modulation

2.1. Adrenaline stimulates I_f

Catecholamines increase the amount of I_f current by shifting its activation curve to more positive U-values; this causes a faster rate of development of the pacemaker depolarization phase and leads to cardiac acceleration (HAUSWIRTH, NOBLE and TSIEN, [7]; DI FRANCESCO *et al.*, [2]). This action of catecholamines is due to a positive displacement of the I_f activation curve, and is mediated by β–adrenergic stimulation of adenylyl-cyclase and an increase of intracellular cAMP.

Although the position of the I_f activation curve is modified by catecholamines, the I_f fully activated $I-U$ relation is not. The action of β–stimulation is thus to affects the I_f kinetics, but not the fully-activated current magnitude. At the single-channel level this means that in any cell the maximal number of channels and the channel conductance are unaffected, whereas the probability of channel opening on hyperpolarization increases during β–stimulation after an increase of cAMP levels. This is indeed what experimentally observed in single-channel recordings (DI FRANCESCO, [8]). Single-channel measurements indicate that the single I_f- channel conductance is fairly small (about 1 pS with 70 mM KCl, 70 mM NaCl in the extracellular solution).

2.2. Acetylcholine inhibits I_f

Muscarinic receptor stimulation by ACh has on I_f an effect opposite to that of catecholamines, which consists of a shift of the I_f activation curve to the negative direction on the U–axis (DI FRANCESCO and TROMBA, [3]); this slows rhythm by reducing the degree of I_f by muscarinic-induced inhibition of adenylyl-cyclase (DI FRANCESCO and TROMBA, [4]) and a reduction of the intracellular cAMP levels, effects which are opposite to those caused by β–adrenergic stimulation. As with catecholamines, ACh modifies I_f–kinetics without altering the fully-activated current intensity.

An important feature of the muscarinic-induced I_f-inhibition is that it occurs at fairly low concentrations, in a range (up to about 10–30 nM) where activation of the ACh-dependent K$^+$-conductance is not observed. Thus, I_f is responsible for modulating rhythm under conditions of moderate vagal activity (DI FRANCESCO, DUCOURET and ROBINSON, [5]).

2.3. cAMP activates I_f directly

As mentioned above, cAMP is the second messenger involved in both the adrenergic and cholinergic regulation of I_f channels. β–adrenergic stimulation and muscarinic inhibition of adenylyl-cyclase lead to an increase and a decrease, respectively, of the intracellular cAMP level. This in turn controls I_f availability to U-dependent activation by modulating the position of the current activation curve on the U–axis (DI FRANCESCO and TROMBA [4]).

It is known that cAMP modulates various ionic channels, including the cardiac L-type Ca^{2+}-channel and delayed K$^+$-channel (OSTERRIEDER *et al.*, [9], WALSH and KASS, [10]). This action is performed through PKA-dependent phosphorylation of the channel protein. Phosphorylation by the cAMP-dependent protein kinase is indeed a widespread pathway for biochemical regulation of a variety of cellular processes. It could be expected, therefore, that I_f channels, too, are regulated through a phosphorylation mechanism. Instead, we have found that in the SA node pacemaker cells, cAMP activates I_f channels directly, *via* a mechanism independent of phosphorylation (DI FRANCESCO and TORTORA, [6]). Activation can be shown to occur when the intracellular side of inside-out macro-patches excised from the membrane of isolated SA node myocytes is perfused with cAMP in the absence of ATP and Mg^{2+}.

cAMP shifts the I_f activation curve to more positive U–values without modifying the fully-activated $I-U$ relation. This is in accordance with data obtained in whole-cell conditions during stimulation of cAMP production by β–adrenergic activation (Di Francesco *et al.*, [2]; Di Francesco and Tromba [3, 4].

The dose-response relation of I_f activation (measured as the positive shift of I_f activation curve) as a function of cAMP concentration can be fitted with a Hill equation yielding a half-maximal concentration K_d = 0.211 μM and a slope factor n = 0.85 (Di Francesco and Tortora, [6]).

I_f channels are activated by other cyclic nucleotides (cGMP and cCMP), although with a lower efficiency. The cAMP-induced I_f regulation is similar to that observed in other cyclic-nucleotide-activated channels of sensory cells (Fesenko *et al.*, [11]; Haynes and Yau, [12]; Johnson *et al.*, [13] and Nakamura and Gold, [14]). So far, however, I_f appears to be the only current exhibiting kinetic regulation by both membrane potential and direct cAMP binding.

Acknowledgements

This work was supported by the CNR.

References

[1] D. Di Francesco, *Trends Cardiovascular Med.*, **1**, 250 (1991).

[2] D. Di Francesco, A. Ferroni, M. Mazzanti and C. Tromba, *J. Physiol.*, **377**, 61 (1986).

[3] D. Di Francesco and C. Tromba, *J. Physiol.*, **405**, 477 (1988).

[4] D. Di Francesco and C. Tromba, *J. Physiol.*, **405**, 493 (1988).

[5] D. Di Francesco, P. Ducouret and R.B. Robinson, *Science*, **243**, 669 (1989).

[6] D. Di Francesco and P. Tortora, *Nature* (*London*), in the press.

[7] O. Hauswirth, D. Noble and R.W. Tsien, *Science*, **162**, 916 (1968).

[8] D. Di Francesco, *Nature* (*London*), **344**, 470 (1991).

[9] W. Osterrieder, G. Brum, J. Hescheler, W. Trautwein, V. Flockerzi and F. Hofman, *Nature* (*London*), **298**, 576 (1982).

[10] K.B. Walsh and R.S. Kass, *Science*, **242**, 67 (1988).

[11] E.E. Fesenko, S.S. Kolesnikov and A.L. Lyubarsky, *Nature* (*London*), **313**, 310 (1984).

[12] L.W. Haynes and K.W. Yau, *Nature* (*London*), **317**, 61 (1986).

[13] E.C. Johnson, P.R. Robinson and J.E. Linsman, *Nature* (*London*), **324**, 468 (1986).

[14] T. Nakamura and G.H. Gold, *Nature* (*London*), **325,** 442 (1987).

EXCITATION-CONTRACTION COUPLING IN SKELETAL MUSCLE

EDUARDO RÍOS, ADOM GONZÁLEZ, MILOSLAV KARHANEK, JIANJIE MA,
ROMAN SHIROKOV, GONZALO PIZARRO[1], LASZLO CSERNOCH[2], ROBERT FITTS[3],
ISMAEL URIBE[4] and M. MARLENE HOSEY[5]

Departement of Physiology
Rush University School of Medicine
1750 W. Harrison St. Chicago, Illinois, USA.

1 Departamento de Biofísica, Facultad de Medicina, Montevideo, Uruguay.
2 Department of Physiology, Medical University of Debrecen, Hungary.
3 Departament of Biology, Marquette University, Milwaukee, Wisconsin, USA.
4 Departamento de Fisiología, Universitad de Colima, México.
5 Department of Pharmacology, Northwestern University, Chicago, Illinois, USA.

Contents

Bioelectrochemistry IV
Edited by B.A. Melandri *et al.*, Plenum Press, New York, 1994

1.   Historical introduction

A historical approach is a simple way for reviewing the background necessary for discussions of the current concerns. 1947 is a convenient starting point; in this field, as well as in others, a number of contributions appeared two to three years after the war, that defined the field for years to come. HEILBRUNN and WIERCINSKI [1] essentially demonstrated Ca^{2+} to be the second messenger that controls contraction, and proposed that Ca^{2+} entered the myoplasm from outside; A.V. HILL [2] demonstrated, with a simple diffusion calculation, the need for a propagation mechanism of activation, other than diffusion from the outside. This system was identified in the fifties: in 1958 HUXLEY and TAYLOR [3] suceeded in depolarizing localized regions of the membrane of a fiber with an extracellular electrode placed very close to the fiber, a predecessor of patch clamp. The depolarization only induced contractions (localized to the underlying zone) when the pipette was placed over the I band, but not over the A band. It is interesting that C. FRANZINI - ARMSTRONG, when reviewing the field, always starts at this time. This result obviously motivated her, as it did other morphologists, to search for a structure underlying these localized effects. Her celebrated studies have a predecessor, closer to home, E. VERATTI [4] who had described the T tubular system in 1902, using the GOLGI staining technique. Apparently his contemporaries, when presented with the pictures, only said *se non è vero è ben trovato,* and ignored the result, wich was only exploited in the sixties and seventies by FRANZINI - ARMSTRONG [4 bis], B. EISENBERG [5], L. PEACHEY [6] and others. Figure 1 shows a view of a thick transversal cut of a fiber bundle, with GOLGI stain (PEACHEY and EISENBERG [7]), demonstrating how the T system effectively reduces the (diffusion) distances in the transversal direction to under a micrometer. As the morphological description of the T system progressed, so did that of the sarcoplasmic reticulum (SR).

HODGKIN and HOROWICZ [8] demonstrated the existence of selective, inwardly rectifying permeability to K^+ in the resting muscle membrane, and used this permeability to induce steady, graded changes in membrane potential* by changing $[K^+]_{ext}$. This demonstrated the existence of a monotonically increasing relationship between membrane potential and muscle tension, and identified the transmembrane potential difference as the key independent variable in the control of muscle tension, a finding that was of course similar to the more celebrated finding of A. HODGKIN of a similar role for potential difference (U) in the control of Na^+ and K^+ permeabilities. It is only now that the ultimate unity of these two discoveries has become evident; we'll come back to this. In these studies HODGKIN and HOROWICZ [8] generated a new tool for the study of excitation-contraction (EC) coupling, the K contracture, which was later elegantly used by C. CAPUTO [9] and C. LÜTTGAU (LÜTTGAU and SPIECKER [10]) to demonstrate

---

* This is actually a potential difference, for easiness this quantity will be called in the following shortly *membrane potential*, and given the symbol U_m

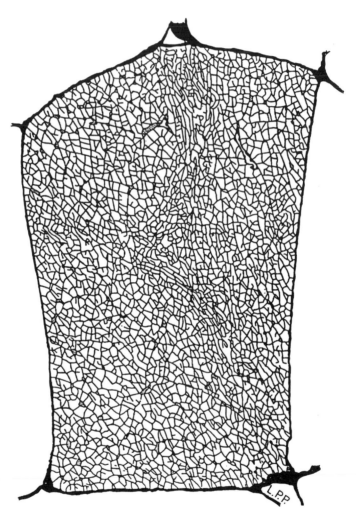

FIG. 1. The transverse tubular system. This is a reconstruction that includes a complete T system *disc,* not entirely planar. The image was reconstructed starting from thick cuts, examined with the GOLGI stain. In this frog fiber, the T tubule disc follows approximatly the Z band. The larger diameter of the fiber shown is about 100 μm. From PEACHEY and EISENBERG, [7].

a modulatory effect of extracellular Ca^{2+} (an effect studied recently by G. BRUM, RÍOS and other ([11]) by other methods).

In the late sixties EBASHI [12] and others clarified the role of Ca^{2+} as key to the mechanochemical reactions of contraction, while HASSELBACH and others [13] established the preponderance of SR as organelle of Ca^{2+} storage and controlled release. After the studies of EBASHI the field split in two, activation of contraction and control of contractile proteins by Ca^{2+}, and the field now termed EC coupling, which was restricted theretofore to the questions of translation of action potential to Ca^{2+} release. In the seventies these studies were dominated by the need to understand propagation, which was clarified by, among others, H. GONZÁLEZ - SERRATOS [14]. His cinemicrography approach helped measure the speed of inward propagation, with the end result that the propagation down the tubules was *active*, as an action potential, although substantial depolarization and contractile activation was still possible if action potentials were impeded.

In the meantime, the researchers of the SR were laboriously convincing themselves that the ubiquitous SR pump could not be the site of physiological Ca^{2+} release, and when thinking about the whole process before coming to Europe, I could not help associating it with the Eurotunnel, where the French burrow from one side, and the British from the other, narrowing the field as they go. In the present stage, we are faced with the last un- known, all the attention of the field is focused on the T-SR junction, somewhere under the English Channel.

2.   The T-SR junction

The tubules are devices specialized for propagation and communication. Most of its membrane is devoted to communication, forming the triadic or T-SR junction (Fig. 2). The section of the T tubule is flattened, about 100 nm wide. Most of the area of the lateral sides participates in the junctions, where both membranes come within 15 nm but do not touch. The well known *feet* described by FRANZINI-ARMSTRONG and NUNZI [15] in 1970, have recently been identified with the large *spanning protein* described by CADWELL and CASWELL [16]. Later SUTKO and KENYON [17] and PESSAH *et al.* [18] identified ryanodine as a specific antagonist of Ca^{2+} release, with complex actions. FLEISCHER *et al.* [19] showed that ryanodine binding and pharmacological effect are restricted to the terminal cisternae, and work in the labs of FLEISCHER, (INUI *et al.*. [20]), CAMPBELL [21] and MEISSNER, LAI *et al.* [22,23]) resulted in purification of the receptor and its identification with the feet structure, the spanning protein and the Ca^{2+} release channel. WAGENKNECHT *et al.*. [24] have produced 3-d reconstructions of the molecular structure showing intriguing hints of multiple and complex pathways within the channel (Fig. 3).

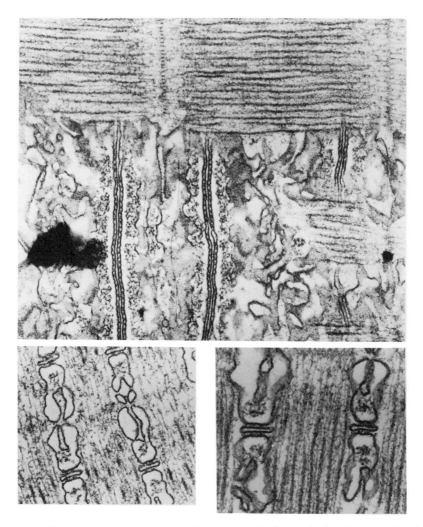

FIG. 2. Triads. The lower panels show sections that cut the fiber longitudinally, thereby cutting the triads transversally. The T tubule appears flattened (0.1 μm large diameter). The space between tubule and cisternae, of circa 0.01 μm, is crossed by feet. Feet are arranged in pairs in the transversal section, in rows in the longitudinal sections (top panel). From FRANZINI - ARMSTRONG and NUNZI [15]

On the other side of the junction, FRANZINI - ARMSTRONG and NUNZI [26] and BLOCK *et al.* [27] described formations of intramembrane particles that they termed j (junctional) T tetrads (Fig. 4), which are natural candidates for the role of sensors of U - changes in the T membrane. BLOCK *et al.* [27] demonstrated in the toadfish swimbladder a striking correspondence between the dimensions of the jT tetrads and the feet, as well as their orientation, and proposed a specific alignment of jT particles and ryanodine receptor protomers (Fig. 4).

There are other chemical components in the triadic junction, including a 95

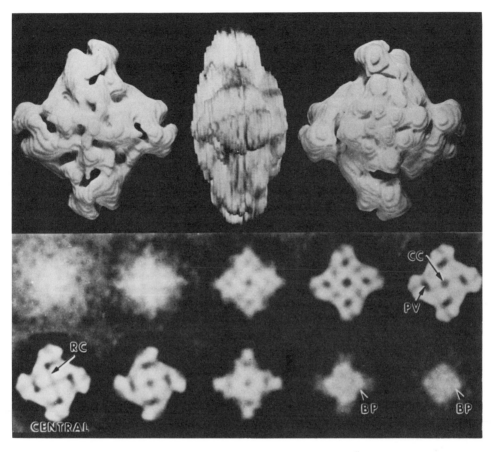

FIG. 3. Computer-generated surface representations and sections of the three-dimensionally reconstructed release channel. BP, base platform (intra SR membrane); PV, peripheral vestibules; CC central channel; RC radial channels. Approx. diagonal dimension: 25 nm. From WAGENKNECHT *et al.* [24].

kDa protein of the SR membrane, termed triadin (BRANDT *et al.* [28]) which binds to ryanodine receptors and DHP receptors, and the protein TS28 desribed by JORGENSEN *et al.* [29]. We will come back to the issue of possible roles of all these particles in the last part of this chapter, let me point out that these detailed studies essentially rule out other forms of communication proposed earlier, like the existence of gap junction-style continuity between the T and SR compartments.

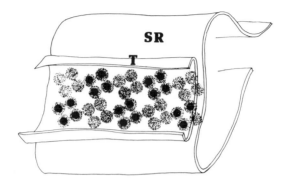

FIG. 4. A) Schematic arrangement of intramembrane particles in T-SR junction. This diagram, modified from re-constructions by FRANZINI-ARMSTRONG, [26] is a lateral perspective of a half-triad: one terminal cisterna and the proximal half of the T tubule were omitted for clarity. The larger circles, stippled, represent protomers of the ryanodine receptor, occupying the thickness of the SR membrane and spanning the T-SR gap. The feet are the homotetramers, arranged along the tubule with their diagonal at a 20 degree angle with the axis. The filled circles represent the intra-T membrane particles, forming jT tetrads. From RIOS and PIZARRO, [25].

3.   The *U*-sensor and calcium release

At this point I will abandon the pretense of objectivity and describe in more detail electrophysiological studies with which I am familiar. I will give first a brief description of methods.

3.1.   *Methods for studying the U-sensor*

These are essentially an outgrowth of the techniques of potential difference clamp of HODGKIN and HUXLEY [30] (in fact ADRIAN, CHANDLER and HODGKIN [31] were the first to set or *clamp* the electric potential difference of a muscle fiber to a constant value) with the addition of Vaseline gap techiques, in which the ends of a fiber segment are used for passing current and measuring potential difference (HILLE and CAMPBELL [32]; KOVACS, RIOS and SCHNEIDER [33]) and more recently of patch clamp techniques. In the double Vaseline gap technique the amplifier is connected in an adequate manner to maintain the membrane in the central pool at a desired potential and the current is measured. RIOS and PIZARRO [34] recently reviewed the various gap methods and discussed some of their intrinsic errors; in brief, the Vaseline gap affords a fast clamp, with low noise in the measurement of current (these advantages are due to the very low resistance of the connections to the intracellular medium), the ability to equilibrate the internal medium of the fiber to the solutions in the cut ends, and hours of stable measurement per fiber. The main disadvantage is the contamination of the measured current with currents generated in the membrane regions under the Vaseline seals, which are inhomogeneously polarized. In recent work we have seen that the saponin permeabilization procedure of IRVING *et al.* [35] substantially reduces this contamination (A. GONZÁLEZ and E. RÍOS [36], unpublished observations).

The charge movement and the gating current of nerve, since first described in 1973 (SCHNEIDER and CHANDLER [37], ARMSTRONG and BEZANILLA [38]) are measured in the presence of impermeant ions, or blockers of all ionic channels, as the main part of the *asymmetric current* (difference between the current in a test pulse and the current in a control pulse). When this is done well, one gets records as in Fig. 5. It is relatively straightforward to obtain good kinetic records of charge movement current.

An important concept here, much more than in the measurement of ionic currents, is that of *charge moved at a given potential difference*; this is more difficult to evaluate because it is the time integral of the current, and it may be affected greatly by small ionic currents. Since these intramembrane charge movements are purported to be movement of charged molecules trapped within the membrane, the amount of charge moved is an important quantity. For instance, it should reach a maximum at very high U-values, and it should be equal at all U-values for ON and OFF. This is just a special case of a more general rule, that the amount of charge that moves in the transition to a certain final value of U should be a function of U only, and not of the pathway or the history of potential differences that the system went through in order to get to U (in thermodynamics parlance charge movement *current* is an exact differential, whose integral does not depend on the path, but only on the initial and final values of the state variable U).

As shown in Fig. 5B the $Q (U)$ distribution is a sigmoidal function. Thermodynamic considerations lead, in the simplest hypothesis of a two state system, to the BOLTZMANN equation

$$Q (U) = Q_{max} \; \frac{1}{1+ \exp \left(-\dfrac{U - \bar{U}}{K} \right)}$$

whose derivation and extensions can be found in RÍOS and PIZARRO [34]. As shown by the dotted line in the figure, the BOLTZMANN function is adequate for description, but in this and other cases the distribution is better described as the sum of two terms of the same BOLTZMANN form (continous line), a point recently studied carefully by HUI and CHANDLER [41].

A less known property of charge movement is that it should only have the direction of the field change, outward for a positive change in potential, inward for a hyperpopolarizing pulse. Another way of saying this is that the charge movement current *has no reversal potential.*

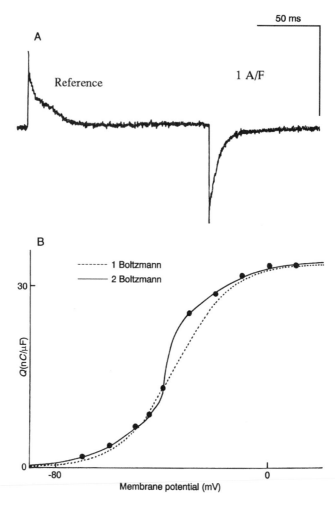

FIG. 5. Intramembrane charge movement. Example of asymmetric current, obtained as difference between the total measured current during a pulse to -40 mV and a pulse from -120 mV to -100 mV appropriately scaled to match *U*-values. B) *Q vs.* test pulse *U* - values. The dotted line corresponds to a best fit single BOLTZMANN. The continuous line is a sum of two BOLTZMANN terms. The internal solution contained Cs glutamate, 15 m*M* EGTA, ATP, creatin phosphate, tris maleate buffer and no added Ca^{2+}. The external solution contained TEA methanesulfonate, tris maleate buffer and the channel blockers Cd^{2+}, La^{3+}, anthracene-9-carboxylic acid and TTX, with no other ion (see CSERNOCH *et al.* [39] for details) (from unpublished work of GONZÀLEZ and RIOS)

3.2. *Methods for studying Ca^{2+} release*

Intracellular $[Ca^{2+}]$ can be measured now with many different techniques, but only the many available optical techniques have adequate sensitivity and temporal resolution. In our laboratory we sometimes combine an optical absorption technique with a fluorescent technique. The former, using the dye Antipyrylazo III (RÍOS and SCHNEIDER, [42]) has a temporal resolution better than a millisecond and relatively low sensitivity. The latter, using the *fourth generation* fluorescent dye Calcium green, has high sensitivity and is essentially immune to artifacts caused by contraction.

Taking into account all fluxes per liter of fiber-water the simple equation applies:

$$\frac{d[Ca^{2+}]_{in}}{dt} = \text{input flux} - \text{output flux}$$

In the situation of functional EC coupling the contribution of the SR release to input flux is so overwhelming that there is no error in neglecting all others and writing:

$$\frac{d[Ca^{2+}]_{in}}{dt} = \dot{R}(t) - \text{output flux} \qquad (1)$$

with release flux $\dot{R}(t)$ instead of input flux. If we are to understand the control of calcium release, a necessary goal is to calculate $\dot{R}(t)$. Since we are measuring $[Ca^{2+}]_{in}(t)$ with the optical techniques, the problem is reduced to calculating output or removal flux.

We have done this in two ways: in MELZER *et al.* [43] we have described a method based on measuring the rate of decay of free Ca^{2+} after the pulse as an experimental way of estimating the magnitude of removal fluxes. Once we have that estimate of removal flux we insert it in equation (1) together with $[Ca^{2+}]_{in}(t)$ and derive the release flux. A current version of this method is described in BRUM *et al.* ([44], appendix) and Fig. 6 summarizes the waveforms of $[Ca^{2+}]_{in}$, $d[Ca^{2+}]_{in}/dt$, removal and release flux.

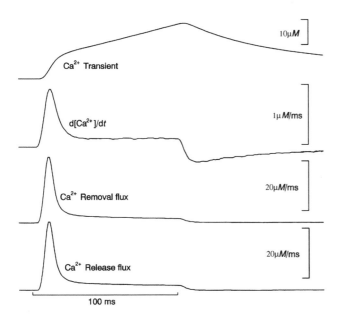

FIG. 6. Calcium fluxes underlying a transient. Top, Ca^{2+} transient, recorded with Antipyrylazo III, and its time derivative. Third record from top, removal flux estimated by the method of MELZER *et al.* [43] modified as described by BRUM *et al.* [11] (appendix). The release flux, bottom record, is the sum of $d[Ca^{2+}]/dt$ and the removal flux. The removal flux is by far the dominant term. (From RÍOS and PIZARRO, [34]).

Another method that we are currently using is based on the fortunate fact that the Ca²⁺ buffer EGTA is very slow, with an ON rate constant that is about 100-fold slower than in a binding process limited by diffusion. The method requires the simultaneous presence of the calcium indicator and a large concentration of EGTA (as much as 30 mM) in the cell.

To understand the method we rewrite equation (1) as

$$\frac{d\,[Ca^{2+}]_{in}}{dt} = \dot{R}\,(t) - \text{removal flux}_{intrinsic} - \frac{d\,([EGTA:\,Ca^{2+}])}{dt} \quad (2)$$

where we have neglected a dye:Ca²⁺ term as the EGTA term is very large. If it is large enough, all other removal fluxes can also be neglected, as well as d [Ca²⁺]/dt, which as seen in the figure is usually much smaller than the other terms, even without EGTA. Thus:

$$\dot{R}\,(t) = \frac{d\,[EGTA:\,Ca^{2+}]}{dt} \quad (3)$$

The right side of this equation is the net rate of the binding reaction:

$$\frac{d\,[EGTA:\,Ca^{2+}]}{dt} = [Ca^{2+}]\,[EGTA]\,k_{on} - [EGTA:\,Ca^{2+}]\,k_{off} \quad (4)$$

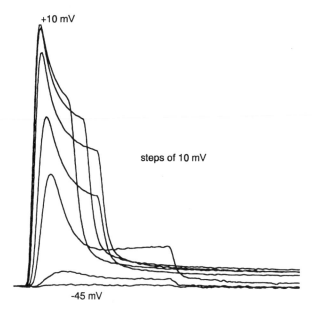

FIG. 7. Ca²⁺ transients recorded simultaneously with the current records of Fig. 5. Since the internal solution contained a high [EGTA], The transients have a shape similar to records of release flux (see text). (Unpublished work by A. GONZÀLEZ and E. RÍOS).

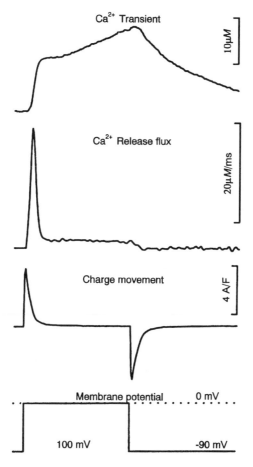

FIG. 8. Summary of simultaneous waveforms, recorded or computed during a pulse to 0 mV (represented schematically at a bottom). From RÍOS and PIZARRO, [25].

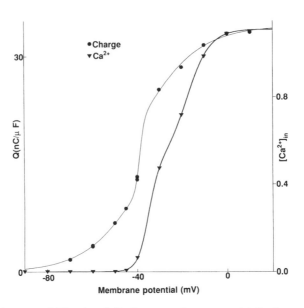

FIG. 9. Charge displacement $Q(U)$ and peak $[Ca^{2+}]_{in}$ *vs.* membrane potential U_m. Same experiment as Fig. 5 and 7. The peak $[Ca^{2+}]_{in}$ is approximately proportional to the peak value of Ca^{2+} release flux during the transients. Note the high steepness of the U-dependence of release near threshold, and how it is consistent with the Q_γ component in the distribution of charge.

If the buffer is sufficiently concentrated and sufficiently slow, then the amount of free buffer can be considered constant and the back reaction can be neglected during the pulse, so that, substituting equation (4) in equation (3), equation (5) results:

$$\dot{R}\ (t) = [Ca^{2+}]\ [EGTA]\ k_{on} \cong const\ [Ca^{2+}] \tag{5}$$

Thus, if [EGTA] is large, free Ca^{2+} concentration follows the time course of the release flux (an example of Ca^{2+} transients measured in the presence of high EGTA is in Fig. 7).

Both the MELZER method and the EGTA method give a similar waveform of release flux, with an early rise to a peak followed by inactivation to a lower level. We will get back to this waveform later.

With these techniques, some of the most basic results are demonstrated in Figs. 8 and 9. First, the charge movement has the right timing to control Ca^{2+} release (essentially the charge has the right U-dependence). In this sense a qualification is necessary, as shown in Fig. 9 a good portion of charge movement takes place at U-values that are subthreshold for Ca^{2+} release, and the overall steepness of dependence is greater for release, especially at or near the minimun U-value that elicits release. As shown in Fig. 5, in

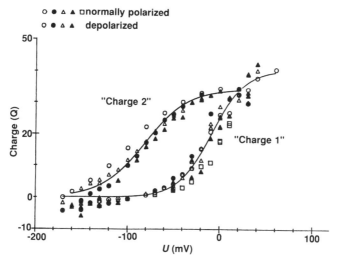

FIG. 10. The effect of prolonged depolarization. The distribution labeled *charge 1* was obtained from a holding *U*- value of -100 mV. The distribution labeled *charge 2* from a holding *U*-value of 0 mV. The extracellular medium contained 100 m*M* Co²⁺ as sole cation. Therefore the *U*– dependence is shifted to higher *U*-values than in more conventional solutions. Fibers stretched to 4.2 μm/sarcomere. Other solutions as in BRUM *et al.* [11]. Unpublished work of G. PIZARRO, R. FITTS, I. URIBE and E. RÍOS [47].

many circumstances a second kinetic component of charge movement is visible, termed the *hump* or Q_γ (ADRIAN and PERES, [45]) and this component is visible in the Q (*U*) distribution as a second BOLTZMANN component, with greater steepness, usually centered at about – 50 mV. This has generated the consensus that Q_γ is intimately related to Ca²⁺ release, and in some researchers the opinion that Q_γ is an independent species of charge, carried by a subset of *U*– sensitive molecules, which are the only sensors controlling Ca²⁺ release (HUANG, [46]).

With this methodology of measuring simultaneously events at the T side of the junction, presumably monitoring the *U*– sensor, and events at the SR side, namely release flux, we carried out a number of experiments that can be conveniently divided in two major sets. These experiments have largely determined the way in which I see this field and I will summarize some of the main results. In a first series of experiments we interfered in multiple ways with the *U*-sensor; that took until 1988 or so. In a second series we tried to interfere specifically and primarily with the release channel. In both cases, by observing what happened to the other side we were able to derive useful conclusions about their interaction.

3.3. Interfering with the U-sensor

Under this heading we must consider three types of studies:

i) prolonged depolarization, which causes *U*-dependent inactivation, much in the same way as a Na$^+$ channel is inactivated;

ii) elimination of Ca^{2+} in the extracellular medium, which leads to a state similar to inactivation; and

iii) extracellular application of Ca^{2+} antagonists like dihydropyridines and phenylalkylamines.

3.3.1. U-dependent inactivation.

When the cell is depolarized for a long time the well known phenomenon of contractile inactivation ensues, that is for instance the reason that a K$^+$ contracture relaxes after a few seconds. Together with this the distribution of mobile charge *Q* (*U*) shifts to the left (Fig.10). This is an important result, the BOLTZMANN parameter *(U)* shifts from about $-$ 30 mV to about $-$ 110 mV. Surprisingly, the BOLTZMANN parameter Q_{max} does not change much.. Thus, even though the effect of prolonged depolarization on charge movement has always been termed charge immobilization, such is not the effect in skeletal muscle. The term obeys to the fact that much less charge moves after inactivation in the range of *U*$-$ values near 0 mV. The charge that moves in the inactivated fiber is termed charge 2 or Q_α A similar phenomenon has been described for gating current of squid giant axon (TAYLOR, BEZANILLA and FERNANDEZ, [48]). In both squid axon and skeletal muscle it is understood with a 4 state model including a resting state R, an active state A and two inactivated states.

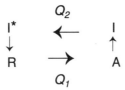

The four-state model assumes that the transitions represented horizontally are *U*dependent, generating charge movements, whereas the vertical transitions are *U*-independent. When a fiber held at the resting potential difference is depolarized, the *U*-sensors go from R to A, and that underlies charge 1, then slowly go into I. Once the system is inactivated, changes in *U* will only move the system between I and I*, causing charge 2.

3.3.2. Effects of extracellular metallic ions

Another intervention that was found to interfere with the *U*-sensor is the elimination of extracellular Ca^{2+}. When this was done without other metal ions being present (Figs. 11 and 12), release was eliminated. Since charge movement then acquired the characteristic shifted distribution of charge 2 (Fig. 12) we believe that the effects on release were secondary, and that the primary effect was the induction of inactivation of the *U*-sensor. The effect could be prevented by having high concentrations of alkali ions like Na+ (Fig. 11) or Li+, or low concentrations of divalents replacing Ca²⁺. We interpreted this as evidence of the existence of a binding site for Ca²⁺ on the *U*-sensor, that has to be occupied in order to prevent the sensor from inactivating at the resting *U*-value. This *priming site* has affinity for other ions, and both PIZARRO *et al.* [47] and LÜTTGAU [50] and collaboratos, in results not yet published, found a suggestive similarity between the profile of affinities of the priming site and the relative permeabilities of the cardiac L-type Ca²⁺-channel (HESS *et al.* [51]) attributed also to binding affininities to an intrapore site.

3.3.3. Channel blockers

These similarities with *U*-sensitive channels were of course causing a profound impression in our laboratory by 1986, but it was the effect of nifedipine on EC coupling which finally opened our eyes to the fact that the *U*-sensor is a Ca²⁺ channel or similar molecule. The result, now well known (Fig. 13), is that low concentrations of nifedipine reduced charge movement (of the charge 1 type, while insreasing charge 2) in a strictly *U*-dependent fashion, analogous to the effect on Ca²⁺ channels (BEAN, [53]) and in some ways consistent with the *U*-dependence of specific DHP binding to muscle (SCHWARTZ *et al.* [54]). When we found this effects, EISENBERG and coworkers [55] had described a curious *U*-dependent paralysis caused by the phenylalkylamine D600, later shown to be mediated by effects on the *U*-sensor (HUI *et al.* [56]). GLOSSMAN's and LAZDUNSKI's laboratories had demonstrated that the phenylalkylamines and DHPs bound to the same molecule (reviewed by GLOSSMANN and STRIESSNIG, [57]), and KURT BEAM and collaborators [58] had shown that his now famous myodysgenic mice lacked the slow membrane Ca²⁺ current. Thus, even though it took some gall at the time to hypothesize that the DHP receptors of the T membrane were the *U*-sensors of EC coupling (RIOS and BRUM, [52]), the proposal was never controversial. A few months later SHOSAKU NUMA had the DHP receptor α_1 subunit cloned (TANABE *et al.* [59] and subsequent years saw the spectacular results of expression of this cDNA in cells from myodysgenic mice, (TANABE *et al.*, ADAMS *et al* . [60]).

In summary, we now know that the *U*-sensor is in every respect a *U*-sensitive Ca²⁺ channel of the L religion, with the only peculiarity that it appears to be doing something else; *U*-sensing, yes, but for the benefit of another channel in a nearby membrane. Up to this date we really do not know whether the *U*-sensor actually passes I_{Ca}, and this is a relevant current question (DE JONGH *et al.*, [61] PIZARRO *et al.* [25]).

3.4. Interfering with the release process

In work that involved collaborations with the laboratories of KOVACS in Debrecen and STEFANI in Houston, we have recently taken the opposite approach, of applying interventions intended to cause primarily and directly alterations of release. As in the previous case, there were many different ways in which this was done:

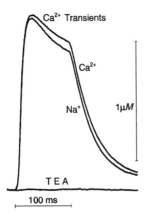

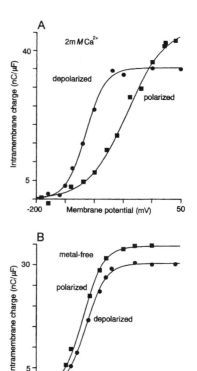

▲ FIG. 11. Effect of extracellular cations. When all metal cations are replaced by TEA, EC coupling fails. The Ca²⁺ transient recovers when 125 mM Na⁺ are admitted in the extracellular solution, replacing TEA. From RIOS *et al.* [49]. Experimental conditions as in BRUM *et al.* [11].

▶ FIG. 12. $Q(U)$ in the absence of metal cations. A) the effect of changing the holding potential from 0 (depolarized) to -100 mV (polarized) is a shift to higher U-values of the $Q(U)$ as described for Fig. 10. B) In the absence of extracellular metal cations, however, polarization of the holding potential to -100 mV does not cause a significant change in $Q(U)$, which remains with the same properties as in the depolarized case (that is, it remains as charge 2). Unpublished work of G. PIZARRO, R. FITTS, I. URIBE and E. RÍIOS [47].

i) by inactivating Ca²⁺ release with a conditioning pulse,

ii) by depleting the SR of calcium,

iii) by applying known Ca²⁺ release channel blockers.

A total of ten interventions applied so far, all of which fall within one of the above types, had similar effects: in all cases the abolition or reduction of release flux was accompanied by a reduction in amplitude and in some cases an elimination of the component of charge movement termed Q_γ.

This is illustrated with just two experiments. One is to deplete the SR by repetitive pulsing in the presence of a high concentration of the high affinity, fast-equilibrating Ca²⁺ buffer BAPTA. Fig. 14 A shows records of charge movement current in a reference situatuation and twenty minutes after changing the solution in the end pools (that is, the internal solution) to one containig 20 mM of BAPTA, and pulsing the fiber repeatedly to deplete the SR. In panel B the solutions are the same, but they were applied in the reverse order. In both cases the reference record contained a substantial amount of Q_γ current, which was abolished in the depleted situation. As we said before, Q_γ is a hump component in the records, that appears in the potential difference distri-

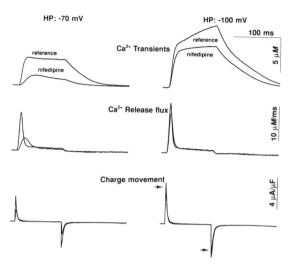

FIG. 13. Effects of a dihydropyridine on EC coupling. The effects of 0.5 μM nifedipine are different at a well polarized holding potential difference (right side panel) and at a relatively depolarized holding potential (–70 mV). In the less polarized situation Ca²⁺ transients, release fluxes and charge movement currents are smaller due to *U*-dependent inactivation. In this situation the drug has a marked effect, interpreted as due to state-dependent binding to the *U*-sensor (RIOS and BRUM, [52]).

bution as a second BOLTZMANN term of high steepness. The graph at the bottom of the figure (C) shows the potential difference distribution of charge in experiment A. It is obvious from inspection (and corroborated by the best fit with two BOLTZMANNS, continuous curves) that in reference solution there is a conspicuous high steepness component, centered near – 50 mV, that disappears in BAPTA. Surprisingly, even though Q_γ disappears, the maximum charge in this experiment did not change, and in several experiments was reduced only slightly.

As said before, there are many more examples, one to mention is shown in Figs. 15 and 16, the effect of low concentration of tetracaine. Fig. 15 shows the effect of tetracaine at 20 μM on Ca²⁺ release flux and Fig. 16 the corresponding effect on charge movement. Tetracaine reduces the peak of the release flux, surprisingly without affecting the final level of release (PIZARRO *et al.* [65]). This important observation will be taken up later. On the charge movement current the effect is the elimination of Q_γ.

A number of other interventions have been used with the same result. CSERNOCH *et al.* have recently reviewed this work [66]. The conclusion is that Q_γ, the hump component in intramembrane current, the high steepness component of the charge distibution, is a consequence of release, and is reduced, abolished or modified secondarily to the reduction or abolition of Ca release.

The manner in which we have explained the generation of Q_γ is illustrated in Fig. 17. The intramembrane microscopic potential difference (U_m) is made up of several components, including the bulk transmembrane potential difference and a negative surface potential difference presumably on both sides of the membrane, but represented in the diagram for only the intracellular face. We assume the existence near the inner face of the *U*-

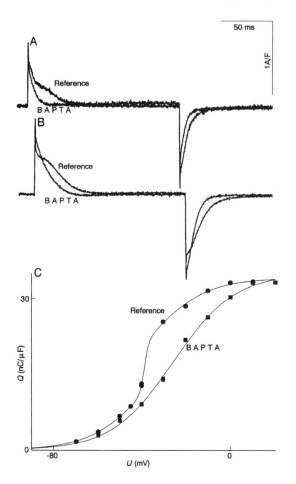

FIG. 14. BAPTA eliminates Q_y. A) charge movement current during a pulse to -40 mV in reference (same conditions as in Fig. 5) and after 20 minutes of exposure of the cut fiber ends to an internal solution with 20 mM BAPTA and no Ca^{2+}. B) In a different fiber, the internal solutions were applied in reverse order. C) $Q(U)$ distributions in reference and BAPTA, for the experiment represented by records in A. From GONZÀLEZ and RÍOS, [62].

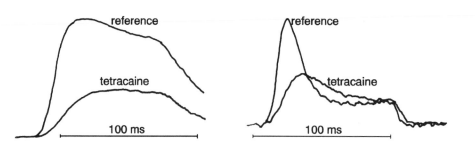

FIG. 15. Tetracaine selectively blocks the inactivating component of release. Ca^{2+} transients (top) and Ca^{2+} release flux (bottom) during a pulse to -60 mV in reference and in the presence of 20 µM tetracaine. The effect is a reduction in the amplitude of the inactivating component, but not in the maintained component of release. From PIZARRO *et al.* [63].

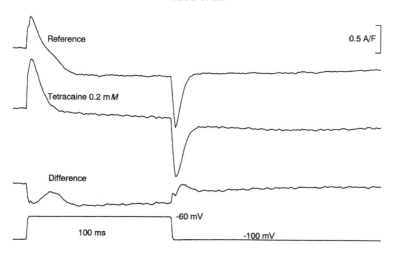

FIG. 16. Tetracaine blocks Q_γ Asymmetric currents in reference, and after exposure to 0.2 mM tetracaine. The difference record at bottom shows that the drug eliminated a hump component. A similar, though sometimes less complete, block is observed with 20 μM tetracaine. CSERNOCH, PIZARRO, URIBE, RODRÍGUEZ and RÍOS [64].

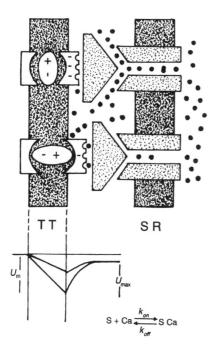

FIG. 17. A Ca^{2+}-binding model of Q_γ. See text. From PIZARRO *et al.* [65].

sensors of Ca^{2+}-binding sites, which are free of Ca^{2+} at the low $[Ca^{2+}]_{in}$ of the resting situation.When Ca^{2+} is released and increases locally, it occupies those sites and makes the surface potential difference less negative (or more positive) in fact increasing the intramembrane potential difference beyond the level determined by the U-clamp pulse. Therefore the initial charge movement, caused by the applied pulse, and presumably exponential, is followed by a second kinetic component of delayed onset, *humpy* because it is a secondary rise determined by the local increase in $[Ca^{2+}]$. Simultations with this model account reasonably well for the experimental properties of Q_γ (PIZARRO et al. [67]).

As shown in Fig. 18, the effect of this mechanism is to establish a positive feedback loop between release and charge movement. Indeed, the model explains Q_β and Q_γ with a single class of U-sensors. Whether they move as Q_β, initially, or as Q_γ, these U-sensors are going to open Ca^{2+} channels. Thus, when a small pulse is applied, capable of moving some sensors and opening some channels, Ca^{2+} flows from the SR, increases locally, binds to the sensors and in fact causes some that had not moved to move, thereby opening more release channels and constituting a positive feedback loop. We have demonstrated quantitatively, in work that is now submitted for publication, that one of the consequences of this positive feedback is to increase substantially the response to a just suprathreshold pulse. That has the effect of increasing the steepness of both the $Q(U)$ and the $\dot{R}(U)$ dependences slightly above threshold.

This helps to understand one of the puzzling questions in control of Ca^{2+} release, namely the extraordinary steepness of the $\dot{R}(U)$-dependence, only matched by the steepness of the dependence of Q_γ on U. For instance, the steepness factor can be as low as 2 mV, which from simple BOLTZMANN statistics would require the complete transfer of 12 electronic charges per particle across the membrane. Since this is unheard of in the study of other U-sensitive channels and as we now know, the U-

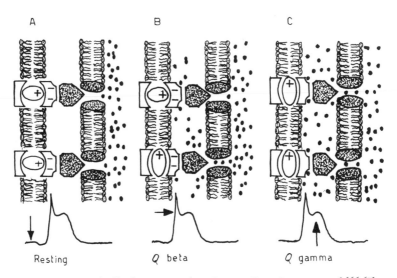

FIG. 18. Positive feedback in EC coupling. See text. From CSERNOCH *et al.* [66,67].

sensor of EC coupling should not be very different from *U*-sensitive channels in its structure, the positive feedback mechanism is a much better explanation of the high steepness than assuming intrinsic high valence in the mobile particle.

Finally, let me stress the significance of another result. As shown in Fig. 15, tetracaine blocks the inactivating component of release, without touching the non-inactivating component. Tetracaine is, as defined by the researchers that study fractionated SR, perhaps the best inhibitor of Ca^{2+}- induced Ca^{2+} release, with for instance 10 times higher effectiveness than procaine. That it selectively blocks one of the component is entirely consistent with the proposal that PIZARRO and RIOS made [25] that these two components of Ca^{2+} release correspond to two different mechanisms: the peak component mediated by Ca^{2+}, the non-inactivating component directly controlled by potential difference. Recently JACQUEMOND *et al.* [68] reported that microinjection of

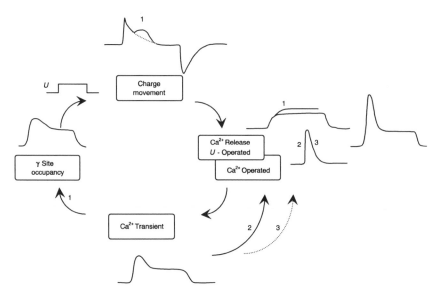

FIG. 19. A collective view of events in EC coupling. Two positive feedback mechanisms (1,2) and one negative feedback mechanism (3) are shown.

high concentrations of BAPTA eliminated selectively the peak of release, without touching the maintained component, very much like the tetracaine effect. Since BAPTA is a very effective buffer, the result is again consistent with the idea that this component is due to a local increase in $[Ca^{2+}]_{in}$. In turn, this increase is secondary to the opening of channels directly under control by the T membrane potential. The alternating structure of feet with and without *U*-sensor described by BLOCK *et al.*. [27] provides the structural scenario that makes this explanation feasible.

As a summary then, let me give a hypothetical view of the sequence of events that are involved in a normal cycle of Ca^{2+} release (Fig. 19): a depolarization starts to move charged moieties in the *U*-sensors (as we shall see, these are likely to reside in the S_4 transmembrane segments of the α_1 subunit), some of these open underlying Ca^{2+} release channels, by a mechanism that we will address later (these channels will remain open for as long as the sensors remain activated). As a consequence of channel opening Ca^{2+} diffuses from the SR (probably without generating major potential gradients) and increases the local $[Ca^{2+}]_{in}$. This increase has, in addition to the second messenger role, at least three feedback roles: first it binds to sites on the U-sensor (mechanism labeled 1 in the figure), boosting the microscopic potential that acts on the sensors and determining more transitions in sensors that had not activated (and causing Q_γ). Also, the local $[Ca^{2+}]_{in}$ activates directly other release channels, by the mechanism of Ca^{2+}-induced Ca^{2+} release, this results in the rapid rise of release (mechanism labeled 2 in the figure). Finally, one has to account for the relaxation of release that occurs after the peak even if the depolarizing pulse continues. M. SCHNEIDER and B. SIMON [69] have convincingly explained this inactivation as due to local Ca^{2+}, that is, as a Ca^{2+}-dependent inactivation, on the basis of its dependence on the history of previous pulses and the bulk $[Ca^{2+}]_{in}$ reached during the pulse. This mechanism (3 in Fig. 19) is also consistent with inactivating effects of Ca^{2+} on release channels reconsituted in bilayers. The explanation, however, was brought into question because the sensitivity to $[Ca^{2+}]$ required is much higher than that of Ca^{2+}-dependent inactivation of channels in bilayers, and because SIMON failed to observe inactivation of release in experiments in which $[Ca^{2+}]_{in}$ was increased by release from photolysable Ca^{2+} buffers (HILL and SIMON, [70]).

4. Mechanisms of transmission

I will now discuss succinctly three candidate mechanisms of trasmission from T depolarization to channel opening (Fig. 20).

4.1. Ca^{2+} - induced Ca^{2+} release

Since ARMSTRONG *et al.* [71] it is known that skeletal muscle works without extracellular Ca^{2+}. The concept of the priming site (*vide supra*) serves to explain many observations, for instance by J. B. FRANK [72] on the inhibitory effects of lowering extracellular $[Ca^{2+}]$. If extracellular Ca^{2+} is replaced by other metals, known not to cross the membrane (like Cd^{2+} or La^{3+}) EC coupling is restored, which really rules out Ca^{2+} entry as a transmission mechanism.

As we have seen, it is likely that Ca^{2+}-induced Ca^{2+} release plays an amplifying role, but only *after* channel openings elicited by a different transmission mechanism have occurred. The question that ensues is how the system remains graded on potential difference, and how can it be stopped rapidly by repolarization, even during the inactivating portion of release, given the intrinsically self-sustaining nature of Ca^{2+}-induced Ca^{2+} release. The answer to this question will require improving our understanding of the whole mechanism and will be an area of interest in the future.

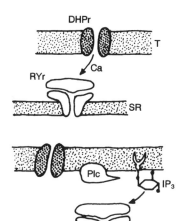

Chemical
transmission

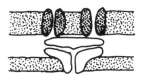

Mechanical
transmission

FIG. 20. Three possible mechanisms of T-SR transmission. Among them, only the mechanical transmission model predicts necessary *reverse* effects of the interaction on the *U*-sensor.

4.2. *Inositol trisphosphate*

The hypothesis that InsP$_3$ mediates T-SR transmission (VERGARA *et al.* [73] VOLPE *et al.* [74]) has been tested repeatedly, and even though there are very many observations suggesting a role, we do not believe that this is the primary transmitter. We have discussed this extensively (RIOS and PIZARRO, [34]) finding three especially troublesome problems:

i) the density of precursor (PIP$_2$) available to be hydrolyzed in the T membrane is 60/μm^2 (HIDALGO and JAIMOVICH, [75] or an order of magnitude less than that of the putative receptors (feet structures);

ii) the sensitivity of the Ca^{2+} release system to InsP$_3$ is modulated by the state of the sensor, with only fibers with depolarized T tubules or with inactivated *U*-sensors being highly sensitive to the putative transmitter. This implies that the putative receptor has an independent line of communication with the *U*-sensor, thus, InsP$_3$ would be redundant;

iii) finally, the sensitivity to InsP$_3$ of the release system decays drastically between [Ca^{2+}]$_{in}$ 100 nM and 10 nM and no such dependence has been found in the Vaseline gap studies of Ca^{2+} release.

4.3. *Mechanical coupling*

According to this hypothesis (CHANDLER *et. al.* [40]), the *U*-sensor in the T membrane would be mechanically linked to the SR channel. Perhaps the strongest evidences in its favor are the newly revealed structure of the release molecule, the molecular nature of the *U*-sensor, its arrangement in the T membrane, and new demonstrations of the association of *U*-sensors and release channels during development. We will briefly review all these arguments.

The release channel has a large cytoplasmic moiety, sufficiently tall to reach the T membrane, that gives a structural basis to the postulated mechanical link. The *U*-sensor is sort of a *U*-sensitive channel, and there is a limited repertoire of functions that can be assigned to such a molecule. Since it is not performing its transmitter function by transporting ions, it could be simply changing conformation and modifying a molecule(s) in close contact. It could conceivably be activating a phospholipase, or a G protein. It is more parsimonious, however, to think that it is acting on the release channel.

This is especially likely given the strict alignment and stoichiometry in their association, as revealed by the microscopic images of BLOCK *et al.*[27]. The strict steric association is brought home in a compelling way in recent studies of JORGENSEN [29], who demonstrated with immunofluorescence microscopy that during development the DHP receptor α_1 subunit and the ryanodine receptor are associated since very early, even before the formation of T tubules. Thus, if there is a fixed stoichiometry of association, if the spatial arrangement of the tetrads of *U*-sensors is suitable to fall just on top of the homotetramer of release channel subunits, with a on-to-one correspondence between individual sensors and release promoters, and if this association is programmed since the earliest stages of ontogenesis, it is almost unthinkable that the functional interaction between the proteins would still be mediated by, say, a diffusible messenger. Even a bridge protein, as proposed by CASWELL and coworkers (BRAND *et al.*[28]) would seem difficult to justify.

4.4. *Ongoing functional studies favor the mechanical hypothesis*

One definite prediction of the mechanical hypothesis is that *there should be reciprocal effects* of the release channel on the *U*-sensor, due to the hypothesized contact. To test this prediction we have started in our laboratory a systematic comparison of pharmacology, kinetics and other properties of the native *U*-sensor, in the single fiber, and the corresponding properties of *U*-sensors in non interacting situations. By this we mean three other preparations or situations (Fig. 21):

i) the inactivated *U*-sensor,

ii) the DHP receptor from T membrane vesicle fractions, reconstituted in planar bilayers (MA *et al.* [76]), and

iii) the *U*-sensor of the L type cardiac Ca^{2+} channel (SHIROKOV *et al.* [77]).

The inactivated *U*-sensor presumably interacts very differently or does not interact with the release channel. The *U*-sensor from the T tubule reconstituited in bilayers is devoid of the EC coupling interaction. The L type channel of the heart inter-

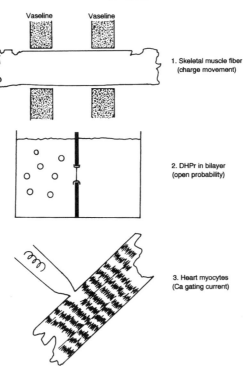

Vaseline   Vaseline

1. Skeletal muscle fiber
(charge movement)

2. DHPr in bilayer
(open probability)

3. Heart myocytes
(Ca gating current)

FIG. 21. An empirical approach to test the prediction of reverse effects. Three preparations in which DHP receptors can be studied. Only in the active DHP receptor of the skeletal fiber (top) there should be reverse effects.

acts with the release channel but only trough Ca^{2+}-induced Ca^{2+} release (NABAUER *et al.* [78]). In initial stages we studied the effects of the anion perchlorate on all systems. ClO_4^- is the best studied, and perhaps the most effective agonist of EC coupling (LÜTTGAU *et al.* [79]), causing a large shift of $Q(U)$ to lower U-values, an increase in steepness of the distribution, and a general slowing of the kinetics of charge movent, especially at the OFF. It greatly shifted to lower U-values the activation of release flux and reduced substantially the amount of charge movement needed to start opening release channels (GONZÁLEZ *et al.* [36]).

By contrast, ClO_4^- had essentially no effects on any of the other situations of non-interacting U-sensors. It did not shift the U-dependence of the DHP-sensitive channel of skeletal muscle in bilayers (RIOS *et al.* [80]), and it shifted only slightly but did not change the kinetics of the charge of both inactivated skeletal U-sensor and the L-type Ca^{2+} channel of ventricular myocytes (RIOS *et al.* [81]). These results are consistent with the idea that the pharmacological action requires the interaction of the two molecules, and are thus consistent with the mechanical model. Of course, there are other interpretations, however, the quantitative aspects of the changes due to ClO_4^- are best explained assuming an allosteric interaction between four U-sensors and one tetrameric release channel (RIOS *et al.* [81]).

References

[1]   L.V. HEILBRUNN and F. WIERCINSKI, *J. Cell. Comp. Physical,* **29**, 15 (1947).

[2]   A.V. HILL, *Proc. Roy . Soc. (London),* **B 135**, 446 (1948).

[3]   A. HUXLEY and R.E.TAYLOR, *J. Physiol. (London) ,* **144**, 426 (1958).

[4]   E. VERATTI, *Men. Ist. Lombardo Cl. Sci. Matt. Nat.***19**, 87 (1902).
         translated in *J. Biophys Biochem. Cyttol.,, 10* (4), Supp. 3-59 (1961).

[4 bis] C. FRANZINI - ARMSTRONG, *J. Cell Biol,* **47**, 488 (1970).

[5]   B. EISENBERG, in *Handbook of Physiology, Skeletal Muscle.* Am-Physiol. Soc.
         Bethesda, Md. (1983) Chapt. 3 pp. 73.

[6]   L. PEACHEY, *J. Cell Biol.,* **25**, 209 (1965).

[7]   L. PEACHEY and B. EISENBERG, *Biophys. J.,* **22**, 145 (1978).

[8]   A. L. HODGKIN and P. HOROWICZ, *J. Physiol. (London),* **153**, 386 (1960).

[9]   C. CAPUTO, *J. Physiol. (London),* **223**, 483 (1972).

[10]   H. C. LÜTTGAU and W. SPIECKER, *J. Physiol. (London),* **296**, 411 (1979).

[11]   G. BRUM and E. RÍOS, *J. Physiol. (London),* **387**, 489, (1987), G. BRUM, R. FITTS,
         G. PIZARRO and E. RÍOS, *J. Physiol. (London),* **398**, 475 (1988).

[12]   S. EBASHI, M. ENDO and I. OHTSUKI, *Q. Rev. Biophys.,* **2**, 351 (1969).

[13]   W. HASSELBACH, *Prog. Biophys.,* **14**, 167 (1964).

[14]   H. GONZÁLEZ SERRATOS, *J. Physiol. (London),* **212**, 777 (1971).

[15]   C. FRANZINI - ARMSTRONG and G. NUNZI, *J. Muscle Res. Cell Motil.,* **4**, 233 (1983).

[16]   J.J. S. CADWELL and A. H. CASWELL, *J. Cell, Biol.,* **93**, 543 (1982).

[17]   J.L. SUTKO and J. L. KENYON, *J. Gen. Physiol.,* **82**, 385 (1983).

[18]   I.N. PESSAH in A. O. FRANCINI, D. J. SCALES, A. L. WATERHOUSE and J. E. CASIDA, *J.
         Biol. Chem.,* **261**, 8643 (1986).

[19]   S. FLEISCHER, E. M. OGUNBUNMI, M. C. DIXON and E. A. FLEER, *Pro. Natl., Acad.
         Sci. USA,* **82**, 7256 (1985).

[20]   M. INUI, A. SAITO and S. FLEISCHER, *J. Biol. Chem.,* **262**, 1740 (1987).

[21]   K. CAMPBELL, *et al., J. Biol. Chem.,* **262**, 6460 (1987).

[22]   F. A. LAI, H. ERICKSON, B. BLOCK and G. MEISSNER, *Biochem. Biophis. Res.
         Commun.,* **143**, 704 (1987).

[23]   F. LAI, H. P. ERICKSON, E. ROUSSEAU, Q.Y. LIU and G. MEISSNER, *Nature
         (London),* **331**, 315 (1988).

[24]   T. WAGENKNECHT, R. GRASSUCCI, J. FRANK, A. SAITO, M. INUI and S. FLEISCHER,
         Nature (London), **338**, 167 (1989).

[25]   E. RÍOS and G. PIZARRO, *N.I. P.S.,* **3**, 223(1988).

[26]   C. FRANZINI - ARMSTRONG and G. NUNZI, *J. Muscle Res. Cell. Motil.,* **4**, 233 (1983).

[27]   B. A. BLOCK, T. IMAGAWA, K. CAMPBELL and C. FRANZINI - ARMSTRONG, *J. Cell.
         Biol.,* **107**, 2587 (1988).

[28]   N. R. BRANDT, A. H. CASWELL, S. R. WEN and J.A. TALVENNEIMO, *J. Membr. Biol.,*

113, 237 (1990).

[29] A. O. JORGENSEN, S. ACY, S. H. YUAN, M. GAVER. and P. K. CAMPBELL, *J. Cell. Biol.*, **110**, 1173 (1990).

[30] A. L. HODGKIN and A. HUXLEY, *J. Physiol. (London)*, **117**, 500 (1952).

[31] R. ADRIAN, W. CHANDLER and A. L. HODGKIN, *J. Physiol. (London)*, **208**, 607 (1970).

[32] B. HILLE and D. T. CAMPBELL, *J. Gen. Physiol.*, **67**, 265 (1976).

[33] L. KOVACS, E. RÍOS and M. F. SCHNEIDER, *Nature (London)*, **279**, 391 (1979).

[34] E. RIOS and G. PIZARRO, *Physiol. Rev.*, **71**, 849 (1991).

[35] M. IRVING, J. MAYLIE, N. SIZTO and K. CHANDLER, *J. Gen. Physiol.*, **89**, 1 (1987).

[36] A. GONZÁLEZ and R. RÍOS, *unpublished observations.*

[37] M. F. SCHNEIDER and W. K. CHANDLER, *Nature (London)*, **242**, 244 (1973).

[38] C. M. ARMSTRONG and F. BEZANILLA, *Nature (London)*, **242**, 459 (1973).

[39] L. CSERNOCH, G. PIZARRO, I. URIBE, M. RODRIGUEZ and E. RÍOS, *J. Gen. Physiol.*, **97**, 845 (1991).

[40] W.K. CHANDLER, R.F.RAKOWKSI and M.F. SCHNEIDER, *J. Physicol (London)*, **254**, 285 (1976).

[41] C. S. HUI and W. K. CHANDLER, *J. Gen. Physiol.*, **96**, 257 (1990).

[42] E. RIOS and M. F. SCHNEIDER, *Biophys. J.*, **36**, 607 (1981).

[43] W. MELZER, E. RÍOS and M. F. SCHNEIDER, *Biophys. J.*, **45**, 637 (1984); **51**, 849 (1987).

[44] G. BRUM, E. RÍOS, and E. STEFANI, *J. Physiol. (London)*, **398**, 441 (1988).

[45] R. H. ADRIAN and A. PERES, *J. Physiol. (London)*, **398**, 83 (1979).

[46] C. H. L. HUANG, *Physiol. Rev.*, **68**, 1197 (1979).

[47] G. PIZARRO, R. FITTS, I. URIBE and E. RÍOS, *J. Gen. Physiol.*, **94**, 405 (1989).

[48] F. BEZANILLA R. E. TAYLOR, and J. FERNANDEZ, *J. Gen. Physiol.*, **79**, 21 (1982).

[49] E. RIOS, G. BRUM, G. PIZARRO and M. RODRIGUEZ, *Biophys. J.*, **57**, 341a (1990).

[50] H. C. LÜTTGAU, *personal communication.*

[51] P. HESS, G. B. LANSMAN and R. W. TSIEN, *J. Gen. Physiol.*, **88**, 293 (1986).

[52] E. RÍOS and G. BRUM, *Nature (London)*, **325**, 717 (1987).

[53] B. P. BEAN, *Proc. Natl. Acad. Sci. USA*, **81**, 6388 (1984).

[54] L. SCHWARTZ, E. MC CLESKEY and W. ALMERS, *Nature (London)*, **314**, 747 (1985).

[55] R. S. EISENBERG, R. T. MC CARTHY and R. L. MILTON, *J. Physiol. (London)*, **341**, 459 (1983).

[56] C. S. HUI, R. L. MILTON and R. S. EISENBERG, *Proc. Natl. Acad. Sci. USA*, **81**, 2582 (1984).

[57] M. GLOSSMANN and J. STRIESSNIG, *Vitamins and Hormones*, **44**, 155 (1988).

[58] K. G. BEAM, C. M. KNUDSON and J. A. POWELL, *Nature (London)*, **320**, 168 (1986).

[59] T. TANABE, H. TAKESHIMA, A. MIKAMI, V. FLOCKERZI, M. TAKAMASI, K. KANGAWA, M. KOJIMA, H. MATSUO, T. HIROSE and S. NUMA, *Nature (London)*, **328**, 313 (1987).

[60] T. TANABE, K. G. BEAM, J. A. POWELL and S. NUMA, *Nature (London)*, **336**, 134 (1988); B. ADAMS, T. TANABE, A. MIKAMI, S. NUMA and K. BEAM, *Nature*

(London), **346**, 589 (1990).

[61]   K. S. De Jongh, D. K. Merrick and W. A. Catterall, *Proc. Natl. Acad. Sci USA,* **86**, 8585 (1989); G. Pizarro, R. Fitts, M. Rodriguez, I. Uribe and E. Ríos, in *The Calcium Channel,* M. Morad (Editor), Springer Verlag (1988), p. 138.

[62]   A. González and E. Rios, *Biophys. J.,* **61**, 130 a (1992).

[63]   G. Pizarro, L. Csernoch and E. Ríos, *J. Physiol. (London),* **452**, 525 (1992).

[64]   L. Csernoch, G. Pizarro, I. Uribe, M. Rodriguez and E. Ríos, *J. Gen. Physiol.,* **97,** 845 (1991).

[65]   G. Pizarro, L. Csernoch, I. Uribe and Rios, *Biophys J.,* **55**, 237 (1989).

[66]   L. Csernoch, E. Stefani, G. Pizarro, G. Szucs, J. Garcia and E. Ríos, *In E C coupling in Skeletal, Cardiac and Smooth muscle.* Ed. G.B. Frank. p. 137-148. Plenum Press, New York, 1992.

[67]   G. Pizarro L. Csernoch, I. Uribe M. Rodriguez and E. Ríos, *J. Gen. Physiol.,* **97**, 913 (1991).

[68]   V. Jacquemond, L. Csernoch, M. Klein and M. F. Schneider, *Biophys. J.,* **60**, 902 (1991).

[69]   M. Schneider and B.J. Simon, *J. Physiol. (London),* **405**, 727 (1988).

[70]   B.J. Simon and D. Hill, *Biophys. J.,* **61**, 1109 (1992).

[71]   C. M. Armstrong, F. Bezanilla and P. Horowicz, *Blochem. Biophys. Acta.,* **267**, 605 (1985).

[72]   G. B. Franck, *Biochem. Pharmacol.,* **29**, 2399 (1980).

[73]   J. Vergara, R. Y. Tsien and M. Delay, *Proc. Natl. Acad. Sci. USA,* **82,** 6352 (1985).

[74]   P. Volpe, G. Salviati, F. Divirgilio and T. Pozzan, *Nature (London),* **316**, 347 (1985).

[75]   C. Hidalgo and E. Jaimovich, *J. Bioenerg. Biomembr.,* **21**, 267 (1989).

[76]   J. Ma, C. Mundiña - Weilenmann, M.M. Hosey and E. Ríos, *Biophys. J.,* **60**, 890 (1991).

[77]   R. Shirokov, R. Levis, N. Shirokova and E. Ríos, *J. Gen. Physiol.,* **99**, 863 (1992).

[78]   M. Nabauer, L. Callenwaert, L. Cleeman and M. Morad, *Science,* **244**, 800 (1989).

[79]   M. C. Lüttgau, G. Gottschalk, L. Kovacs and M. Fuxreiter, *Biophys. J.,* **43**, 247 (1983).

[80]   E. Ríos, R. Shirokov, R. Levis, A. González, I. Stavrosky, J. Ma, C. Mundiña Weilenmann and M. M. Hosey, *Biophys J.,* **59**, 20 a (1991).

[81]   E. Ríos, M. Karhanek, A. González and J. Ma, *Biophys. J.,* **61**, 131 a (1992).

MOLECULAR BIOLOGY OF
SARCOPLASMIC RETICULUM Ca^{2+} CHANNELS

FRANCESCO ZORZATO

Istituto di Patologia Generale, Università di Ferrara
Via Borsari, 44100 Ferrara, Italy

Contents

Bioelectrochemistry IV
Edited by B.A. Melandri *et al.*, Plenum Press, New York, 1994

1. Introduction

The sarcoplasmic reticulum (SR) is an intracellular membrane compartment that controls the myoplasmic Ca^{2+} concentration, thereby playing an important role in the excitation-contraction coupling mechanism (ENDO, and RIOS and PIZARRO [1]). Skeletal muscle contraction is initiated by Ca^{2+} release from terminal cisternae (SOMLYO *et al.*, [2]), the portion of the sarcoplasmic reticulum junctionally associated, *via* feet structures, to invaginations of the plasmalemma called the transverse tubules (FRANZINI-ARMSTRONG; KAWAMOTO *et al.*, [3]). Electrophysiological studies have shown the existence of two types of Ca^{2+} release channels in sarcoplasmic reticulum membranes:

1) a low conductance channel evenly distributed throughout the sarcoplasmic reticulum membrane;

2) a high conductance channel that appears to be selectively localized in the junctional sarcoplasmic reticulum (SMITH *et al.*, [4]).

The high conductance channel is modulated by a variety of agents (SMITH *et al.*, [4]; PALADE *et al.*, [5]) including Ca^{2+}, ATP, Mg^{2+}, doxorubicin and ryanodine. Recently, SALAMA and his coworkers have reported the existence of another SR Ca^{2+} channel, which exhibits properties similar to the high conductance channel but differs in being of lower molecular mass (106 *vs* 450-564 kDa) and gated by sulphydryl reagents (ZAIDI *et al.*, [6]). Another molecule that may act as a Ca^{2+} channel in the sarcoplasmic reticulum is the InsP$_3$ receptor (VOLPE *et al.*, [7]). Hybridization probe analysis of skeletal muscle mRNA with cDNA from the cerebellum InsP$_3$ receptor clone has indicated that skeletal muscle expresses the InsP$_3$ receptor (FURUICHI *et al.*, [8]). To date however, no information regarding its exact intracellular membrane topology is available. This chapter deals with recent data obtained on the structural properties and pathological alterations of the molecular component of the high conductance sarcoplasmic reticulum Ca^{2+} channel (ryanodine receptor) from striated muscle.

2. Molecular structure of the ryanodine receptor Ca^{2+} channel

2.1. *Identification of the protein component of the Ca^{2+} release channel*

In the last few years a tremendous amount of work has been carried out in order to define useful probes to purify the molecular components of the Ca^{2+} release channel. The most popular one is ryanodine a plant alkaloid that modulates Ca^{2+} release from isolated terminal cisternae and binds to its receptor with high affinity (PESSAH *et al.*; FLEISCHER *et al.*, [9]). The ryanodine receptor (RYR) is comprised of four 450 kDa subunits which form:

1) a large homotetrameric complex morphologically identical to the feet structures (INUI *et al.,* [10]) and;

2) a cation channel with a pharmacological profile, conductance and ion selectivity matching those of the native channel (SMITH *et al.;* LAI *et al.;* HYMEL *et al.,* [11]).

In addition, other drugs have been used to characterize the molecular composition of the Ca^{2+} channel. ZORZATO *et al.* [12]) found that doxorubicin selectively activates Ca^{2+} release from skeletal muscle terminal cisternae fractions. By using [^{14}C]–doxorubicin as a natural photoligand, it has been reported that three minor protein components of junctional sarcoplasmic reticulum with an *m.w.* of 350, 170 and 80 kDa are specifically photolabeled by doxorubicin. The most convincing evidence that these proteins might be part of the molecular complex of the Ca^{2+} release channel has been obtained by experiments showing that known modulators of Ca^{2+} release (caffeine, Ca^{2+} and Ag^+) were able to inhibit [^{14}C] –doxorubicin labelling of these proteins. On a molar basis, the 350 and 170 kDa proteins exhibited the highest incorporation of [^{14}C] –doxorubicin.

Subsequently, the high molecular weight 350 kDa doxorubicin-binding protein appeared to be identical to the junctional protein that has been shown to constitute the ryanodine receptor (LAI *et al.,* [11]). In order to further investigate, the structural and functional role of the 350 kDa doxorubicin-binding protein in the native Ca^{2+} release channel molecular complex, polyclonal Ab were raised, and their effect on Ca^{2+} release from isolated terminal cisternae was there studied. ZORZATO *et al.* [13] have shown that anti–(RYR) Ab are able to affect activation of Ca^{2+} release by doxorubicin and Ca^{2+}. FILL *et al.* [14] examined the effect of the same Ab at the single channel level and demonstrated that the anti–(RYR) Ab decreased the open probability of the Ca^{2+} channel upon stimulation with Ca^{2+}. On the other hand, in the presence of ATP the Ab-bound Ca^{2+} channel displayed no decrease in the open probability. These results indicate that the Ab-dependent abnormalities of the Ca^{2+} gating of the channel may be due to alterations of the regulatory mechanism of the channel rather than to a modification of the channel pore.

A preliminary step towards the elucidation of the structural and functional domain(s) of the Ca^{2+} channel is the definition of its primary structure. Thus, these anti –(RYR) Ab were used as probes to screen λ gt11 cDNA expression libraries to isolate the full length cDNA clone encoding the rabbit and human skeletal muscle RYR.

2.2. Primary structure of the skeletal muscle ryanodine receptor Ca²⁺ release channel

The human skeletal muscle ryanodine receptor monomer is encoded by an mRNA of about 16 kb containing one open reading frame encoding 5032 amino acids with a predicted molecular weight of 564 kDa (Table 1). Comparison of the primary structure of the skeletal RYR between different species (human and rabbit) showed that overall the two sequences are highly conserved (>90%). However, the degree of homology is not evenly distributed along the molecule. As a result of several non-conservative substitutions, the polypeptide defined by residue 4394-4434 displays 77,5% identity. Few deletions were also observed in correspondence with an extremely acidic region (residue 1872-1923) (Figure 1). However, there are long stretches of complete identity of the amino acid sequences between the species examined. Human and rabbit RYR are identical in the region encompassing:

a) transmembrane segments M1 to M2 and M5 to M10), and

b) potential calmodulin binding sites (residues 2948-3293) (Figure 2).

Skeletal rabbit RYR sequences that have been determined so far display one significant discrepancy. TAKESHIMA et al. [15] reported a Ala-Gly-Asp-Ala-Glu sequence at position 3481-3485; this was not found in both rabbit and human cDNA clones isolated by ZORZATO et al. [16]. These results may be consistent with an alternative splicing of an exon of the skeletal RYR gene, which may account for the existence of different RYR isoforms. Analysis of the RYR aminoacid sequence deduced from the cDNA led to the identification of potential domain boundaries (Figure 3).

TABLE 1. CHARATERISTICS OF THE CARDIAC AND SKELETAL RYNODINE RECEPTORS cDNA AND PROTEIN

	Skeletal	Cardiac
cDNA size	15,346 bp	16,732 bp
5'-untraslated	106 bp	307 bp
3'-untraslated	148 bp	1,318 bp
Amino acids	5,032	4,969
Molecular mass	564,073 Da	564,711 Da
Isoelectric point	5.0	5.65

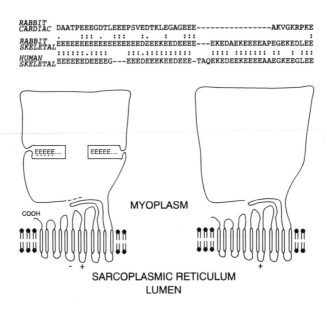

```
RABBIT
CARDIAC   DAATPEEEGDTLEEEPSVEDTKLEGAGEE----------------AKVGKRPKE
          .    ::: .  :::   :.  :   :::                        :
RABBIT    EEEEEEEEEEEEEEEEEEEEEDEEKEEDEEEE---EKEDAEKEEEEAPEGEKEDLEE
SKELETAL
          ::::::.::::    :::.:::::::::    ::::  :::::::  ::  :  :::
HUMAN     EEEEEEDEEEG---EEEDEEEKEEDEEE-TAQEKEDEEKEEEEAAEGKEEGLEE
SKELETAL
```

MYOPLASM

COOH

SARCOPLASMIC RETICULUM
LUMEN

SKELETAL   CARDIAC

FIG. 1. Comparison of acidic region of skeletal and cardiac ryanodine receptors. Alignment of amino acid sequences was carried out by Genepro computer program.

PUTATIVE CALMODULIN BINDING SITE

SKELETAL KEKEMITSLFCKLAALVRHR 3018
:::::.:::: ::. :::::
CARDIAC KEKEMVTSLFLKLGVLVRHR 3052

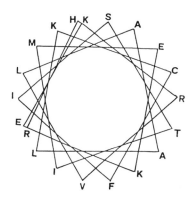

FIG. 2. Calmodulin binding site: Amino acid numbering for cardiac and skeletal RYR isoform is as described by OTSU *et al.* [28] and ZORZATO *et al.* [16], respectively. Helix wheel representation of the skeletal RYR calmodulin consensus sequence shows that the majority of basic residues are facing one side of the helix.

The skeletal muscle RYR monomer may contain up to 12 transmembrane segments, ten (M1 to M10) of which are restricted to the 1000 carboxy-terminal residues, two transmembrane sements (M' and M") are localized in the central portion of the molecule (residues 3123–3143 and 3187–3205). This is consistent with the suggestion that the 4000 amino–terminal residues comprise protein domains which make up the hydrophylic (cytoplasmic) portion of the RYR, while the channel is mainly formed by the carboxy–terminus. According to the three–dimension al reconstruction of the RYR tetramer proposed by WAGENKNECHT *et al.* [17], the size of the hydrophobic portion (basal plate) of the receptor is about 140 x 140 Å. Such a structure presumably contains the transmembrane portion of the RYR and it may accomodate up to 150 helices of 11 Å in diameter, a much higher than that of the maximal number (48) of the putative transmembrane segments of the RYR oligomer. If the estimate of the basal plate is correct, it may be speculated that either, the domains of the protein embedded in the hydrophobic helices or lipids contribute to the formation of the basal plate. Provided that there are 12 transmembrane segments for each RYR subunit, the resulting model would have six lumenal loops, five of which lie in the carboxy–terminal fifth of the molecule. The third and fourth lumenal loops contain clusters of positively (RRR-VRRLRR, residues 4307–4314) and negatively (EEAGDEDE, residues 4612–4620) charged residues, respectively (Figure 4). The presence of charged residues at the lumenal mouth of the channel might form an ionic screen at the channel entrance. These resi-

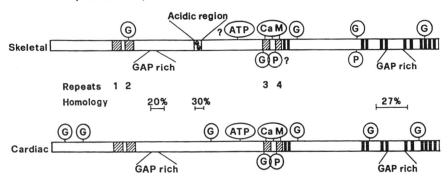

Fig. 3. Planar model of ryanodine receptor: transmembrane segments are indicated by closed rectangles (M′, M″; M1 to M10 from left to right); G,P, Ca M and ATP indicate putative glycosylation, calmodulin and ATP–binding sites, respectively. Hatched boxes indicate two tandem pairs of repeated sequences.

dues might also be involved in the protein–protein interaction between the RYR and lumenal proteins such as calsequestrin and/or calsequestrin binding protein (MITCHELL et al., DAMIANI and MARGRETH, [18]).

Apart from a pair of putative transmembrane segments, the 4000 amino –terminal residues of the RYR monomer are also predicted to contain (Figure 3):

(i) four repeats arranged in two tandem pairs;

(ii) an acidic region which is located within a 1600 residue long polypeptide separating two pairs of repeated sequences;

(iii) an unstructured region enriched in glycine, alanine and proline (GAP); a second GAP–rich region is also present in the myoplasmic loop between putative transmembrane segments M4 and M5.

The functional role of these repeated sequences is unknown. However, it is interesting to notice that the third and fourth repeated sequence are located near the first two putative transmembrane segments M′ and M″ within a region containin consensus sequences for calmodulin, a modulator of Ca²⁺ release. The effect of calmodulin on the channel activity might be mediated by a modification of the function of these repeated sequences. Another interesting feature of the RYR sequence concerns the region defined by residues 1872 – 1923 which is highly enriched in glutamates. The acidic region appears to be located within protease–insensitive sequences (MARKS et al., [19]) and it has therefore been suggested that it may be either buried in the glubular portions of the molecule or associated with the channel wall.

Ca²⁺ binding sites, with low and high affinity, have been proposed (MEISSNER et al., [20]). The high–affinity sites are involved in the activation of the Ca²⁺ release by a mechanism known as Ca²⁺ –induced Ca²⁺ release; the low–affinity sites appear to be involved in the channel inactivation. In view of the similarity between the acidic region

```
SKELETAL  EPEGEPEADEDEGMGEAAAEGAEEGAAGAEG-AAGTVAAGATARLAAAAARALRGLSYR
          :  . : .   :     : .      ::  .  :  :   :    : :: .
CARDIAC   ESDLNERSANKEESEKERPE---EQGPKMGFFSVLTVRSALFALLRYNILTLMRMLSLK

SKELETAL  SLRRRVRRLRRLTAREAATAL-AALWAVVARAGAAGAGAAAGALRLLWGSLFGGGLVEGA
          ::..  .. .. :.. ::       :  .  :  : :: :::::
CARDIAC   SLKKQMKKMKKMTVKDMVTAFFSSYWSIFMTLLHFVASVFRGFFRIVCSLLLGGSLVEGA

SKELETAL  KKVTVTELLAGMPDPTSDEVHGEQPAGPGGDADGAGEGEGEGDAAEGDGDEEVAGHEAGP
          ::. : :::: :::: ::: :.      :  ...         .       :.  :
CARDIAC   KKIKVAELLANMPDPTQDEVRGD-----GEEGERKPMETTLLLPSEDLTDLKELTTEESDL

SKELETAL  -GGAEGVVAVADGGPFRPEGAGGLGDMGDTTPAEPPTPEGSPILKRKLGVDGEEELVPEP
           :.   .:: ..    .  :      : :   :   :       :  :
CARDIAC   LSDIFGLDLKREGGQYKLIPHNPNAGLSDLM----SNPVLIPEEQEKFQEQKTKEEEKEE

SKELETAL  EPEPEPEPEKADEENGEKEEVPEAPPEPPKKAPPSPPAKKEEAGGAGMEFWGELEVQRVK
          :  :::::. : ::::::        :    . :       ::  .    :
CARDIAC   KEETKSEPEKAEGEDGEKEEKVKE-DKGKQKLRQLHTHRYGEPEVPESAFWKKIIAYQQK

SKELETAL  FLNYLSRNFYTLRFLALFLAFAINFILLFYKVSDSPPGED-DMEGSAAGDLAGAGSGGGS
          :::: :::: : ::::.:::::::::::::: :   :   .    .: .   . :
CARDIAC   LLNYLARNFYNMRMLALFVAFAINFILLFYKVSTSSVVEGKELPSRSTSENAKVTTSLDS

SKELETAL  GWGSGAGEEAEGDEDENMVYYFFLE
                              ::
CARDIAC   SSSHRII-----------AVHYVLE
```

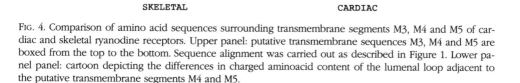

MYOPLASM

SARCOPLASTIC RETICULUM
LUMEN

SKELETAL   CARDIAC

FIG. 4. Comparison of amino acid sequences surrounding transmembrane segments M3, M4 and M5 of cardiac and skeletal ryanodine receptors. Upper panel: putative transmembrane sequences M3, M4 and M5 are boxed from the top to the bottom. Sequence alignment was carried out as described in Figure 1. Lower panel: nel panel: cartoon depicting the differences in charged aminoacid content of the lumenal loop adjacent to the putative transmembrane segments M4 and M5.

```
CARDIAC   IECAEVFSKTVPPGGLPGAGLFGP-KNDLEDYDADSDFEVLMKTAHGHLVPDRVDKDK--   1374
          .      :   : :.:   :: :     :  :   : :   ...: .     .
SKELETAL  VHFHQFHRCTAGATPLAPPGLQPPAEDEARAAEPDPDYEN-RRSAGGWGEAEGGKEGTPG   1352

CARDIAC   ETTKAEFNNH----KDYAQEKPSRLKQR-FLLRRTKPDYSTSHSA               1413
          :              . : : ::  .   :        :   :
SKELETAL  GTPQPGVEAQPVRAENEKDATTEKNKKRGFLFKAKKAAMMTQPPATPAL            1400
```

FIG. 5. Comparison of the amino acid sequence of the ryanodine receptor isoforms: amino acid numbering as described in Figure 2.

TABLE 2. TENTATIVE LIGAND BINDING SITES OF THE CARDIAC AND SKELETAL RYANODINE RECEPTORS

	Skeletal	Cardiac
1) Ca2+ binging sites		
–High affinity		
EF hand type	aa 4248 – 4259 ?	?
	aa 4402 – 4411 ?	
	aa 4484 – 4494 ?	
–Low affinity	acidic region?	?
aa 1873 – 1924		
2) ATP binding	aa 4444 – 4449 ?	aa 2619 – 2652
sites	aa 4447 – 24452	
	aa 1194 – 1199	
3) calmodulin		
binding sites	aa 2807 – 2840	aa 2775 – 2898
	aa 2909 –2930	aa 2877 – 2898
	aa 3033 – 3052	aa 2999 – 3018
4) Mg2+ binging	?	?
sites		
5) cAMP and calmodulin	aa 3940 – 3945	aa 2806 – 2809
dependent protein kinase	aa 4314 – 4317	
phosphorilation sites		

and the carboxy–terminus of calsequestrin, a low affinity Ca^{2+} binding protein
(MacLennan and Wong, Fliegel et al., [21]), it is tempting to speculate that the acidic
region may constitute a low–affinity Ca^{2+} binding site of the RYR (Table 2). Prediction
of high affinity Ca^{2+} binding sites   within newly determined protein sequences has
been achieved by alignment with E–F hand–type consensus sequences: the human
and rabbit RYR amino acid sequences do not contain sequences that fully satisfy the
requirements of the E–F hand–type of Ca^{2+} binding sites. This is not surprising since
there are proteins (Treves et al., Smith and Koch, [22]) with high–affinity Ca^{2+} - binding sites whose primary structure do not exhibit consensus sequence for the E–F
hand motif. As to other Ca^{2+} release modulator such as ATP and Mg^{2+}, there are no
clear sequences that can be predicted to form binding sites for these modulators. Most
known nucleotide binding sites are defined by a Gly rich sequence which is located at
the end of a β–strand and followed by an α-helix (Branden and Tooze, [23]). As noted
by Takeshima et al. [15], skeletal RYR contains Gly–rich sequences within the myopla-

smic loop connecting M4 to M5. However, these sequences would not give a high nucleotide binding site score because they are localized in an unappropriate structural context. The structural information relative to the nucleotide binding sites derives almost exclusively from studies on proteins having high affinity for nucleotides (FRY et al., [24]; BRANDEN and TOOZE, [23]). The Ca^{2+} release channel is activated by mM ATP either in the palanar bilayer and in the isolated terminal cisternae fractions (ROUSSEAU et al., [25]), indicating that the activation of Ca^{2+} release occurs via interaction of ATP with low–affinity binding sites which may be different from those forming high–affinity binding sites.

Skeletal RYR contains several post–translational modification consensus sequences. Potential glycosylation sites are located in region that are predicted to be cytoplasmic. Since glycosylation of proteins occurs in the lumen of ER (LENNARZ [26]), it is likely that none of these site are glycosylated.

There are two potential cAMP and calmodulin–dependent protein kinase phosphorylation sites. One is localized a few residues upstream the putative transmembrane segment M1, the second is located near potential calmodulin binding sites and the third repeated sequence. CHU et al., WITCHER et al., TAKASAGO et al. [27]), and DAMIANI and MARGRETH [27]), reported phosphorylation of the RYR, but to date it is not known whether skeletal muscle RYR channel activity is modulated by phosphorylation.

2.3. Tissue distribution of RYR isoforms

Cardiac muscle contractility is initiated by release of Ca^{2+} from sarcoplasmic reticulum terminal cisternae through a channel which exhibits conductance, modulation by Ca^{2+}, ATP Mg^{2+}, ryanodine and calmodulin similar but not identical to those of the skeletal muscle RYR (Rosseau et al. [25]). Major advances in the characterization of the differences between cardiac and skeletal RYR have been achieved by the identification and comparison of the primary structure of the two RYR iso-forms (Fig. 3). Cardiac and skeletal RYR are the product of two different genes which are localized in human chromosome numbers 1 and 19, respectively (OTSU et al., MCKENZIE et al., [28]). The mRNA of the cardiac RYR monomer is about 16 kb long with one open reading frame encoding 4969 amino acids with a predicted m.w. of 564 kDa. Analysis of the cardiac RYR deduced amino acid sequence showed that the predicted structure of the molecule is similar to that described for the skeletal muscle isoform: the putative channel–forming hydrophobic domain is localized at the carboxy –terminus and the remaining sequence forms the hydrophylic portion of the molecule (Fig. 3). Overall the two RYR sequences are 66 % identical and as indicated by Figure 3, the differences are not evenly distributed throughout the molecule. The highest degree of similarity between the two isoforms was observed in correspondence with the region encompassing the transmembrane segments M6 to M10, including the carboxy –terminal end. Interestingly, the same region appears to be quite similar to that of the InsP3 receptor, indicating that these conserved regions may be crucial to the function of all three endocellular Ca^{2+} channels.

The most remarkable differences between the two RYR isoforms were observed in three regions defined by residues 1319 – 1352, 1873 – 1824 and 4210 – 4562 (see Figure 3).

i) The amino –terminus of the first GAP rich region (residues 1319 – 1352) of cardiac RYR differs significantly with respect to the corresponding portion of the skeletal RYR (Fig. 3 and 5); a deletion occurs at the carboxy –terminal half of the portion.

ii) The second portion that displays pronounced sequence divergence concerns the acidic region (Fig. 1 and 3). At variance with the skeletal RYR, the cardiac isoform contains several non–conservative substitutions and a large deletion at the carboxy –terminus of the acidic region.

iii) The second GAP rich sequence (residues 4352 – 4532) shows several substitutions that mainly involve residues that have been proposed to make up the modulatory domain of skeletal Ca^{2+} release channel (TAKESHIMA *et al.*, [15]). Major differences have also been observed in the sequence (residues 4210 – 4352; Fig. 3 and 4) which contains the putative transmembrane segments M3 and M4 including part of the myoplasmic loops between M2 and M3, and M4 and M5. Another interesting feature that distinguished the cardiac RYR from the skeletal counterpart is the lack of a cluster of acidic residues in the lumenal loop following M5.

As indicated by morphological studies, all these sequence modifications hardly affect the oligomeric structure of the ryanodine receptors. However, it is not known whether these changes may account for the well described functional differences existing between the two RYR isoforms. RYR chimera molecules or sequence specific probes against these regions may be useful to elucidate their role in the functional differentiation of the RYR.

The cardiac RYR deduced amino acid sequence contains consensus sequences for Ca^{2+} release modulators (Table 2). In particular, the sequence GWGNFG at position 2619 – 2652 gives a high score for a nucleotide binding site. Such a potential ATP binding site is close to a phosphorylation consensus sequence RRIS (2806 – 2809). WITCHER *et al.* [29] have shown that Ser 2809 of cardiac RYR is a good substrate of cAMP-and calmodulin–dependent protein kinase. On the contrary, the skeletal isoform appears to be phosphorylated to a lower extent, a result that might be explained by the observation that the phosphorylation consensus sequence (residues 2809) plus its flanking sequence are poorly conserved in the skeletal RYR.

Phosphorylation of the cardiac RAR by calmodulin –dependent kinase causes a reversal in the inhibitory effect of calmodulin on the channel activity (WITCHER *et al.*, [29]). This result indicates that at least for the cardiac RYR the modulatory region is separated from the channel – forming domain at the carboxy –terminus.

Hybrydization probe analysis of mRNA from a variety of tissues showed that the skeletal muscle RYR gene is expressed in fast– and slow–twitch muscle, but not in other tissues such as cardiac and smooth muscles, brain, liver and kidney, wheras the cardiac isoform is the only gene expressed in the heart and in the brain (OTSU et al., [28], NAKAI et al., [30]). The exact structural properties of the RYR expressed in the brain are unknown, and await molecular cloning studies.

3.   Malignant hyperthermia and ryanodine receptor Ca²⁺ channel: from the disease to the molecule?

Malignant hyperthermia (MH) is a potentially lethal autosomal dominant disorder which causes rapid increase of body temperature (approx. 1°C every 5 minutes in response to inhalational anesthetics such as halothane and enfluorane (GRONNERT, [31]). The most common symptom of MH is skeletal muscle rigidity. This led to the sugges- tion that an abnormal control of $[Ca^{2+}]_{int}$ in muscle cells may be the cause of the dis-ease. This hypothesis was further supported by the observation that the $[Ca^{2+}]_{int}$ of muscle fibres of patients with MH is increased (LOPEZ et al., [32]). Most of the advances on the ethiopathogenesis of MH derive from studies on sarcoplasmic reticulum from MH susceptible pigs, the latter being an animal model of the human disease. Electrophysiologycal studies of single channel have shown a shift to higher $[Ca^{2+}]$ for the calcium dependent inactivation of the RYR channels (FILL et al., [33]). An insufficient inactivation of MH Ca^{2+} channels may result in the inability of the sarcoplasmic reticulum to mantain a low $[Ca^{2+}]_{int}$ at rest and may also lead to prolonged elevation of free $[Ca^{2+}]$ during twitch. A defect of the RYR has also been suggested by genetic studies. The human RYR gene has been mapped to the q13.1 region of chromosome 19, very close to the GPI locus which was previously shown to be linked to halothane sensitivity in MHS pigs (DAVIES et al., [34]). Thus, human RYR cDNA probes were used to carry out restriction fragment polymorphism length (RLPF) analysis to define the linkage between the MH locus and the human skeletal muscle RYR gene. These studies show that polymorphic markers of the human skeletal RYR always cosegregate with the MH phenotype (MACLENNAN et al., [35]). Altogether these data strongly suggested that alteration(s) of the skeletal RYR is (are) responsible for the MH. The molecular basis of the RYR defect has been recently investigated by comparing the deduced amino acid sequences of RYR from both normal and MHS pigs (FUJII et al., [36]). Twenty-five polymorphisms between RYR cDNA sequences from normal and MHS pigs were found. One of these nucleotide changes results in a substitution of Arg 615 with Cys. Such an amino acid replacement was found in six different breeds of MHS pigs, suggesting that the Arg 615 Cys point mutation is causally correlated with the disease. However, further genetic reconstitution studies are required to determine whether the Arg 615 Cys mutation is the only pathogenetic factor of MH. Arginine 615 is distant from the channel–forming domain; therefore, a complicated mechanism underlies transmission of the modification caused by the mutation from the amino –terminus to the carboxy –terminal portion of the molecule.

Acknowledgements

I am grateful to Prof. A. MARGRETH, P. VOLPE and D.H. MACLENNAN for stimulating supervision on many of the experiments described in this article.

References

[1] M. ENDO, *Curr. Topics Membr. Transp.*, **25**, 181 (1985); E. RÍOS and G. PIZARRO, *Physiol. Rev.*, **71**, 849 (1991).

[2] A.V. SOMLYO, G. MCLELLAN, H. GONZALES–SERRATOS and A.P. SOMPLUO, *J. Biol. Chem.*, **260**, 680 (1985).

[3] C. FRANZINI–ARMSTRONG, FED. PROC. *FED. AM. SOC. EXPER. BIOL.*, **39**, 2403 (1980; R.M. KAMAMOTO, J.P. BRUNSCHWIG, K. C. KIM and A.H. CASWELL, *J. Cell Biol.*, **103**, 1405 (1986).

[4] J.S. SMITH, R. CORONADO and G. MEISSNER, *Biophys. J.*, 50, 921 (1986).

[5] P. PALADE *J. Biol. Chem.* 262, 614-6148

[6] N.F. ZAIDI, C.F. LAGENNAUR, R.J. HILKERT, H. XIONG, J.J. ABRAMSON and G. SALAMA, *J. Biol. Chem.*, **36**, 21737 (1989).

[7] P. VOLPE, G. SALVIATI, F.D. VIRGILIO and T. POZZAN, *Nature (London)*, **316**, 347 (1985).

[8] T. FURUICHI, C. SHISTA and K. MIKOSHIBA, *FEBS. Lett.*, **267**, 85 (1990).

[9] I.N. PESSAH, A.O. FRANCINI, D.J. SCALES, A.L. WATERHOUSE and J.E. CASIDA, *J. Biol. Chem.*, **261**, 8643 (1986); S. FLEISCHER., E.M. OGUNBUNI, M.C DIXON., E.A.M. FLEER *Proc. Natl. Acad. Sci.* U.S.A **82**, 7256-7259 (1985).

[10] M.INUI, A. SAITO and S. FLEISCHER, *J. Biol. Chem.*, **262**, 1740 (1987).

[11] J.S. SMITH, T. IMAGAWA, J. MA, M. FILL, K.P.CAMPBELL, and R. CORONADO, *J. Gen. Physiol.*, **92**, 1 (1988); F.A.LAI, H.P. ERICKSON, E. ROSSEAU, Q.Y. LIU and G. MEISSNER, *Nature (London)*, **331**, 315 (1988); L. HYMEL, M. INUI, S. FLEISCHER and H. SCHINDLER, *Proc. Natl. Acad. Sci.* U.S.A., **85**, 441 (1988).

[12] F. ZORZATO, A. MARGRETH and P. VOLPE, *J. Biol. Chem.*, **261**, 7349 (1986).

[13] F. ZORZATO, A.CHU, and P. VOLPE, *Biochem. J.*, **163**, 863 (1989).

[14] M. FILL, R. MEJIA –ALVAREZ, F. ZORZATO, P. VOLPE and E. STEFANI, *Biochem. J.* **273**, 449 (1991).

[15] H. TAKESHIMA, S. NISHIMURA, H. MATSUMOTO, K. ISHIDA, N. KANGAWA, H.

Minamino, M. Matsuo, M. Ueda, M. Hanakoa, T. Hirose and S. Numa, *Nature (London)*, **328,** 313 (1989).

[16] F. Zorzato, J. Fujii, K. Otsu, M. Phillips, N. M. Green, F.A. Lai, G. Meissner, and D.R. MacLennan, *J. Biol. Chem.*, **265,** 2244 (1990).

[17] T. Wagenkencht, R. Grassucci, J. Frank, A. Sait, M. Inui and S. Fleisher, *Nature (London)*, **302,** 842 (1989).

[18] R.D. Mitchell, H.K.B. Simmerman and R.L. Jones, *J. Biol. Chem.*, **263,** 1376 (1988); E. Damiani and A. Margreth, *Biochem. J.*, **172,** 1253 (1990).

[19] A.R. Marks, S. Fleisher and P. Tempst, *J. Biol. Chem.*, **265,** 13143 (1990).

[20] Meissner G?, Darling E, Eveleth *J. Biochemistry* 25: 236-244 (1986).

[21] D.H. MacLennan and P.T. Wong, *Proc. Natl. Acad. Sci.* U.S.A., **68,** 1231 (1970); L.Fliegel, M. Ohmishir, M.R. Carpenter, V.J. Khanna, R.A.F. Reithmeier and D.H. Machennan, *Proc. Natl, Acad. Sci.* U.S.A., **84,** 1167 (1987).

[22] S. Treves, M. De Mattei, M. Lanfredi, A. Villa. N. M. Green, D.H. MacLennan, D.H. Meldolesi and T. Pozzan, *Biochem. J.*, **271,** 473 (1991); M.J. Smith and G.L.E. Koch, *EMBO J.*, **8,** 3581 (1989).

[23] C. Branden and J. Tooze, *Introduction to Protein Structure*, Garland Publ. Inc., New York and London (1991).

[24] D.C. Fry, S. Kuby and A.S. Mildvan, *Proc. Natl. Acad. Sci.* U.S.A., **83,** 907 (1986).

[25] E. Rosseau, J.S. Smith, J.S. Henderson and G. Meissner, *Biophys. J.,* **80,** 1009 (1986).

[26] W. Lennarz, *Biochemistry,* **26,** 7205 (1987).

[27] A.Chu, C.Sumbilla, G. Inesi, D.S. Jay and K.P. Campbell, *Biochemistry,* **29,** 5899 (1990); D.R. Witcher, R.Jidovacs, H. Scherlman, D.C. Cefali and L.R. Jones, *J. Biol. Chem.*, **266,** 11144 (1991); T. Takasago, T. Imagawa, K. Furukawa, T. Ogurusus, M. Shigekawa, *J. Biochem. (Tokio),* **109,** 163 (1991); E. Damiani and A. Margretyth, *Biochem. J.,* **277,** 825 (1991).

[28] K. Otsu, H.F. Willard, V.K. Khanna, F. Zorzato, N.M. Green, D.M. Machennan, *J. Biol. Chem.*, **265,** 13472 (1990); A.E. Mckenzie, R. Kornebuk, F. Zorzato, J. Fujii, M. Phillips, D. Iles, S. Leblond. J. VBailly, H. F. Willard, C. Duff, R. Wonton, D.H. MacLennan, *Am. J. Hum. Gen.*, **46,** 1082 (1990).

[29] D.R. Witcher et al. sec. Ref. **27.**

[30] J. Nakai, T. Imagawa, Y. Hakamata, M. Shikegawa, H. Takeshima and S. Numa, *FEBS Lett.*, **271,** 169 (1990).

[31] G.A. Gronnert, *Anesthesiology,* **53,** 395 (1980).

[32] J.R. Lopez, L. Alamo, C. Caputo, Kikinski, D. Ledzema, *Muscle and Nerve,* **8,** 355 (1985).

[33] M. Fill, R. Coronado, J.R. Mickelson, J. VIlvex, J. Ma, B.A. Jacabson and C. F. Louis, *Biophys. J.*, 50: 471-475 (1990).

[34] W. Davies, I. Harbitz, R. Fries, G. Stranzinger and J. Hange, *Anim. Genetics,* **19,** 203 (1988).

[35] D.H. MacLennan, C. Duff, F. Zorzato, J. Jujii, M. Phillips, M. Korneluk, R.G. Frodis, B.A. Britt and R. Worton, *Nature (London),* **343,** 559 (1990).

[36] J. Fujii, K. Otsu, F. Zorzato, S. De Leon, V. K. Khanna, J.E. Weiler, P.J. O'Brien and D.H. MecLennan, *Science,* **253,** 448 (1991).

CONTRACTILE PROTEIN ISOFORMS IN SARCOMERIC MUSCLES: DISTRIBUTION, FUNCTION AND CONTROL OF GENE EXPRESSION

STEFANO SCHIAFFINO and PAOLO MORETTI

*CNR Unit for Muscle Biology
and Physiopathology, Department of Biomedical Sciences,
University of Padova, Italy*

Contents

Bioelectrochemistry IV
Edited by B.A. Melandri *et al.*, Plenum Press, New York, 1994

1. Introduction

This chapter aims at reviewing the polymorphism of contractile proteins, with particular reference to sarcomeric proteins in striated muscles of vertebrates. It is divided in three parts dealing with:

1) the heterogeneity of sarcomeric proteins,

2) the functional properties of the different isoforms, and

3) the regulation of isogene expression. The objective is not to provide a comprehensive review, but rather to introduce students to the subject, illustrate critical advances, discuss open issues, and show the direction of ongoing research. Special attention is given to new techniques used to study contractile protein isoforms, their expression and their function. A discussion of the basic molecular mechanism of contraction and its regulation is outside the scope of this presentation.

2. Molecular anatomy of the sarcomere: heterogeneity of contractile proteins

2.1 Diversity of contractile systems and motor molecules. The sarcomeric machine.

Two general classes of motility mechanisms can be defined according to the *track* on which the *molecular motor* works. Actin filament-based systems use myosins as motor, whereas microtubule-based systems use dyneins and kinesins as motors. The sarcomere is a specialized type of actin-myosin system found in striated muscles. Variation in the molecular composition of the sarcomere should be viewed in the context of the wide heterogeneity of actin-myosin systems, including those present in smooth muscles and in non-muscle cells. *Sarcomeric* myosins and actins differ from myosins and actins present in smooth muscle and non-muscle cells; in addition, sarcomeric muscles contain specific proteins that are not found in other cell types.

Sarcomeric muscles provide a unique model to investigate the mechanism of contraction, because of the exceptional regularity of their structure, which allows X-ray diffraction analysis, and also because of the speed and consistency with which their contractile material can be activated, which allows precise physiological measurements [1]. In sarcomeric machines force and/or movement is generated by thick, myosin filaments sliding past thin, actin filaments. The sliding filament model has become a paradigm for cell motility, a similar sliding mechanism is probably operative in microtubule-based motors. Myosin has a central role in the function of the sarcomeric machine and has become the prototype of a motor molecule, *i.e.* a molecule that is able to transform the chemical energy of ATP into the mechanical energy of movement. Some key steps in the elucidation of the mechanism of contraction and the discovery of the sliding filament model are shown in Table 1.

TABLE 1. Major steps toward the elucidation of the mechanism of muscle contraction in sarcomeric muscles[#]

Date	Observation	Author
1939	*Myosin* is an enzyme that hydrolizes ATP	ENGELHARDT and LYUBIMOWA
1941	Physical properties of *myosin* altered by interaction with ATP	ENGELHARDT *et al.*, NEEDHAM *et al.*
1943	*Myosin* resolved into actin plus what is now called myosin	STRAUB
	Dissociation of actin from myosin requires ATP	SZENT-GYORGYI
1953	Myosin is located in the A band	HASSELBACH; HANSON and H. E. HUXLEY
1953	Overlapping double array of thick and thin filaments shown by electron microscopy	H.E. HUXLEY
1954	Breadth of A bands stays constant during stretch and contraction (implies constant thick filament length)	A.F. HUXLEY and NIEDERGERKE; H.E. HUXLEY and HANSON

[#] Modified from A.F. HUXLEY [1].

A subsequent major progress has been the discovery of the mechanism of the Ca^{2+}-dependent activation of actin-myosin interaction, *i.e.* the regulation of the contractile activity by troponin and tropomyosin [2]. Troponin and tropomyosin, the regulatory proteins of the thin filament, represent a kind of Ca^{2+}-sensitive molecular swich, which is responsible for the Ca^{2+}-dependent regulation of actin-myosin interaction in sarcomeric muscles. Troponin consists of three subunits: troponin C, the Ca^{2+}-binding unit, troponin T, the tropomyosin-binding unit, and troponin I, the inhibitory unit which blocks actin-myosin interaction.

2.2   Sarcomeric protein isoforms.

The sarcomere can be viewed as a system of interdigitating thick and thin filaments integrated into a cytoskeletal framework made of connectin (titin) and nebulin filaments and associated Z disks. A scheme of the main protein components of the sarcomere is shown in Fig. 1.

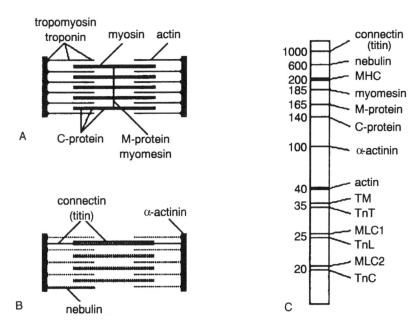

FIG. 1. A schematic view of the sarcomere. A. Distribution of the major contractile proteins in the sarcomere. B. Components of the endosarcomeric cytoskeleton. C. Electrophoretic separation of sarcomeric proteins in sodium dodecyl sulphate polyacrylamide gels. The approximate molecular weight of the different components is indicated on the left. Abbreviations: MHC, myosin heavy chain; MLC, myosin light chain; TnT, troponin T; TnC, troponin C; TnI, troponin I; TM, tropomyosin.

There are numerous variants of each sarcomeric protein and each variant shows a specific tissue distribution [3]. The primary structure of each different isotype is highly conserved between species, whereas marked sequence divergence can be demonstrated between different isoforms in the same organism. The inventory of sarcomeric protein isoforms is very long and still incomplete. It is appropriate to classify these isoforms on the basis of their tissue distribution into three major groups: fast skeletal, slow skeletal and cardiac isoforms. A list of the major actin, myosin and troponin isoforms found in mammalian striated muscle is shown in Table 2. Note that in many cases the same isoform is expressed in cardiac and slow skeletal muscle.

TABLE 2. ACTIN, MYOSIN AND TROPONIN SUBUNITS: MAJOR ISOFORMS IN MAMMALIAN STRIATED MUSCLE[#]

	Fast skeletal	*Slow skeletal*	*Cardiac*
Actin	α-skeletal actin	α-skeletal actin	α-cardiac actin
MHC	2A-MHC, 2X-MHC 2B-MHC, emb-MHC, neo-MHC	β-MHC	α-MHC, β-MHC
MLC (alk)	MLC1 fast, MLC3 fast MLC 1 emb	MLC 1 slow/ventricular	MLC 1 atrial (=MLC1 emb) MLC 1 slow/ventricular
MLC (P)	MLC 2 fast	MLC 2 slow/cardiac	MLC 2 slow/cardiac
TnC	TnC fast	TnC slow/cardiac	TnC slow/cardiac
TnI	TnI fast	TnI slow	TnI cardiac
TnT	TnT fast (several)	TnT slow (several)	TnT cardiac (several)

[#] MLC (alk): alkali MLC; MLC (P): phosphorylatable MLC.
For other abbreviations see Fig. 1.

Let us first consider actin and myosin. There are only two major forms of actin in sarcomeric muscles, one present in skeletal muscles, α-skeletal actin, and one present in cardiac muscle, α-cardiac actin. On the other hand there are several forms of myosin heavy chain (MHC) and myosin light chain (MLC). There are also additional isoforms, not indicated in Table 2, with restricted distribution in specialized muscles or in certain animal species, *e.g.* the MHC isoform present in extraocular muscles and the MHC found in the jaw-closing muscles of carnivores.

The polymorphism of troponin varies according to the subunit. There are only two known forms of troponin C, one present in fast skeletal muscles and one present in slow skeletal muscles and cardiac muscle. There are three forms of troponin I, present in fast skeletal, slow skeletal and cardiac muscles, respectively. The situation is even more complicated with troponin T, since there are several variants for each of the three major isoforms.

2.3. Generation of contractile protein diversity. Multigenic families and alternative splicing

Two mechanisms are responsible for the generation of contractile protein diversity. Different isoforms are either coded by different genes (one gene → one protein), or they derive from a single gene that undergoes alternative splicing (one gene → many proteins). Actin and myosin heavy chains exist as multigenic families and in sarcomeric muscles of vertebrates each isoform is coded by a distinct gene. Six actin genes have been identified in mammals [4]: two are expressed in sarcomeric muscles (skeletal and cardiac α-actins), two in smooth muscles (aortic type α-actin and enteric type α-actin), and two in nonmuscle cells (cytoplasmic β- and α-actins). The six genes are located on different chromosomes and display a high degree of sequence conservation, thus probably arise from one common ancestral gene.

The list of myosin heavy chain (MHC) genes is much longer and still incomplete: at least eight MHC genes have been identified in mammalian sarcomeric muscles, additional MHC genes are expressed in smooth muscle and nonmuscle cells [5]. A novel MHC form expressed in type 2X fibers in different mammalian species has been recently identified (our unpublished observations). Most sarcomeric MHC genes are located in two gene clusters, one comprising the two genes expressed in cardiac muscle, the α-MHC and β-MHC genes that are arranged in tandem in chromosome 14, and one comprising the embryonic, neonatal, type 2A, 2X and 2B MHC genes, expressed in skeletal muscles, which are clustered on chromosome 11 (mouse) or 17 (human). This is the only instance of gene clustering so far recognized for contractile protein genes: all other sarcomeric isogenes are spread in different chromosomes.

Myosin light chain isoforms, as well as troponin and tropomyosin isoforms, are generated by multiple genes, some of which contain alternative promoters and/or exhibit alternative splicing of pre-mRNA transcripts. As shown in Fig. 2, two myosin light chain isoforms expressed in fast skeletal muscles, MLC1 and MLC3, originate from a single gene: this gene has two transcription initiation sites from which two precursor mRNAs are transcribed [6]. The two primary transcripts are further processed by different modes of splicing, so that the two proteins have different aminoterminal portions whereas the rest of the molecule is identical. A combination of alternative promoters and alternative splicing of pre-mRNA transcripts has been demonstrated in the two major tropomyosin genes, α-tropomyosin and β-tropomyosin, both of which give rise to different products in muscle and non-muscle tissues [7]. A particularly intricate combinatorial pattern of alternative splicing has been found in the fast skeletal troponin T gene: a minimum of 10 and potentially 64 distinct mRNAs originate from this gene [8]. Different forms of slow skeletal TnT [9] and cardiac TnT [10] are produced by alternative splicing of the respective pre-mRNAs.

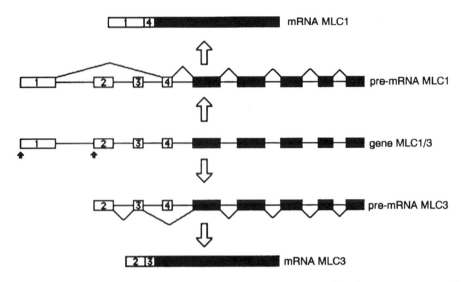

FIG. 2. The MLC1/3 gene (exons represented by boxes, introns by horizontal lines) can generate two different primary transcripts (MLC1 pre-mRNA or MLC3 pre-mRNA) from two different promoters (small arrows). The primary transcripts undergo alternative splicing producing two different mRNAs: MLC1 mRNA contains exon 1 and 4 but not exons 2 and 3, whereas MLC3 mRNA contains exons 2 and 3 but not exon 4. Exon 5 to 9 (in black) are present in both mRNAs.

2.4 Tissue-specific and developmental stage-specific distribution of sarcomeric protein isoforms.

Contractile protein isoforms show a developmental-stage-specific and cell-type-specific pattern of expression. For instance, MHC isoforms can be used as markers for distinguishing fiber types in skeletal and cardiac muscle and for defining various developmental stages. As shown in Fig. 3, four major fiber types can be identified in rat skeletal muscle using specific monoclonal antibodies to MHC isoforms [11]. Type 1 or slow fibers contain β-MHC, and the three subsets of type 2 or fast fibers, called 2A, 2X and 2B fibers, express 2A-, 2X- and 2B-MHC, respectively.

Additional MHCs, called embryonic and neonatal (or perinatal) MHC, are sequentially expressed during muscle development [12]. These MHCs are expressed even in adult animals in extraocular muscles [13, 14] and can be re-expressed during muscle regeneration [15], in denervated and paralyzed muscle fibers [16] or in certain muscles after hypothyroidism [17]. MHC switching varies in different fiber populations during embryonic development [18]. One subset of type 1 fibers that arise early during embryonic development coexpress initially embryonic MHC and low levels of β-MHC, but never express neonatal MHC; embryonic MHC disappears during subsequent developmental stages. In contrast, most type 2 fibers, that are formed at later developmental stages, express sequentially embryonic, then neonatal, then either 2A- or 2X- or 2B-MHC (our unpublished observations).

In avian and mammalian heart at least three distinct MHC isoforms can be detected in atrial myocardium, ventricular myocardium and conduction tissue, respectively [19-22]. Both MHC and other contractile proteins undergo developmental changes in cardiac muscle that can be detected by analyses at the mRNA and protein level (Table 3).

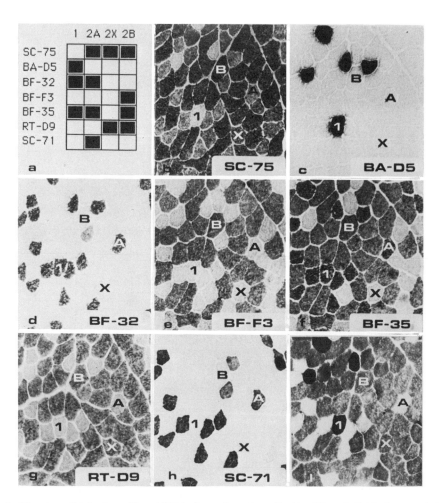

FIG. 3. Fiber type distribution of four MHC isoforms (type 1- or β-, 2A-, 2X- and 2B-MHC) in rat skeletal muscle. *a*, scheme showing the pattern of reactivity of seven anti-MHC monoclonal antibodies; a positive reaction is indicated by a black box. *b-h*, immunoperoxidase staining with anti-MHC antibodies. *i*, myosin ATPase histochemical staining after pretreatment at pH 4.6. Note that type 2B and 2X fibers are not distinguishable by this histochemical procedure. (Modified from SCHIAFFINO et al., ref. 11).

TABLE 3. DEVELOPMENTAL TRANSITIONS OF CONTRACTILE PROTEIN GENE EXPRESSION IN THE RAT
AND MOUSE VENTRICULAR MYOCARDIUM[o]

Gene families	mRNA expressed in	
	Early embryo	Young adult
actin	α-cardiac actin α-skeletal actin α-smooth actin	a-cardiac actin
MHC	β- MHC α-MHC	a-MHC
MLC_1	MLC 1 ventricular MLC 1 atrial	MLC 1 ventricular
TM	α-TM β-TM	α-TM
TnT	cardiac TnT_{emb}[**]	cardiac TnT_{adult}[***]
TnI	slow skeletal TnI	cardiac TnI

* Modified from Ausoni et al. (ref. 23). For abbreviations, see Fig. 1.
** Embryonic splicing products of the cardiac TnT gene
***Adult splicing products of the cardiac TnT gene

3. Physiology of the sarcomere: functional significance of contractile protein isoforms

3.1 Correlation between isoform composition and physiological properties of striated muscles.

On the basis of a comparative study of different types of muscles in different species, BARANY [24] first reported that the actin-activated ATPase activity of myosin is correlated with the maximal speed of muscle shortening. The ATPase activity is mainly determined by the heavy subunits (MHC), as shown by studies on myosin hybrids containing light chains and heavy chains from different muscles [25]. A strong correlation was also found between MHC composition and velocity of shortening of single fibers [26], suggesting that the MHC composition of an individual fiber is a primary determinant of the maximal velocity of shortening.

I will illustrate three examples from our own work aimed at correlating physiological properties and MHC composition in rat skeletal muscles at the whole muscle, single motor unit and single fiber level. Force-velocity properties and MHC composition were investigated in fast-twitch (*extensor digitorum longus,* EDL) and slow-twitch *(soleus)* rat muscles, and in the soleus muscles denervated and stimulated with a high-frequency pattern of impulses [27]. As shown in Fig. 4, the stimulated soleus muscle contained mainly (about 85 %) 2X-MHC and displayed force-velocity properties intermediate between those of normal soleus (about 97 % type 1- or β-MHC) and normal EDL (about 75 % type 2B and 25 % type 2A-2X-MHC).

Interpretation of studies at the whole muscle level is complicated by the mixed fiber type composition of most skeletal muscles. More direct correlations can be established from studies of single fibers or single motor units, that are homogeneous with respect to fiber types. Physiological properties of single motor units from rat skeletal muscle have been correlated with the type of MHC expressed in the motor unit fibers. Repetitive stimulation of a single motor axon leads to glycogen depletion in the muscle fibers belonging to the same motor unit. PAS staining allows to identify the muscle fibers in the units, and their MHC composition can be determined by immunofluorescence or immunoperoxidase staining of serial sections with specific anti-MHC antibodies [28]. Motor units composed of type 2X fibers were found to display contraction and half-relaxation times similar to those of type 2A and 2B motor units and a resistance to fatigue intermediate between that of type 2A and 2B units.

Force-velocity relations were determined in skinned muscle fibers and the MHC isoform composition of the same fibers was subsequently analyzed by staining with anti-MHC antibodies [29]. Type 1, 2A, 2B and 2X fibers were found to differ in maximal velocity of shortening and in the shape of force-velocity curves, however marked variation was also observed within each fiber type population.

Contractile protein composition can be correlated with other physiological properties. For instance, troponin and tropomyosin isoforms were correlated with Ca^{2+} sensitivity in skinned rabbit muscle fibers [30]. The Ca^{2+} sensitivity can be determined from force/pCa curves, *i.e.* by measuring the force produced by permeabilized fibers at different Ca^{2+} concentration (Fig. 5).

However, a major drawback of correlative studies is that one compares muscle fiber types that differ with respect to a wide variety of contractile, membrane and soluble proteins, thus it is impossible to identify a single component as responsible for a specific functional property. This limitation can be overcome by substituting only one component in single fibers and comparing contractile properties before and after the substitution, as described in the next section.

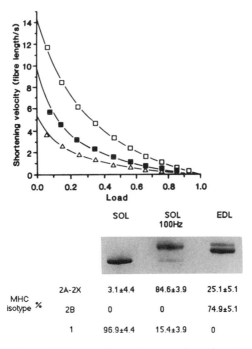

FIG. 4. A. Force-velocity relations of the normal rat soleus muscle, a slow-twitch muscle (triangles), the normal *extensor digitorum longus* (EDL) muscle, a fast-twitch muscle (open squares), and the soleus muscle after long-term stimulation with a high frequency pattern of impulses (black squares). B. MHC composition of the muscles used for physiological studies, as determined by sodium dodecyl sulphate- 6 % polyacrylamide gel electrophoresis. Densitometric measurements of the relative amounts of type 2A/2X-, 2B- and type 1 (B)-MHC bands are indicated. (From SCHIAFFINO *et al.*, ref. 27).

3.2. Selective extraction and replacement of sarcomeric proteins: isoform exchange in skinned fibers.

The effects of partial or total extraction of myosin light chains, troponin C and C-protein on mechanical properties of skinned muscle fibers have been investigated in several studies. Moss [31] has recently published a comprehensive review of extraction and exchange techniques in permeabilized fibers and their relevance to studies on Ca^{2+} regulation of tension development and shortening velocity.

Troponin C isoform exchange has been used to explore the role of troponin C isoforms in modulating the length dependence of Ca^{2+} sensitivity. Ca^{2+} sensitivity of the myocardium is higher at a sarcomere length of 2.3-2.4 mm than at 1.8-1.9 mm: the length dependence is twofold greater in cardiac and slow skeletal muscle fibers than in fast skeletal fibers. Replacement of endogenous cardiac troponin C with fast skeletal troponin C in skinned cardiac trabeculae was found to reduce the length depen-

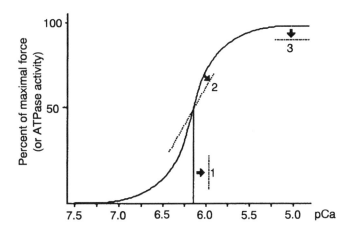

FIG. 5. The relationship between Ca^{2+} concentration, expressed as pCa (-log molar concentration), and force or ATPase activity, expressed as percentage of the maximal force or ATPase activity, can be analyzed in permeabilized (*skinned*) muscle fibers. Three variables of this sigmoidal curve can be found to vary in different preparations or as a result of different stimuli: 1) the Ca^{2+} concentration that produces half-maximal activation (pCa_{50}), 2) the slope of the curve at the midpoint or Hill coefficient, and 3) the maximal force or ATPase activity. These variables can be affected by changes in troponin isoform composition, as well as post-translational changes (*e.g.* phosphorylation) of troponin isoforms.

dence of Ca^{2+} sensitivity of tension [32]. Conversely, replacement of endogenous fast troponin C with cardiac troponin C in skinned fast skeletal fibers was found to increase the length dependence of Ca^{2+} sensitivity of tension [33]. However, interpretation of these results is complicated by the fact that stoichiometric readdition of TnC to TnC-extracted preparations may be incomplete: other studies indicate that replacement of fast skeletal TnC with cardiac TnC in fast fibers has apparently no effect on the length dependence of Ca^{2+}-activated tension [34]. Extraction and replacement experiments have also been performed with genetically engineered troponin C molecules (see below, Section 2.4.4).

Purified contractile protein isoforms can be used to perform *in vitro* reconstitution experiments. A reconstituted system of contractile proteins, made of actin, myosin, tropomyosin and troponin, shows a Ca^{2+}-dependent activation of ATPase activity. TOBACMAN and LEE [35] isolated two isoforms of bovine cardiac troponin T, which differ in sequence in the amino-terminal region, and found that they confer slightly different Ca^{2+} sensitivities (different pCa_{50}) in this reconstituted system.

3.3.   In vitro motility assays.

In vitro motility assays allow to directly visualize motion and force using purified contractile proteins. In a typical assay a glass slide is coated with actin filaments and tiny plastic beads coated with myosin slide over the surface when ATP is added. Alternatively, myosin is absorbed onto the glass surface and the ATP-dependent movement of fluorescently labeled individual actin filaments can be observed by fluorescence microscopy. These assays have been used to analyze the properties of different motor molecules, such as myosin, dynein and kinesin, and to dissect the portion of the molecule responsible for movement. Quantitative analyses of velocity and force have been performed in this system. However, *in vitro* motility assays have not yet been applied systematically to compare the properties of different contractile protein isoforms, *e.g.* different MHCs.

3.4.   Analysis of contractile protein function by molecular genetic techniques.

3.4.1 Insertion of extra genes (transfection in vitro and in vivo, transgenic animals).

DNA transfection techniques can be used to reintroduce in cells or organisms cloned sequences following appropriate genetic engineering *in vitro;* this allows the construction and the analysis of *artificial* mutants and the dissection of the molecular details of the process of interest.

Cloned DNA sequences (cDNA or genomic) can be introduced in cultured animal cells (transfection *in vitro*) or in organisms (transgenic animals and transfection *in vivo*). To reintroduce cloned DNA into cultured animal cells, two general methods have been widely utilized: transient DNA transfection and stable DNA transfection. In the first of these, the DNA of interest is introduced into the host cell nucleus and the cells are analyzed within 1 to 3 days post-transfection. The foreign DNA is lost as the cells replicate, since it is not integrated into the host cell chromosomes and does not carry sequences that allow it to be replicated into the transfected cell nucleus. In the second method the DNA of interest is transfected along with a gene which confers a selectable phenotype (drug resistance). The very low efficiency integration process can then be selected for and the surviving cells contain the transfected DNA integrated into the host cell chromosomes (the site of integration being random). In neither case can the expression of the endogenous gene be eliminated and this is in clear contrast to classical genetic mutations, in which the mutation has disrupted the wild-type (endogenous) copy of the gene. This problem can be only solved by gene disruption experiments (homologous recombination, see next section).

In vitro transfection experiments have been mostly used for studies of gene regulation, i.e. to identify regulatory sequences, often located in regions 5' upstream of the transcription initiation site of the gene. In a typical experiment, a portion of the 5' upstream sequence of a gene, say a skeletal muscle myosin heavy chain, is fused with the coding region of a gene not expressed in vertebrate tissues, e.g. the bacterial chloroamphenicol acetyl transferase (CAT) or the firefly luciferase gene, and transfected in fibroblasts and in myogenic cells. The finding that CAT activity is detected in myogenic cells but not in fibroblast, and that the activity increases with muscle cell differentiation in vitro, can be taken as evidence that the regulatory MHC sequence used for transfection contains elements recognized by trans-acting factors present in muscle cells but not in fibroblasts. The DNA motifs responsible for regulation can be identified by progressive deletions of this region and transfection analyses with deleted constructs, in combination with other techniques such as DNAse footprinting and gel retardation.

More rarely, DNA transfection experiments have been used to analyze the function of different isoforms. The molecular dissection of the tubulin gene family is a good example of this approach [36]. In higher eukaryotes both α and β tubulin are encoded by multigene families: in vertebrates there are at least six functional genes that encode each subunit. The functional significance of multiple tubulin genes is unknown as are the mechanisms responsible for the segregation of individual isotypes into specific classes of microtubules and for the morphological and functional differences among microtubule arrays. Transient DNA transfection experiments have been used to force the inappropriate expression of the chicken β-tubulin isotype 6 (specifically expressed only in erythroid cells where it is assembled into marginal band microtubules) in monkey fibroblasts. The cβ6 polipeptyde is by far the most divergent of the chicken tubulin isotypes, having 17 % of the residue positions that differ from the other isoforms. The assembly of cβ6 into microtubule arrays, analyzed with a monoclonal antibody, demonstrated the efficient incorporation of the transfected isoform in microtubules of interphase as well mitotic cells. Stable transformants have been utilized to express a chimeric protein comprised of the amino-terminal 344 aminoacids of the chicken β2-tubulin isotype fused to the carboxy-terminal 113 residues of a yeast β-tubulin. Using an antibody that recognised the yeast portion of the protein, it has been shown that the chimeric protein coassembled with the endogenous tubulin into interfase and mitotic microtubules, without any detectable effect upon cell growth properties. As previously mentioned, a serious limitation of these experiments is the presence of the wild-type proteins and the fact that only the compatibility of different isoforms can be tested. In addition, if the transfected sequence is expressed at a low level relative to the endogenous gene(s), then the transfected product could simply act as a tracer for assembly and one cannot conclude that they are functionally indistiguishable from the endogenous protein [36].

The reintroduction of cloned DNA sequences into a whole organism is not different in principle from transfection *in vitro*. There are three main routes: direct microinjection of DNA into the male pronucleus of fertilized eggs, infection of pre- or post-implantation embryos with retroviral vectors and via transfected multipotent embryonic stem (ES) cells [37]. The transfected cells or embryos are then implanted in foster mothers and the progeny is screened for germ line transformation. Animals having the sequence integrated into the chromosomes of germ line cells are used in crosses and heterozygous and homozygous organisms (or even chimerae) can be studied. Also in this case the expression of the endogenous gene is not eliminated.

Transgenic animals can be used to study the regulation of contractile protein gene expression during development and the functional significance of the different isoforms. An interesting application of transgenic animals to the study of gene regulation has been the analysis of the expression of rat α-skeletal actin in the heart of mice with a mutation of the α-cardiac actin gene [37]. Two strains of inbred mice (BALB/c and DBA/2) contain a duplication of the promoter and the first three exons of α-cardiac actin that significantly down-regulate the expression of this gene [38]. This decrease is partially compensated in living animals by an increase in the level of expression of α-skeletal actin in the heart. By analizing the level of transcription of a modified rat α-skeletal actin gene inserted into the genome of BALB/c mice, it was shown that the transgene responded in a manner very similar to the endogenous α-skeletal actin gene in mice carrying the wild-type allele and in homozygous for the duplicated allele. In wild-type mice the expression of a-skeletal actin gene in the heart was 2-5 % of that in skeletal muscle; in contrast, in the mutant strain the level of transcription in cardiac cells was as high as in skeletal muscle. These results demonstrate a coregulation of the expression of striated muscle actin genes in the heart and the functional substitution of the cardiac actin by the correspondig skeletal isoform without evident physiological perturbations.

Recently, the possibility to transfect striated muscle cells of living adult animals *in situ* by direct injection of DNA has been demonstrated (transfection *in vivo*). An alternative approach is the transplantation *in vivo* of *in vitro* transfected myoblasts. Transfection *in vivo* has been used to analyze the regulation of the expression of the slow/cardiac troponin C gene in the heart [39]. By directly injecting a reporter CAT gene fused to promoter/enhancer sequences of the slow/cardiac TnC gene into the left ventricular wall of adult rats, the authors have confirmed the results obtained by transfection *in vitro* experiments. A serious limitation of these techniques is the very low efficiency, *i.e.* poor gene expression. Their major interest is in the potential for gene therapy and the possibility of using myoblasts as a gene delivery vehicle.

3.4.2 Selective inactivation of endogenous genes by homologous recombination or antisense RNA.

Two general methods have been used to eliminate (knock out) the expression of selected genes in living cells: homologous recombination and antisense RNA. In the first method an homologous recombination event between a specific gene in its genomic locus and newly introduced foreign DNA sequences (gene targeting) allows the inactivation of the gene of interest in a living cell [40]. The second method consists in the introduction in living cells (or the production by them) of RNA or DNA complementary to a specific mRNA, whose expression is blocked by the formation of duplexes *in vivo* [41, 42].

An interesting application of these methods has been the analysis of the function of myosin in nonmuscle cells, using as a model the *Dictyostelium discoideum* myosin heavy chain gene. *Dictyostelium discoideum* is a eukaryote that displays many forms of cell motility found in higher eukariotic cells. Its major myosin consisting of two heavy chains (MHC A) and two pairs of light chains, assembles into thick filaments, interacts with actin filaments and displays an actin activated ATPase activity.

In the gene targeting experiments shown in Fig. 6, the single genomic copy of the MHC A gene has been eliminated and substituted by a fragment corresponding to the amino terminal half of the same gene [43]. The resulting cells do not express the native myosin but express a myosin fragment corresponding to the muscle heavy meromyosin (HMM). These mutants are viable demonstrating that myosin is not essential for growth and exhibit many forms of cell movement, including membrane ruffling, phagocytosis and chemotaxis. These cells are defective in cytokinesis but not in karyokinesis and display a delayed and incomplete developmental cycle (they do not form the fruiting bodies that contain the spores required to complete the cycle). Both the cytokinesis block and the developmental block have been eliminated in spontaneus revertants in which excision of the transforming plasmid from the genome restores the endogenous MHC gene function; the same result has been obtained by transformation of the mutant cells with an extrachromosomal copy of the MHC A gene.

The results obtained by gene targeting has been confirmed by antisense RNA experiments. In this case the inactivation of MHC gene expression has been obtained by stable transformation of *Dictyostelium discoideum* cells with a vector encoding a RNA molecule complementary to MHC A mRNA [44]. The 250-fold reduction of MHC A protein product caused the cells to growth slowly and to display the same phenotype as the mutants created by homologous recombination. Also in this case the restoration of the expression of the MHC A gene reverted the cells to the wild-type phenotype.

As an alternative way to abolish or interfere with the functions of proteins in living

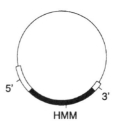

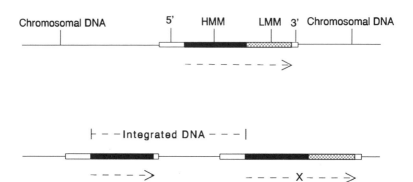

FIG. 6. Integration of the amino terminal half of the myosin heavy chain gene into the MHC A gene. The circle represents the plasmid used for transformation: the white boxes correspond to the 5' and 3' flanking sequences, the black box to the heavy meromyosin (HMM) coding portion; the thin line represents the plasmid sequences. In the center of the diagram is the chromosomal region of the MHC A gene: the thin line represents chromosomal DNA, the white boxes the 5' and 3' flanking sequences, the black and dotted boxes the HMM and light meromyosin (LMM) portions of the MHC A gene respectively; the direction of transcription is indicated by the arrow. In the bottom part is represented the product of the homologous recombination event. The integrated DNA blocks the transcription of the endogenous gene.

cells, the method used by the group of H. Holtzer with intermediate filaments has to be mentioned [45]. In this case, the authors took advantage of the need for intermediate filament proteins to interact with each other to build the intermediate filament network. Expression by living cells of a truncated version of the intermediate filament protein desmin, which lacks a portion of the highly conserved α-helical rod region and the entire nonhelical carboxy-terminal domain, blocks the assembly of desmin and vimentin intermediate filaments and completely dismantles the preexisting intermediate filament network; in the transformed cells vimentin and desmin are packaged into spheroid bodies scattered throughout the cytoplasm. However myogenic cells as -sembled normally aligned striated myofibrils that contracted spontaneosly; myotubes also appeared morphologically indistinguishable from control myogenic cells. This experiment demonstrates that desmin is not necessary for myofibrillogenesis.

3.4.3 Analysis of mutants in Drosophila and Caenorhabditis elegans.

Most higher eukaryotic systems are not readily amenable to the traditional genetic analysis of microorganisms. Two main reasons are the genome complexity and the long generation time. Both of these problems are partially overcome by simpler animal models, e.g.Caenorhabditis elegans and Drosophila melanogaster. Their relatively simple and small genome (about 1/30 the size of tipical mammalian genomes) and a short life cycle allow the dissection by genetic methods of complex biological structures, such as the sarcomere, which involves the assembly of multiple gene products interacting in precisely defined and controlled ways. In addition, site-directed mutagenesis and DNA transfection techniques can be used to complement the analysis of the function of a protein of interest. Null mutants of selected genes are available as transgenic hosts to study the expression of the modified protein in the absence of the wild-type endogenous gene product [46-48].

Mutants with altered motility have been described in the nematode C. elegans, mutants characterized by flight muscle defects have also been described in the fruit fly D. melanogaster. The contractile protein genes responsible for some of these mutations have been identified [49]. In C. elegans four genes encode MHCs of muscle cells [50]. MHC A and B are primarily expressed in muscles of the body wall and are required for motility of the animals; MHC C and D are specifically expressed in the pharynx and are essential for nutrition. MHC A and B coexist in the same muscle cells but are located in different positions in the thick filaments: isoform A is present in the central region of the filaments whereas isoform B is incorporated into the peripheral portion. Several mutations of these two genes have been identified and utilized as tools for the analysis of sarcomere assembly. The unique role of MHC A in initiation of thick filament assembly has been demonstrated by suppressors of null mutations of myo-3 gene (encoding MHC A) and unc-54 gene (encoding MHC B). Null mutants of unc-54 gene can revert to a normal phenotype by amplifications of the myo-3 gene and accumulation of MHC A. On the contrary, in myo-3 null mutants, increased levels of MHC B do not compensate for the absence of isoform A [51]. Mutations designated unc-54(d) produce an altered MHC B isoform and are of particular interest in the analysis of sarcomere assembly: in unc-54(d)/+ heterozygotes, the mutated protein not only fails to assemble into thick filaments, but also interferes with the assembly of the wild-type isoform (as little as 2 % of the wild-type level is sufficient to prevent myosin assembly). The analysis of a series of independent unc-54(d) mutations reveals a cluster of aminoacid substitutions in the MHC head region, most often in a highly conserved glycine-rich loop near to the ATP-binding site or in the actin binding site, but never in the rod [52].

The correct assembly and function of the sarcomere requires the presence of several components in precise stoichiometric proportions. Flight muscles of *Drosophila melanogaster* offer a good model to study this problem. *D. melanogaster*, in contrast to *C. elegans* and all vertebrates investigated, has only one skeletal muscle MHC gene. This gene generates by differential pre-mRNA splicing two isoforms that differ in their carboxy-terminal region: one is specifically expressed in the highly specialized indirect flight muscles, the second in all other muscles. A null mutation of this gene is letal in the homozigous state but in the heterozygotes only the non-essential muscles with highly organized myofilament arrays (indirect flight muscles and jump muscles of adult flies) are paralized. Normally these muscles have a regular myofilament organization and a ratio of thin to thick filaments of 3 to 1 (in contrast to ratios of 9 to 12 thin filaments per thick filament of other *D. melanogaster* muscles). In the mutants the reduction of the amount of MHC and a halving of the ratio of thick to thin filaments is critical for the more specialized muscles, while muscles essential for viability are still capable of functioning [53].

3.4.4 Generation of mutant proteins by site-directed mutagenesis

The power of *in situ* mutagenesis approaches for the molecular dissection of muscle protein function is well illustrated by studies on troponin C [54]. Troponin C (TnC) belongs to a multigene family of Ca^{2+} binding proteins characterized by helix-loop-helix Ca^{2+} binding sites. The fast TnC isoform contains four Ca^{2+} binding sites, two carboxy-terminal high-affinity sites and two amino-terminal low-affinity Ca^{2+} binding sites: the latter are the regulatory sites that trigger contraction upon binding of Ca^{2+} (Fig. 7A). The slow/cardiac TnC isoform contains only three Ca^{2+} binding sites, since site 1 is inactive due to amino acid substitutions in the Ca^{2+} binding loop.

The sequence of TnC genes is known and it is possible to introduce specific mutations in the nucleotide sequence of TnC genes to investigate the molecular basis of the different function of cardiac and fast TnC, as well as other aspects of structure/function relationships in the TnC molecule [55-57]. The function of various

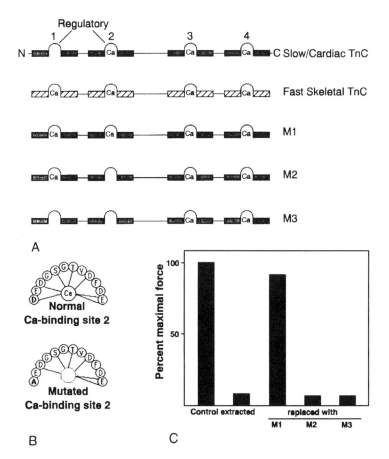

FIG. 7. A. Schematic representation of normal slow/cardiac and fast skeletal TnCs and mutant TnC molecules (M1, M2 and M3) obtained by site-directed mutagenesis of Ca^{2+} binding sites of slow/cardiac TnC. The four helix-loop-helix structures, corresponding to high-affinity (sites 3 and 4) and low-affinity (1 and 2) Ca^{2+} binding sites are indicated. Site 1 is normally inactive in slow/cardiac TnC and is activated by mutation in M1 and M2. Site 2 is inactivated in M2; both site 1 and 2 are inactivated in M3. B. Schematic representation of Ca^{2+} binding site 2: amino acids are indicated by single letter code, coordination bonds are represented by lines radiating from Ca^{2+}. A single aminoacid substitution (D → A) leads to inactivation of this site. C. Force exerted at pCa 4.3 by rabbit soleus muscle fibers after permeabilization and extraction of TnC or exchange of endogenous TnC with mutant TnCs. (B and C, modified from PUTKEY et al., Ref. 55).

mutant molecules can be tested *in vitro* by exchanging the endogenous TnC in skinned fibers (see above, Section *2.2*). Thus it was shown that the ability of TnC to regulate contraction in skinned muscle fibers is abolished by inactivation of site 2 produced
by substitution of a single aminoacid (Fig. 7 B and C). On the other hand, activation of
the normally inactive site 1 in cardiac TnC leads to changes in the Hill coefficient in
force/pCa curves, *i.e.* confers greater cooperativity to Ca^{2+} dependent contraction in
skinned muscle fibers.

TABLE 4. MUSCLE DIFFERENTIATION FACTORS

Transcription factor	Regulatory sequence	Structure
Myo-D Myogenin Myf-5 MRF-4	CANNTG (E box or MEF1 site)	bHLH structure (13 basic aa followed by Helix-Loop-Helix that makes heterodimers with ubiquitous E2A gene products)
Id	not binding DNA	HLH structure without basic region (inhibitory function: competes for E2A products)
MAPF1/2 (muscle actin promoter factor)	CC(AT)$_6$GG (CArG box)	similar to the serum response factor that binds the serum response element (SRE, also containing a CArG motif) in c-fos and cytoskeletal actin genes
M-CAT binding factor	CATTCCT (M-CAT)	?
MEF 2	NTAAAAATAAN (MEF 2 site)	?

4.   Regulation of sarcomeric protein gene expression

4.1   Regulation of muscle gene expression. Muscle-specific differentiation factors.

The first gene coding for a muscle differentiation factor, called MyoD1, was identified in 1987 using a subtraction hybridization procedure [58]. This gene, which is exclusively expressed in skeletal muscle, is present in different vertebrate species; similar genes have been identified in *Drosophyla*. A distinctive property of these genes is that they are able to convert fibroblasts to myoblasts in transfection experiments *in vitro*. This effect has been demonstrated with different cell lines and primary cultures of cells of mesodermic, endodermic and ectodermic origin.

Three additional genes, called myogenin [59]. myf-5 [60] and MRF-4 or myf-6 [61], were subsequently identified by similar approaches, or by screening cDNA libraries at low stringency with MyoD. These genes code for DNA binding proteins, structurally similar to the c-myc protooncogene, which act as trascription factors. Common structural features include the presence of a basic domain (13 aminoacids long) responsible for binding to specific DNA sequences (E box, see Table 4), and an associated helix-loop-helix structure, involved in dimer formation. Heterodimers, made by Myo-D with products of the ubiquitous E2A gene products are especially active as transcription factors and appear to be the physiological activators of muscle genes.

Other transcription factors that bind sequences distinct from the E boxes, such as the factors binding the CArG box [62], the M-CAT site [63] and the MEF-2 site [64], are involved in contractile protein gene activation (Table 4). Regulatory genes with inhibitory function have also been identified. The protein Id, coded by a gene with structure similar to Myo D1 was found to inhibit the activation of muscle genes induced by MyoD1 [65]. Id contains the helix-loop-helix domain and thus can form heterodimers with MyoD1 and E2A products, however it lacks the basic domain required for DNA binding: it is likely that the inhibitory effect of Id is due to competition for E2A products.

4.2   Developmental and tissue-specific regulation of muscle genes.Cell lin-eages and fiber type diversification.

The precise role of the different myogenic regulatory factors described above in muscle differentiation and in the regulation of sarcomeric protein genes is not yet clear. Transfection experiments have shown that the expression of many muscle genes is regulated by MyoD1 and the other helix-loop-helix factors, however the expression of some genes, *e.g.* the b-MHC gene, is apparently regulated by MyoD1-independent pathways [66]. Furthermore, MyoD1 and the other helix-loop-helix factors are not expressed in cardiac muscle.

The pattern of expression of the myogenic regulatory factors is not syncronous during skeletal muscle development [67]. However, it is not known whether the sequential activation of developmentally regulated genes, *e.g.* the transition embryonic MHC → neonatal MHC → type 2A- or 2X- or 2B-MHC, is induced by the changing proportion of the myogenic factors at different developmental stages. Also, it is not known whether the expression pattern of the myogenic regulatory factors is identical in different fiber type populations and whether fiber diversification may be induced by different combinations of the known myogenic regulatory factors or by others still

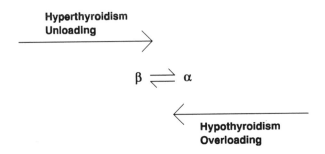

FIG. 8. Switching from β- to α-MHC or from α- to β-MHC gene expression can be induced in the rat ventricular myocardium by changes in thyroid hormone or pressure load.

unknown fiber type-specific factors. The factors reponsible for the differential splicing pattern of the troponin T genes in different muscle cell types and in different developmental stages have not been identified.

The analysis of embryonic development of cardiac and skeletal muscle in mammals and birds shows that different muscle cell populations or muscle cell lineages are present since early developmental stages. In cardiac muscle differences in MHC and troponin isoform composition can be detected very early in development between atrial and ventricular myocardium and conduction tissue. It remains to be established whether there are distinct lineages of cardiac muscle cells since the formation of the heart tube. It has been suggested that the cardiac conduction tissue may not derive from the cardiac mesoderm but from the neural crest (68). In skeletal muscle the existence of different cell lineages (fast vs. slow myoblasts, embryonic vs. satellite cell myoblasts) is supported by clonal studies of cultured embryonic muscle cells [69, 70]. These studies raise the problem of the regulatory factors responsible for the differentiation of the various muscle cell populations.

4.3   Modulation of muscle gene expression by hormonal, mechanical and neural factors.

Contractile protein gene expression in skeletal and cardiac muscle can be modulated by different hormones, by the pattern of activity imposed by neural impulses and by mechanical signals, such as tension overload or unload.

In the rat ventricular myocardium the transition from a *fast-type* phenotype, characterized by the high ATPase activity α-MHC, to a *slow-type* phenotype, characterized by the low ATPase activity β-MHC, can be induced by hyperthyroidism or by pressure overload, *e.g.* after aortic coarctation [71-74] (Fig. 8). This change is reversible and a transition in the opposite direction can be induced by hyperthyroidism or by decrease in pressure load. Thyroid hormone has been shown to have a direct effect on MHC gene expression. In the ventricular myocardium the nuclear thyroid hormone (T3) receptor can up-regulate the α-MHC gene by binding to a specific regulatory sequence in the promoter [75].

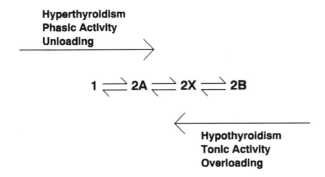

FIG. 9. Switching from type 1 or B-MHC, typical of slow fibers, to the *fast*-specific isoforms 2A-, 2X- and 2B-MHC can be induced in rat skeletal muscle by hyperthyroidism, phasic activity or unloading; changes in the reverse direction is induced by hypothyroidism, tonic activity and overloading.

A more complex transition from the *slow-type* phenotype of type 1 fibers, characterized by β-MHC, to the *fast-type* phenotype of the type 2A-, 2X- and 2B fibers, characterized by the corresponding MHC isoforms, can be induced in skeletal muscle in response to changes in thyroid hormone level, pattern of activity and mechanical tension (Fig. 9). The effect of thyroid hormone is illustrated in Fig. 10. The effect of the pattern of activity is clearly shown by the changes in the rat soleus after denervation and stimulation with a high frequency pattern of impulses. This pattern of activity, typical of fast muscles, was found to induce down-regulation of β-MHC and up-regulation of 2A- and 2X-MHC [23]. Similar isoform transitions are seen after hindlimb suspension, an experimental model of unloading used to simulate weightlessness (M. CAMPIONE and S. SCHIAFFINO, unpublished observation). A remarkable finding is that under these conditions there seems to be an obligatory sequence in MHC switching, in the order 1 ↔ 2A ↔ 2X ↔ 2B [76]. It will be of interest to determine whether the

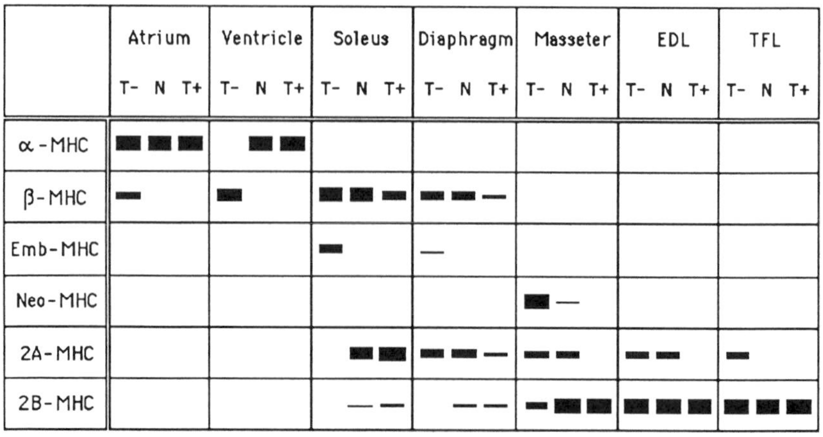

	Atrium T-	Atrium N	Atrium T+	Ventricle T-	Ventricle N	Ventricle T+	Soleus T-	Soleus N	Soleus T+	Diaphragm T-	Diaphragm N	Diaphragm T+	Masseter T-	Masseter N	Masseter T+	EDL T-	EDL N	EDL T+	TFL T-	TFL N	TFL T+
α-MHC	■	■	■		■	■															
β-MHC	■				■		■	■	■	■	■	—									
Emb-MHC							■					—									
Neo-MHC													■	—							
2A-MHC								■	■	■	■	—	■	■		■	■		■		
2B-MHC								—	—		—	—	■	■	■	■	■	■	■	■	■

FIG. 10. The scheme illustrates the changes in the expression of different MHC genes in different muscle tissues in response to changes in thyroid hormone. The thickness of each band reflects the level of the corresponding mRNA, as determined by S1 nuclease mapping analysis. N indicates normal tissues, T- and T+ tissues from hypothyroid and hyperthyroid rats, respectively. (Modified from IZUMO *et al.*, Ref. 17).

coordinated regulation of type 2 MHC genes is related to their physical linkage in the same chromosomal locus.

The same gene may respond in completely opposite directions to the same hormonal stimulus in different muscles, *e.g.* the 2A-MHC gene is up-regulated in soleus but down-regulated in EDL by thyroid hormone [17]. The tissue-specificity in MHC expression is clearly illustrated by the different response of different muscle tissues (atrial *vs.* ventricular myocardium, fast skeletal *vs.* slow skeletal muscles) to variations in thyroid hormone level (Fig. 10). The same difference is found when one compares the response of skeletal muscles to different patterns of electrical stimulation [23]. In conclusion, these studies show that gene expression is variably modulated in different muscle cell populations in response to the same inducing stimulus. These intrinsic differences between various muscle cells may reflect their different origin or their different developmental history.

Acknowledgments

Original work presented in this review was supported in part by institutional grant from CNR and by a grant from the Italian Space Agency (ASI).

References

[1]   A.F. HUXLEY, *Reflections on muscle.* Princeton Univ. Press, Princeton (1980).

[2]   S.EBASHI, and M. ENDO, *Progr. Biophys. Mol. Biol.* **64**, 465 (1972).

[3]   D.PETTE, and R.S. STARON, *Rev. Physiol. Biochem. Pharmacol.* **116**, 1 (1990).

[4] T. MIWA, Y. MANABE, K. KUROKAWA, S. KAMADA, N. KANDA, G. BRUNS, H. UEYAMA and T. KAKUNAGA, *Mol. Cell. Biol.* **11**, 3296 (1991).

[5] V. MAHDAVI, S. IZUMO, and B. NADAL-GINARD, *Circ. Res.* **60**, 804 (1987).

[6] P. BARTON and M.E. BUCKINGHAM, *Biochem. J.,* **231,** 249 (1985).

[7] J.P. LEES-MILLER and D.M. HELFMAN, *BioEssays* **13**, 429 (1991).

[8] R.E. BREITBART, H.T. NGUYEN, R.M. MEDFORT, A.T. DESTREE, V. MAHDAVI and B. NADAL-GINARD *Cell,* **41**, 67 (1985).

[9] R. GAHLMAN, A.B. TROUTT, R.P. WADE, P. GUNNING and L. KEDES, *J. Biol. Chem.,* **262**, 16122 (1987).

[10] J.P. JIN, Q.Q. HUANG, H.I. YEH and J.J.C. LIN, *J. Mol. Biol.,* **227**, 1269 (1992).

[11] S. SCHIAFFINO, L. GORZA, S. SARTORE, L. SAGGIN, M. VIANELLO, K. GUNDERSEN and T. LØMO, *J. Muscle Res. Cell Motil.* **10**, 197 (1989).

[12] R.G. WHALEN, S.M. SELL, G.S. BUTLER-BROWNE, K. SCHWARTZ, P. BOUVERET and I. PINSET-HARSTROM, *Nature (London),* **292,** 805 (1981).

[13] D.F. WIECZORECK, M. PERIASAMY, G.S. BUTLER-BROWNE, R.G. WHALEN and B. NADAL-GINARD, *J. Cell Biol.* **101,** 618 (1985).

[14] S. SARTORE, F. MASCARELLO, A. ROWLERSON, L. GORZA, S. AUSONI, M. VIANELLO and S. SCHIAFFINO, *J. Muscle Res. Cell Motil.* **8**, 161 (1987).

[15] S. SARTORE, L. GORZA and S. SCHIAFFINO, *Nature (London),* **298,** 294-296 (1982).

[16] S. SCHIAFFINO, L.GORZA, G. PITTON, L. SAGGIN, S. AUSONI, S. SARTORE and T. LØMO, *Dev. Biol.,* **127,** 1 (1988).

[17] S. IZUMO , B. NADAL-GINARD and B. MAHDAVI *Science* **231,** 597 (1986).

[18] K. CONDON, L. SILBERSTEIN, H.M. BLAU and W.J. THOMPSON, *Dev. Biol.* **138**, 256 (1990).

[19] S. SARTORE, S. PIEROBON BORMIOLI and S. SCHIAFFINO, *Nature (London),* **274,** 82 (1978).

[20] S. SARTORE, L. GORZA, S. PIEROBON BORMIOLI, L. DALLA LIBERA and S. SCHIAFFINO, *J. Cell Biol.* **88,** 226 (1981).

[21] L. GORZA, S. SARTORE, S. SCHIAFFINO, *J. Cell. Biol.* **95,** 838 (1982).

[22] L. GORZA, S. SARTORE, L.E. THORNELL and S. SCHIAFFINO, *J. Cell Biol.* **102,** 1758 (1986).

[23] S. AUSONI, L. GORZA, S. SCHIAFFINO, K. GUNDERSEN and T. LØMO, *J. Neurosci.* **10,** 153 (1990).

[24] M. BARANY, *J. Gen. Physiol.* **50,** 197 (1967).

[25] P.D. WAGNER, *J. Biol. Chem.* **256,** 2493 (1981).

[26] P.J. REISER, R.L. MOSS, G.G.GIULIAN and M.L.GREASER, *J. Biol. Chem.* **260,** 14403 (1985).

[27] S. SCHIAFFINO, S. AUSONI, L. GORZA, L. SAGGIN, K. GUNDERSEN and T. LØMO, *Acta Physiol. Scand.,* **134,** 565 (1988).

[28] L. LARSSON, L. EDSTROM, B. LINDERGREN, L. GORZA and S. SCHIAFFINO, *Am. J. Physiol.,*

261, C93 (1991).

[29] R. BOTTINELLI, S. SCHIAFFINO, C. REGGIANI, *J. Physiol.* **437,** 655 (1991).

[30] F.H. SCHACHAT, M.S. DIAMOND and P.W. BRANDT, *J. Mol. Biol.,* **198,** 551 (1987).

[31] R.L. MOSS, *Circ. Res.* **70,** 865 (1991).

[32] A. BABU, E. SONNENBLICK, J. GULATI, *Science* **240,** 74 (1988).

[33] J. GULATI, E. SONNENBLICK and A. BABU, *J. Physiol.* **441,** 305 (1991).

[34] R.L. MOSS, L.O. NWOYE and M.L. GREASER, *J. Physiol.* **440,** 273 (1991).

[35] L.S. TOBACMAN and R. LEE, *J. Biol. Chem.* **262,** 4059 (1987).

[36] D.W. CLEVELAND, *Cell Motil. Cytoskel.,* **14,** 147 (1989).

[37] M. SHANI, *Cell Motil. Cytoskel.,* **14,** 156 (1989).

[38] I. GARDNER, A. MINTY, S. ALONSO, P. BARTON and M. BUCKINGHAM, *EMBO J.,* **5,** 2559 (1986).

[39] M.S. PARMACEK, J.A. VORA, T. SHEN, E. BARR, F. JUNG and J.M. LEIDEN, *Mol. Cell. Biol.,* **12,** 1967 (1992).

[40] A. DE LOZANNE, (1989) *Cell Motil. Cytoskel.,* **14,** 62-68.

[41] G.J. IZANT, *Cell Motil. Cytoskel.,* **14,** 81 (1989).

[42] D.A. KNECHT, *Cell Motil. Cytoskel.,* **14,** 92 (1989).

[43] A. DE LOZANNE, J.A. SPUDICH, *Science* **236,** 1081 (1987).

[44] D.A. KNECHT, W.F. LOOMIS, *Science* **236,** 1081 (1987).

[45] T. SCHULTHEISS, L. ZHONGXIANG, H. ISHIKAWA, I. ZAMIR, C.J. STOECKERT and H. HOLT-ZER, *J. Cell Biol.,* **114,** 953 (1991).

[46] E. FRYBERG, *Cell Motil. Cytoskel.,* **14,** 118 (1989).

[47] T.M. FULLER, C.L. REGAN, L.L. GREEN, B. ROBERTSON, R. DEURING and T. HAYS, *Cell Motil. Cytoskel.,* **14,** 128 (1989).

[48] R.H. WATERSTONE, *Cell Motil. Cytoskel.,* **14,** 136 (1989).

[49] G. GERISH, A.A. NOEGEL and M. SCHLEICHER, *Ann. Rev. Physiol.,* **53,** 607 (1991).

[50] N.J. DIBB, I.N. MARUYAMA, M. KRAUSE and J. KARN *J. Mol. Biol.,* **205,** 603 (1989).

[51] R.H. WATERSTON, *EMBO J.,* **8,** 3429 (1989).

[52] A. BEJSOVEC and P. ANDERSON, *Cell,* **60,** 133 (1990).

[53] P.T. O'DONNEL, S.I. BERNSTEIN, *J. Cell. Biol.,* **107,** 2601 (1988).

[54] M.S. PARMACEK, J.M. LEIDEN, *Circulation,* **84,** 991 (1991).

[55] J.A. PUTKEY, H.L. SWEENEY and S.T. CAMPBELL, *J. Biol. Chem.,* **264,** 12370 (1989).

[56] H.L. SWEENEY, R.M.M. BRITO, P.R. ROSEVEAR and J.A. PUTKEY, *Proc. Nat. Acad. Sci. USA,* **87,** 9538 (1990).

[57] J. GULATI, E. SONNENBLICK and A. BABU, *J. Physiol.,* **441,** 305 (1990).

[58] R. L. DAVIS, H. WEINTRAUB and A. B. LASSAR, *Cell,* **51,** 987 (1987).

[59] W. E. WRIGHT, D. A. SASSOON and W. K. LIN, *Cell,* **56,** 607 (1989).

[60] T. BRAUN, G. BUSCHHAUSEN-DENKER, E. BOBER, E. TANNICH and H.H. ARNOLD, *EMBO J.,* **8,** 701 (1989).

[61] S. J. RHODES, S. F. KONIECZNY, *Genes Dev.,* **3,** 2050 (1989).

[62] T. Miwa and L. Kedes, *Mol. Cell. Biol.,* **7,** 2803 (1987).

[63] J.H. Mar and C.P. Ordhal, *Mol. Cell. Biol.,* **10,** 4271 (1990).

[64] L. Gosset, D. Kelvin, E. Sternberg and E.N. Olson, *Mol. Cell. Biol.,* **9,** 5022 (1989).

[65] R. Benezra, R. L. Davis, D. Lockshon, D. L. Turner and H. Weintraub, (1990) *Cell* **61,** 49-59.

[66] W.R. Thompson, B. Nadal-Ginard and V. Mahdavi, *J. Biol. Chem.,* **266,** 22678 (1991).

[67] M. Buckingham, *Trends Genetics,* **8,** 144 (1992).

[68] L. Gorza, S. Schiaffino and M. Vitadello, *Brain Res.,* **457,** 360 (1988).

[69] J.B. Miller and F.E. Stockdale, *Proc. Nat. Acad. Sci. USA* **83,** 3860 (1986).

[70] G. Cossu and M. Molinaro, *Curr. Topics Dev. Biol.,* **23,** 185 (1987).

[71] J.F.Y. Hoh, P.A. McGrath, H.T. Hale, *J. Mol. Cell., Cardiol.,* **10,** 1053 (1978).

[72] A.M. Lompré, K. Schwartz, A. D'Albis, G. Lacombe, N.V. Thiem and B. auw, *Nature (London),* **282,** 105 (1979).

[73] L. Gorza, P. Pauletto, A.C. Pessina, S. Sartore and S. Schiaffino, *Circ. Res,* **49,** 1003 (1981).

[74] A.M. Lompré, B. Nadal-Ginard and V. Mahdavi, *J. Biol. Chem.,* **259,** 6437 (1984).

[75] S. Izumo and V. Mahdavi, *Nature (London),* **334,** 539 (1988).

[76] S. Schiaffino, L. Gorza, S. Ausoni, R. Bottinelli, C. Reggiani, L. Larsson, L. trom, K. Gundersen and T. Lømo, In *The Dynamic State of Muscle Fibers,* D.Pette, ed., Walter de Gruyter, Berlin, (1990) pp. 329 -341.

BIOENERGETICS OF MUSCLE CONTRACTION

JOSEPH TIGYI

Biophysical Institute of the Medical University
Pécs, Hungary

Contents

Bioelectrochemistry IV
Edited by B.A. Melandri *et al.*, Plenum Press, New York, 1994

1.   Introduction

The cross-striated muscle is the mechanical energy converter generally used in the animal in kingdom. (The Nobel Laureate Russian physicist, DR. KAPICZA, has estimated that more than half of the mechanical energy utilized on our globe originates from the animal muscle machine). Muscle research has played a prominent role in biological science, starting with GALVANI's famous experiments in 1974 (The first specific monograph about muscle was published in 1882 by A. FICK) [1]. Today the problem of muscle contraction stands in the frontline of physiology, biophysics, biochemistry and molecular biology; for example at the American Biophysical Society meeting in San Francisco, in February 1991, 15 of the 64 symposia dealt with muscle research. Great progress has been achieved over the last decades, thanks to molecular biological methods. Among the many unanswered problems that still exist in the field of muscle science, that of the energetics of muscle contraction is the most complex one.

This chapter will present a shortened overview under consideration of the following points: status of the energetics of the muscle contraction, important contradictions and criticisms, and unsolved problems to be studied more deeply.

2.   Present status of the energetics of muscle contraction

Many reviews, symposia and monographs have been published over the last few decades about muscle contraction (BRENNER, [2]; COOKE [3]; ERNST, [4]; HOMSHER, [5]; KUSHMERICK, [6]; MOMMAERTS, [7]; POLLACK-SUGI, [8]; POLLACK, [9]; SZENT GYÖRGYI [10] WOLEDGE, [11]; WOLEDGE *et al.* [12]). Because of the large number of important publications, this review will be shortened, but an effort will be done to select the most important experimental facts and the most plausible explanations.

It is generally accepted that muscle is a chemo–mechanical energy converter. The energy source of muscular activity is ultimately derived from the metabolic oxidation of the food ingested by the animal. Note that muscle is not only a mechanical energy converter, but also a metabolizing organ performing part of the total metabolic function of the whole body. Table 1 shows that the O_2 consumption of muscles related to that of the whole organism under resting conditions amounts to 20%, and during activity it can increase to 60%.

Consensus is that ATP is the primary, and phosphocreatine (PCr) the secondary energy source for muscular contraction (Fig. 1).

TABLE 1. O₂ CONSUMPTION (%)

Organ	At rest	During activity
Heart	12	16
Skeletal muscles	20	61
Other organs	68	23

The basal O_2 consumption of a resting frog muscle at 0 °C accounts for 90% of the total energy requirements of the animal. The basal level, 7 nmol per gram (wet weight) per min, is equivalent to 0.8 nmol ATP. The rate of ATP turnover increases by a factor of 10^3 during the first few seconds of an isometric tetanus.

The study of high-energy phosphates termed *phosphogens*, as the fuel for muscle contraction was initiated by FISKE and SUBBA ROW [13], EGGLETON and EGGLETON [14], LOHMANN [15], LUNDSGAARD [16], and MEYERHOF *et al.* [17], but the mechanism of how phosphogen energy is converted to mechanical work remained unknown. The essential step concerning the details of the ATP–mechanical work conversion was made by ENGELHARDT and LJUBIMOVA [18], and later by the SZENT–GYÖGYI group [19, 20] by demonstrating that the muscle protein actomyosin acts as an ATPase enzyme, catalyzing the hydrolysis of ATP. The main goal of muscle research during the last half century was to explore the details of the conversion of ATP energy into mechanical energy.

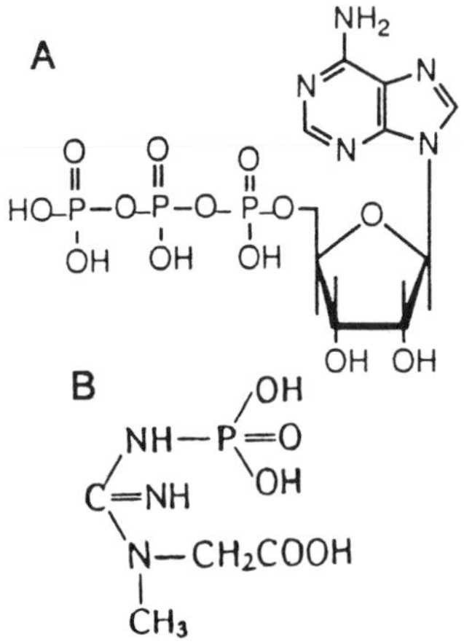

FIG. 1 Adenosine 5'-triphosphate (A) and phosphocreatinc (B) molecules.

After publication of the sliding filament hypothesis in 1954 by HUXLEY and HANSON [21], the main line of research was to understand the coupling mechanism of sliding (cross bridge) action with ATP splitting.

Before discussing the ATP–cross–bridge interactions some experimental details about ATP and the thermodynamical principles of muscle contraction should be mentioned.

2.1. Adenosine triphosphate

The overall equation for ATP hydrolysis is usually written

$$ATP + H_2O \leftrightarrow ADP + P_i^- + \alpha\, H^+ + work$$

In this reaction three processes are always linked together:

i) the performance of mechanical work by the actomyosin;

ii) the Ca^{2+} transport in the sarcoplasmic reticulum (SR);

iii) the Na^+ and K^+ transport through the membranes of the sarcolemma and of the T-system.

2.1.1. Work performance by actomyosin

Actomyosin ATPase activity is the most significant. The concentration of myosin catalytic sites amounts to 0,3 μmol/g wet weight of muscle (frog). The maximum enzymatic turnover rate (k_{cat}) for actin–activated frog myosin subfragment-1 (0 °C, pH = 7, 10 mM KCl) is 4,5 s^{-1}. One can calculate the maximal rate of ATP splitting, which is 1,4 μmol/g s^{-1}. This value is in good agreement with the observed rate of ATP utilization in a frog muscle performing mechanical work, and also harmonizes with the steady utilization rate of an isometric tetanus (0.3 – 0.4 μmol g^{-1} s^{-1}). These figures may vary in different types of muscle, but they can be considered as a good average.

2.1.2. Ca^{2+} transport

The Ca^{2+} transport ATPase concentration is estimated to be 0.1 – 0.3 μmol/g ATP. This value was calculated by the energy requirements of Ca^{2+} transport. The SR in frog muscle, which is 15 % of the volume, contains a maximum of 2 – 3 μmol Ca^{2+} g^{-1} muscle. Normally, a smaller concentration (1 μmol Ca^{2+} g^{-1} can be calculated. In addition, only a fraction of the total Ca^{2+} is realeased in a single twitch (about one fifth of the total). It is assumed that the Ca^{2+} concentration of the myofibrillar and cytoplasmic binding sites increases 100–fold during contraction from 10^{-7} to 10^{-5}. Data gained on skinned fibres showed that for 1 g wet muscle there is 0.1 – 0.2 μmol Ca^{2+} s^{-1} which requires an ATP splitting of 0.1 μmol g^{-1} wet muscle.

2.1.3. Na⁺ – K⁺ transport

A basic phenomenon of muscular activity is the accelerated sodium-potassium exchange, during the initial phase of the contraction. This phenomenon was discovered in our laboratory by Prof. ERNST in 1928 (Fig. 2). HÖBER, the famous German biophysicist, incorporated these results in his book in 1945 and became, thereafter, the authority for Europe and the United States [22].

ERNST, E. und SCHEFFER, L.:
Untersuchungen über Muskelkontraktion
Mitt. VIII. Die Rolle des Kaliums in der Kontraktion.
1928 Pflügers Arch. Ges. Physiol.... Vol. 220 pp. 665-671

FIG. 2 Title of the original ERNST-SCHEFFER article.

2.2. Thermodynamic considerations

The energy cost to restore the initial conditions after a single action potential of

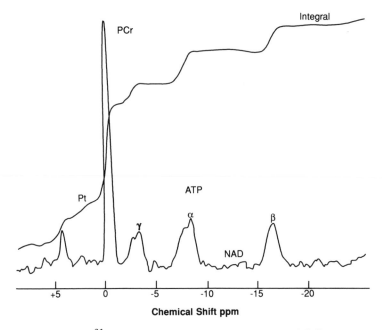

FIG. 3 ³¹P NMR-spectrum of muscle (CHALOVICH et al., [24])

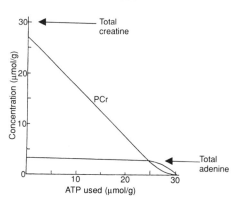

FIG. 4   Change of energy phosphates as a function of ATP used (WOLEDGE *et al.*, [12]).

100 mV can be calculated. If we assume that the capacitance of the muscle surface membranes in 1 g tissue is about 1 mF/g (estimated by THOMAS [23] with the consideration that the capacitance of the membrane is in the order of 1 µF/cm²) and that of the T tubules is 5 mF/g, the total amount of charge necessary to the given capacitance would be 6×10^{-4} C g⁻¹ which would be equivalent to 6 nmol g⁻¹ monovalent ions.

Considering the stoichiometric situation of Na⁺, K⁺ – ATPase which is 3 Na⁺ : 2 K⁺ : 1 ATP, then the necessary energy for this process amount to 2 nmol ATP splitting.

The ATP requirement in the three processes mentioned in section 2.1. (actomyosin ATPase: Ca²⁺–ATPase Na⁺, K⁺ – ATPase), relates as 700 : 50 : 1 (1.4 : 0.1 : 0.002).

Besides the ATP dehydration there is an important reaction in the muscle which produces the ATP from ADP and phosphocreatine by the enzyme creatine phosphokinase (CPK):

$$ATP + Cr \rightleftarrows P\ Cr + ADP + \alpha\ H^+$$

Earlier radioactive tracer studies and recent ³¹P-NMR studies prove that the forward and reverse rate of this reaction are 2-5 times higher than the ATP turnover rate during contraction (Fig. 3).

Table 2 gives an overview of the most important chemical reactions of muscle metabolism. Futher analysis of the molecular changes of muscle is very significantly affected by the mitochondrial content of the muscle. (In a skeletal muscle the mitochondrion content is about 5%, but in heart muscle 40%).

Figure 4 shows a summary of changes of energy–rich phosphate compounds as a function of ATP used (WOLEDGE *et al.* [12]). The whole metabolic process of muscle is summarized in Fig. 5.

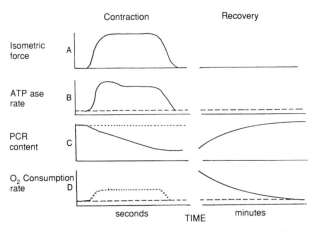

FIG. 5 Overview of the muscle metabolism (KUSHMERICK [6]).

TABLE 2 - THE MOST IMPORTANT CHEMICAL REACTION OF ENERGETICS OF MUSCLE

$MgATP \rightarrow MgADP^- + HPO_4^{2-} + H^+$	(1)
$H^+ + PC^{2+} + MgADP^- \rightarrow Cr + MgATP^{2-}$	(2)
$H^+ + 2MgADP^- \rightarrow AMP^- + MgATP^{2-} + Mg^{2+}$	(3a)
$H^+ + AMP^- \rightarrow IMP^- + NH^+_4$	(3b)
Glycogen unit + 3 $MgADP^-$ + 3 HPO_4^{2-} + H^+ $\rightarrow$ 2 lactate$^-$ + 3 Mg ATP^{2-}	(4)
Glycogen unit + 6O_2 + 36 $MgADP^-$ + 36 HPO_4^{2-} $\rightarrow$ 36H^+ + 6 CO_2 + 6H_2O + 36 $MgATP^{2-}$	(5)

Following the equations given by WALZ in his first chapter, without taking into account, as a first approximation, the dissipation function, and considering the real situation in muscle, figures for the energy balance can be obtained. By direct analysis the following values for [ATP], [P_i], and [ADP] are obtained: 4×10^{-3} M, 2×10^{-3} M and 16×10^{-7} M (by calculation) respectively, at 20 °C, pH = 7 and ionic strength 0.25. Using the figure − 29.9 KJ/mol^{-1} for $\Delta G°$ (WEECH et al. [25]), the actual GIBBS free energy available can be calculated from the formula

$$\Delta G = \Delta G° - RT \ln \frac{[ATP]}{[ADP][Pi]} = -29.9 - 39,8 = -69,7 \text{ KJ/mol}^{-1}$$

The appropriate enthalpy change of PCr hydrolysis is ΔH (PCr) = $-$ 34 KJ/mol⁻¹, (CURTIN et al. [26]). If we block creatin phosphokinase (CPK) by fluorodinitrobenzol (FONB), the appropriate enthalpy change for net ATP hydrolysis ΔH (ATP) = $-$ 46 KJ/mol⁻¹. Since the extent of the reaction PCr $\rightarrow$ Cr + P$_i$ determines the enthalpy change, for *resting muscle* ΔH = $-$ 34 KJ/mol⁻¹, ΔG = $-$ 70 KJ/mol⁻¹ and by calculation $T\Delta S$ = + 36 KJ/mol⁻¹.

In *contracting* muscle, the actual initial and final reactant concentrations are changing, and the free enthalpy change can be determined by integration

$$\Delta G' = \int (dG'/d\xi) \, d\xi$$

Because the muscle is a chemo-mechanical energy converter, it is important to know the efficiency with which it works.

Many textbooks give the efficiency of muscle and values can usually be found for about 0.25–0.33. This value is usually derived as follows

$$\eta = \frac{w}{E} = \frac{\text{the effective work}}{\text{total energy consumption}}$$

As ERNST [27] explained, this is theoretically erroneous, and should be called the *economic efficiency*. The total energy consumption was usually measured by the O_2 consumption, because – as mentioned earlier – the muscle is not only a chemo–mechanical energy converter, but also a significant organ of metabolism.

To give a correct parameter describing the muscle as a mechanical energy converter engine we have to use the mechanical efficiency

$$\eta_m = w/W$$

where w is the effective work and W is the theoretical maximum work (indicated work).

The value of w can be measured directly and the one of W can be determined from the length tension diagram (see Fig. 6).

In an earlier paper TIGYI [28] mathematically analyzed the curve of active tension of the muscle and derived the equation:

$$\frac{T}{T_o} = \exp\left[-K^2\left(\frac{\Delta L}{\Delta L_o}\right)^2\right]$$

Where T is the isometric tension at a given length L of the isometric contraction after an isotonic shortening ($L_o - L = \Delta L$), T_o is the maximal active tension, L_o is the

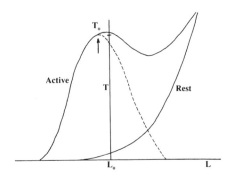

FIG. 6. Length-tension diagram of muscle (TIGYI, [28])

length at rest, ΔL_o is the maximal shortening (Fig. 6).

In the case of frog sartorius the constant was determined and $K = 2.0 \pm 0.2$ was found. The indicated work was:

$$W = \int_o^{\Delta L} T \, \Delta L = 0.45 \, T_o \, \Delta L_o$$

Using the experimental data of the effective work **w** in a case of a human upper arm pulling up the whole body on a horizontal bar, the mechanical efficiency (η_m) can amount to 0.9. This high value is unexpected and not generally known, but nevertheless is a fact.

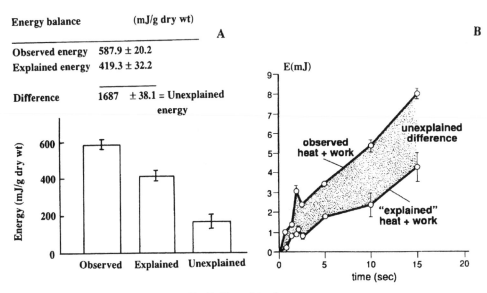

FIG. 7. Unexplained energy.

2.3.   Energy balance experiments

From a practical point of view, the most important information can be gained from *energy balance experiments.*

Over the last decades the question as to whether an energy balance exists during muscle contraction has been investigated intensively. Numerous laboratories and scientists have taken part in these studies and published informative papers. Without going into details, Figures 7A (WOLEDGE, [12] and 7B (POLLACK, [9]) show that a significant gap exists between measured and explained energy levels. The so-called *unexplained* energy stands between 15% and 40% of the observed one, i.e. the energy observed is always larger than the explained one.

At present, this energy gap seems to be the greatest problem in muscle energetics. Obviously further specific myothermic and chemical energy studies are needed to solve this problem.

3.   Relationship between ATP hydrolysis and the cross-bridge function

After publication of the sliding (cross-bridge) hypothesis, muscle scientists spent a great deal of effort in explaining ATP hydrolysis and actomysin joint action (HUXLEY and HANSON [21], HUXLEY [29], HUXLEY and NIEDEREGERKE [30], HUXLEY and GORDON [31].

The basic concept of the sliding hypothesis is shown in Fig. 8. It was supposed that the thick filaments are composed mainly of myosin and are in the anisotropic A-band. The thin filaments are made up of actin and are found on both sides of the Z disk (vertical lines in Fig. 8). The isotropic band (I) contains only thin filaments but part of the thin filaments also exist in the A-band overlapping the thick filaments. The contrac-

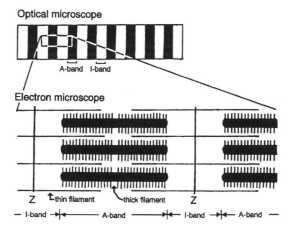

FIG. 8.   The basic concept of the sliding hypthesis (POLLACK, [9]).

tion mechanism consists in the sliding of the thin and thick filaments over each other.

A characteristic supposition of the hypothesis is that the length of the A-band does not change during contraction; it is only the length of the I-band that diminishes.

The main problem of the hypothesis was to ascertain the driving force behind the sliding of the two kind of filaments. Very elegant electron-microscopic pictures demonstrate (Fig. 9) that there are cross–bridges between the filaments.

FIG. 9. ELMI picture of cross–bridges (TROMBITÀS in Ref. 9).

The cross-bridges are made from the heads of the myosin molecules (Fig. 10) and the remaining part of the molecule formes the rod like part of the thick filaments (Fig. 11).

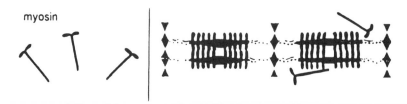

FIG. 10. Myosin molecules in the thick filaments (POLLACK, [9])

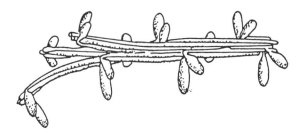

FIG. 11. Steric scheme of thick filaments (POLLACK, [9]).

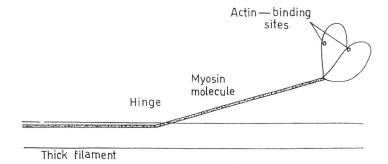

FIG. 12. Scheme of the thick filaments (POLLACK, [9]).

A scheme of the cross-bridge structure can be seen in Fig. 12 and the coupling of the ATP hydrolysis cycle is demonstrated in Fig. 13. A more detailed conception by HUXLEY and KRESS [32] is shown in Fig. 14.

Details of the cross–bridge action have been constantly under study over the last decades and numerous extremely sophisticated methods have been used. The most important experimental approaches are the following ones.

3.1. Elastic properties

BRENNER, SCHOENBERG, CHALOWICH, GREENE and EISENBERG [33] measured the

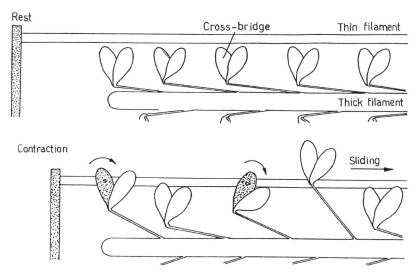

FIG. 13. The cross–bridge cycle (POLLACK, [9]).

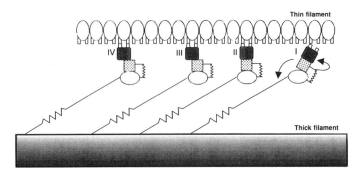

FIG. 14. Cross–bridge model with no rotation during the power stroke (HUXLEY and KRESS, [32]).

change in the stiffness of rigor and resting skinned rabbit psoas using different speed of length changes. They aimed to characterize the kinetics of cross–bridge detachment and reattachment: length changes of 5-10 nm per half sarcomere were applied. It was interesting that in rigor the stiffness changed very little when the speed of the length change was varied over 5-6 orders of magnitude. By contrast, relaxed fibres (at low ionic strength) showed a significant effect of stretch speed on fibre stiffness. From these experiments it can be concluded that a significant fraction (although not all) of the cross–bridges are attached in relaxed fibres and low ionic strengths. Useful arguments can be gained from the *rate constants* of attachment and detachment of cross bridges in such experiments (Fig. 15).

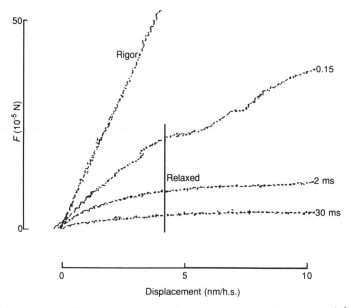

FIG. 15. Change of stiffness as a function of the speed of stretching (BRENNER *et al.*, [33]).

3.2. Fluorescent labelling

YANAGIDA [34] measured the polarized fluorescence from fluorescent ATP or ADP analogs (ε–ATP or ε–ADP 1 N6–Ethano ATP or ADP) bound to myosin heads in a glycerinated muscle and demonstrated that during isometric contraction the bound nucleotides are highly oriented with respect to the axis of the fibre. In addition, the polarization of the tryptophan fluorescence did not change when the fibre was transferred from a rigor to an active state of a low ATP concentration. Such experiments have shown that the rotation of the myosin heads during contraction seems to be extremely limited if at all. Similarly the angle of bound ADP was not changed by passive stretching of the fibre.

Using phalloidin-FITC (phallotoxin) complex bound specifically to actin in glycerinated muscle fibre, the orientation of fibrous molecules changed remarkably when the fibres were activated from relaxation or rigor to contraction. These results suggested that rotation or distortion of actin monomers occur during contraction, which should take part in the process of active tension development (Figs. 16 and 17).

3.3. Spin labels

It is well known that some paramagnetic molecules (spin labels) can be used for detection of oriented molecular movements even in complex biological systems. Many laboratories used this method with interesting results.

An important question of the cross–bridge theory was whether cross–bridges (myosin heads) rotate. COOKE *et al.* [35] used spin labels attached specifically to reactive sulphdryl groups on the myosin heads in glycerinated rabbit muscles. Previously

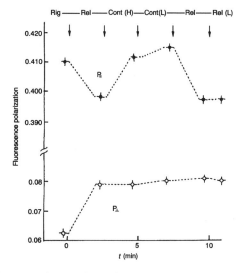

FIG. 16. Fluorescence polarization changes during the contraction-relaxation cycle of the glycerinated fibre (YANAGIDA, [34]).

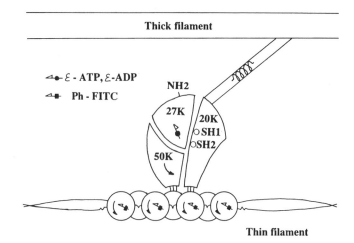

FIG. 17. Scheme of COOKE *et al.*, [35]. No change of angle of nucleotide on myosin heads during tension development.

THOMAS and COOKE [36] had demonstrated that paramagnetic probes are highly ordered in rigor muscle and show random angular distribution in relaxed muscle. COOKE *et al.* [35] observed in their paper that addition of ADP to rigor fibres caused no ESR spectral changes, but addition of AMPPNP or PP_i increased the fraction of this ordered spin probe. The stretching of fibres in the presence of AMPPNP or PP_i did not cause changes in the spectra. Addition of ATP and Ca^{2+} to the fibres produced isometric tension. The EPR spectrum is shown in Fig. 18.

This spectrum is a superimposition of the spectra in relaxation and contraction. When the authors compared the spectrum of rigor fibres (Fig. 19a, dashed line) and the difference spectrum obtained by subtracting 81% of the spectrum of the relaxed fibres from that of contracting (solid line), they obtained a curve that was identical with the rigor spectrum (Fig. 19b).

Using the saturation transfer method as well, they found that the disordered spectral component arises from myosin heads that are attached to the actin. The result that the ordered component displays an identical angular distribution under all condi-

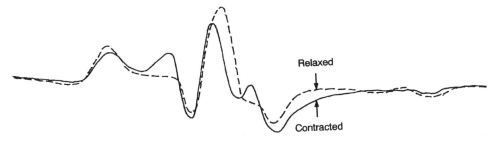

FIG. 18. Change of EPR spectra during contraction [35].

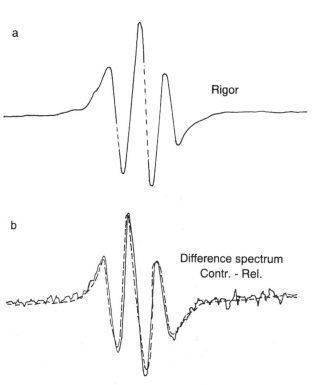

FIG. 19. A. EPR spectrum of rigor; B. difference spectrum: contracted - 81% of relaxed one (solid line; COOKE, et. *al.* [35]).

tions leads to the conclusion that its orientation is not linked to the force generation process. They elaborated a scheme shown in Fig. 20 in which they conceive a cross–bridge power stroke without changing the orientation in the domain of the

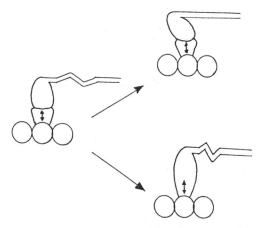

FIG. 20. Power stroke, without change of orientation of the adjacent part of myosin to the actin (COOKE, [38])

myosin head adjacent to the actin.

In our laboratory (LORINCZI *et al.* [37]) the problem has also been studied with spin labels where the label was bound to the reactive thiol site 704 (Cys 704) of the myosin (maleimidopiperidyne-nitroxyl) and the melting behaviour of the myosin was measured by differential scanning calorimetry (DSC). Intact and LC–2 deficient heart muscle myosin were studied. We found a conformational change in LC–2 deficient bovine heart myosin at 18 °C detected by the spin label method. This transition was not observed in intact myosin. Studyng the melting behaviour of myosin with DSC measurement we detected a low temperature endothermic peak at 18 °C in the case of intact myosin. Removal of LC–2 was associated with the disappearence of the 18 °C transition. This example suggests that it is necessary to use a wide spectrum of methods for gaining realistic information about the changes of the parameters of the cross–bridges.

3.4.   X–ray diffraction

X–ray studies have played a decisive role in the development of the sliding filament hypothesis. The X–ray diffraction patterns result from various periodic structures of sarcomeres: the equatorial reflections are due to the filament lattice structure and the meridional reflections and layer lines are due to the periodic structure along the thick and thin filaments. Intensive X–ray studies of meridional reflections performed in the 1960 did not show changes during contraction (ELLIOTT *et al.* [39]; HUXLEY [40]). This result together with studies performed with the new ELMI electron microscope (HUXLEY and HANSON [41]) showed that no coiling or folding occur in the muscle proteins during contraction. This led to the sliding hypothesis.

Nowadays, the X–ray diffraction method has been markedly improved by using position–sensitive proportional counters instead of films. Using high intensity synchroton radiation the new time–resolved X–ray method arrived at the 1 ms time resolution.

Figure 21 shows the redistribution of mass within the myofilament. PODOLSKY'S [43] group assumes that during isometric contraction there is an *overall shift in S1 during the cross-bridge cycle*. H.E. HUXLEY, in spite of many other results, seems to be more pessimistic is saying: *however our results so far provide little information about the actual nature of the structural change in the attached heads* [44].

3.5.   Cross–bridge function

Studying the cross–bridge function with the above-mentioned highly sophisticated methods has not led to the development of a general consensus about the problem. On the contrary, there are many arguments against the cross–bridge theory. In his monograph, G.H. POLLACK [9] summarizes the criticism in an extremely vivid manner.

The most important points of attack against the sliding cross–bridges hypothesis, according to POLLACK, are the following:

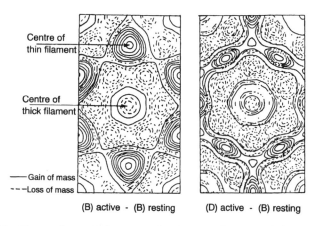

Centre of thin filament

Centre of thick filament

——Gain of mass
- - -Loss of mass

(B) active - (B) resting

(D) active - (B) resting

FIG. 21. Change of mass during contraction (X-ray axial projection) (YU *et al.* [42]).

a) The length of the thick filament is not constant. Table 3 summarizes the results of 40 papers and only 6 of them show that there is no change in the length of thick filaments. There are many brilliant ELMI and light microscopic pictures showing clearly the remarkable shortening of the thick filaments during activity (Figs 22a and b).

b) The tension-length diagram does not correspond to the overlap range of thick and thin filaments.

c) A few papers report contraction beyond the overlap region.

d) No acceptable electron microscopic evidence exists on cross–bridge activity.

e) No evidence of the coupling of ATP hydrolysis and supposed power stroke step.

f) In vitro models of the thin filaments planted cross–bridges have shown surprisingly a reverse gliding.

g) Contradiction between random smooth shortening and wave stepwise shortening (Fig. 23).

TABLE 3. SUMMARIZED DATA ABOUT CHANGE OF THICK FILAMENTS (POLLACK, [9])

Maximum amount of A-band or thick filament shortening	Number of independent Studies	
	Invertebrate	Vertebrate
No Change	1	5
5-15%	2	4
15-40%	5	12
> 40%	6	5

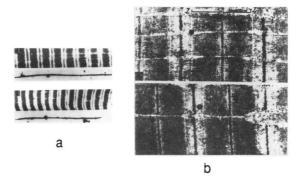

FIG. 22. Shortening of thick filaments. a. Samosudova and Frank; [45] b. Dewey *et al.,* [46].

There are many further contradictions. For example, one of the weakest points of the sliding cross–bridge hypothesis is that it does not have sufficient energetic evidence for tension development and mechanical energy conversion. However, in spite of all these facts, the sliding cross–bridge theory enjoys an overwhelming acceptance and central stance. It can be said that over the last three decades this has been the ruling concept of muscle contraction. It would be difficult to determine the main reasons for its outstanding popularity; perhaps because of its logical beauty, evident internal consistency and the attractive ELMI pictures and convincing schemes that help to teach it persuasively. It may perhaps be due to the great human and scientific personalities of its creators. It is probable, nevertheless, that textbooks will teach it as the basic concept of muscular contraction for many decades in the future.

Perhaps the real reason for its general acceptance is that at present there is no

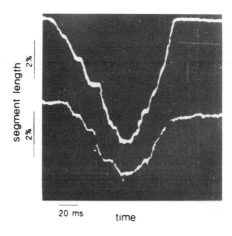

FIG. 23. Stepwise shortening of the sarcomere (Granzier *et al.* [47]).

hypothesis to compete with the sliding cross–bridge model.

Probably POLLACK'S hypothesis of helix-coil transition is noteworthy (Fig. 24), yet needs more study in various different laboratories. At present, the British Muscle Group opposes it very strongly (SIMMONS, [48]).

In his monograph, POLLACK proposed many new possibilities of obtaining experimental evidence to verify the helix coil transition hypothesis, *e.g.* to develop mutants in which the meltable region of the myosin rod is selectively altered. Such kind of myosin may be non-contractile.

4.   Further considerations on muscle energetics

To return to the main problem of muscle energetics, to the unexplained energy fraction, CURTIN *et al.* [26] observed as follows: *The phenomenon of unexplained energy fraction first observed in isometric contraction is also observed in isotonic ones, although the unexplained energy fraction is lower in muscles that do work...* RALL *et al.* [49] reported that the unexplained energy fraction was much greater (0.6-0.85) in a short (0.5–0.75 s) tetanus of rapidly shortening muscle, but in a tetanus of 3 s the unexplained fraction was smaller (0.3).

A working hypothesis has been elaborated that the rapid shortening induces a phase shift between heat production and chemical reaction of a hypothetical phosphorylated compounds (ξ–P). This phase shift between energy production and chemical reaction

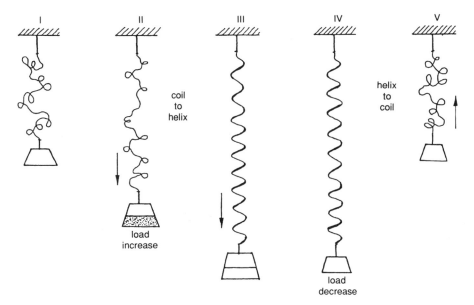

FIG. 24. Scheme of the helix-coil transition (POLLACK, [9]).

was interpreted later in terms of cross–bridge interaction, *i.e.* the relative distribution of myosin states in its kinetic cycle changes during rapid shortening. Therefore, additional ATP splitting is apparently necessary immediately after rapid shortening to restore the initial isometric distribution.

This and similar explanations are not very convincing. Therefore, I should like to explain our hypothesis about the unexplained energy fraction in muscle (see p. 311.). This hypothesis was based on our mechanical– volumetric – microcalorimetric – and light microscopic experiments and was developed during the 1950s. The hypothesis did not receive much attention, perhaps because the papers were published in the local journal *Acta Physiologica* of the Hungarian Academy of Sciences (ERNST and TIGYI, [50, 51], TIGYI [57, 58]).

The basic concept of the hypothesis is that muscle proteins behave like all other polymer structures and have a so–called rubber–like elasticity. It is a well–known effect in polymer chemistry that polymer crystallization is caused by external mechanical stretch. In such a conformational change (a phase transition), the proportion of ordered molecular domains will be increased in relation to the disordered fraction. In our volumetric measurements it was clearly demonstrated that the vo-

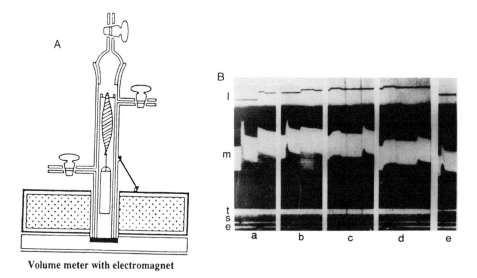

Volume meter with electromagnet

FIG. 25. A. The magnetic volumeter; B. Volume decrease of the muscle during active and passive tension (ERNST and TIGYI, [50, 51]).

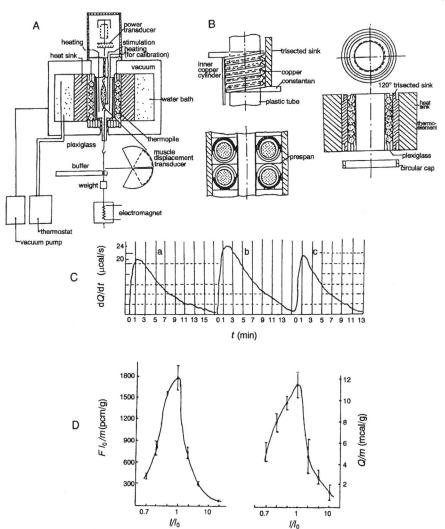

FIG. 26. Muscle heat measurements. A. and B. Microcalorimeter scheme [52]; C. the heat production of isotonic and isometric twitch [53]; D. isometric force and heat production plotted against length [54].

lume decrease is enhanced by external mechanical stretch (Figs. 25, 26a, b, c, d, 27). Similarly, increased heat production was observed with the stretch. It is well known that the birefrigence of the stretched muscle increases, too.

In general: the mechanical stretch (passive or active) evokes this *polymer crystallization* which appears in the microcalorimetric measurements as extra heat production.

This concept may also help to explain many phenomena of muscle mechanics, (EDMAN, [56]).

In energy balance experiments the unexplained energy fraction always showed a lack of chemical energy.

If we assume that the structural extra energy component, the *heat of crystalli-zation*, appears, then we can give an explanation for the unexplained part of the energy balance. This kind of hypothesis belongs to the socalled beaten path, *i.e.* some parts of the results do not necessarily find easy interpretation within the framework of the cross–bridge model. However, if they are well based and experimentally reproducible in many laboratories, they should be considered as precious pieces on the whole treasure of muscle sciences.

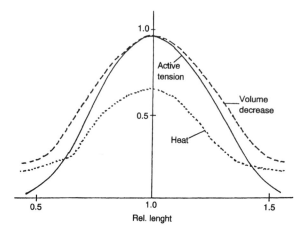

FIG. 27. The relationship between active tension, volume decrease and heat production (TIGYI, [55]).

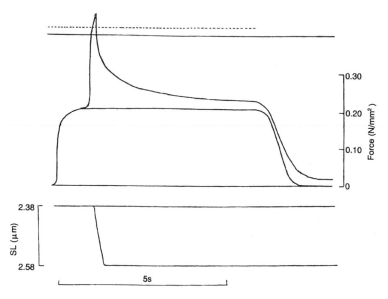

FIG. 28. Increase in isometric tension after a quick stretch (EDMAN, *et al.*, [56]);

References

[1] A. FICK, *Mechanische Arbeit und Wärmeentwicklung bei der Muskeltätigkeit.* Brockhaus, Leipzig (1882).

[2] B. BRENNER, *Annu. Rev. Physiol.,* **49,** 655 (1977).

[3] R. COOKE, *Crc. Crit. Rev. Biochem.,* **21** (1), 53 (1986).

[4] E. ERNST, *Biophysics of Striated Muscle,* Akad. Kiadó, Budapest (1963).

[5] E. HOMSHER, C.J. KEAN, Annu. Rev. Physiol., **40,** 93 (1978).

[6] M.J. KUSHMERICK, *Energetics of Muscle Contraction,* in: L.D. PEACHEY, *Handbook of Physiology* Section **10.** Am. Physiological Society, pp. 189-230. B. (1983).

[7] W.F.H.M. MOMMAERTS, *Physiol. Rev.,* **49,** 427 (1969).

[8] G.H. POLLACK, H. SUGI, (Editors) *Contractile Mechanism in Muscle,* Plenum Press, New York (1984).

[9] G.H. POLLACK, *Muscles and Molecules,* Ebner and Sons, Seattle (1990).

[10] A. SZENT-GYÖRGYI, *Chemical Physiology of Contraction in Body and Heart Muscle.* Acad. Press., New York (1953).

[11] R.C. WOLEDGE, *Heat Production and Chemical Change in Muscle,* In: BUTLER-NOBILE, (Editor), *Progress in Biophysics and Molecular Biology,* **22,** 37 (1971).

[12] R.C. WOLEDGE, N.A. CURTIN, E. HOMSAER, *Energetic Aspects of Muscle Contraction,* Academic Press, London (1985).

[13] C.H. FISKE, Y. SUBBAROW, *The Isolation and Function of Phosphocreative Science,* **67,** 169 (1928).

[14] G.P. EGGLETON, P. EGGLETON, *Biochem. J.,* **21,** 290 (1927).

[15] K. LOHMANN, *Biochem. Z.,* **271,** 264 (1934).

[16] E. LUNDSGAARD, *Proc. R. Soc. London Ser. B.,* **137,** 73 (1938).

[17] O. MEYERHOF, W. SCHULZ, and P. SCHUSTER, *Biochem. Z.,* **293,** 309 (1937).

[18] V.A. ENGELHARDT and M.N. LJUBIMOVA, *Nature London;* **144,** 668 (1939).

[19] A. SZENT-GYÖRGYI, *Studies, Szeged,* Vol. **1** (1942).

[20] A. SZENT-GYÖRGYI, *Introduction to Submolecular Biology,* Acad. Press, New York (1960).

[21] H.E. HUXLEY, J. HANSON, *Nature (London),* **173,** 973 (1954).

[22] R. HÖBER, *Physical Chemistry of Cells and Tissues,* The Blackiston Com. Philadelphia (1945).

[23] R.C. THOMAS, *Physiol. Rev.,* **52,** 563 (1972).

[24] J.M. CHALOVICH, C.T. BURT, M.J. DANON, T. GIONEK, and M. BÀRÀNY, *Ann. N.Y. Acad. Sci.,* **317,** 649 (1979).

[25] R.L. VEECH, J.W.R. LAWSON, N.W. CORNELL, and H.A. KREBS, *J. Biol. Chem.,* **254,** 6538 (1979).

[26] N.A. CURTIN, C. GILBERT, D.M. KRETSCHMAR, and D.R. WILKIE, *J. Physiol. (London),* **238,** 455 (1974).

[27] E. ERNST, *Orvosi Hetilap,* **90,** p. 225 (1949) (Hungarian).

[28] J. TIGYI, *The Role of Mechanical Tension in the Muscular Activity,* Ph. D. Dissertation, Hung. Acad. Sci., Budapest (1952) (Hungarian).

[29] A.F. HUXLEY, *Prog. Biophys. Biophys. Chemistry.,* **7,** 255 (1957).

[30] A.F. HUXLEY, R. NIEDERGERKE, *Nature (London),* **173,** 971 (1954).

[31] A.F. HUXLEY and A.M. GORDON, *Nature (London),* **193,** 28081 (1962).

[32] H.E. HUXLEY, M. KRESS, *J. Mus. Res. Cell Motil.,* **6,** 153 (1985).

[33] B. BRENNER, M. SCHOENBERG, H.M. CHALOVICH, L.E. GREENE, and E. EISENBERG, *Proc. Natl. Acad. Sci. U.S.A.,* **79,** 7288 (1982).

[34] T. YANAGIDA, *Angles of fluorescently labelled myosin heads and actin monomers in contracting and rigor strained muscle fiber,* in: *Contractile Mechanisms in Muscle,* G.H. POLLACK, H. SUGI, (Editors), Plenum Press, New York, London, (1984) pp. 397-412.

[35] R. COOKE, M. CROWDER, C.H. WENDT, V.A. BARNETT, D.D. THOMAS, *Muscle cross–bridges: do they rotate?* in: *Contractile mechanisms in muscle,* G.H. POLLACK, and H. SUGI, (Editors), Plenum Press, New York, London, (1984) pp. 413-428.

[36] D.D. THOMAS, R. COOKE, *Biophys. J.,* **32,** 891 (1980).

[37] D. LORINCZI, U. HOFFMANN, L. POTO, J. BELÀGYI, P. LAGGNER, *Gen. Physiol. Biophys.,* **9,** 589 (1990).

[38] R. COOKE, *Nature (London),* **294,** 570 (1981).

[39] G.F. ELLIOT, J. LOWY, and B.M. MILLMAN, *J. Mol. Biol.,* **25,** 31 (1967).

[40] H.E. HUXLEY, *Time resolved X–ray diffraction studies of cross–bridge movement, and their interpretation,* G.H. POLLACK, and H. SUGI (Editors), Plenum Press, New York, London, (1984) pp. 161-176.

[41] H.E. HUXLEY and J. HANSON, *Symp. Soc. Exp. Biol.,* **9,** 228 (1955).

[42] L.C. YU, T. ARATA, A.C. STEVEN, G.R.S. NAYLOR, R.C. GAMBLE and R.J. PODOLSKY, *Structural studies of muscle during force development in various states,* in: *Contractile Mechanisms in Muscle,* G.H. POLLACK, H. SUGI (Editors), Plenum Press, New York, London (1984) pp. 207-220.

[43] R.J. PODOLSKY, G.R.S. NAYLOR, T. ARATA (1982) *Cross-bridge properties in the rigor state,* in: *Basic Biology of Muscles; A Comparative Approach,* B.M. TWAROG, R.J.C. LEVINE and M.M. DEWEY (Editors), Raven Press, New York (1982) pp. 79-89.

[44] H.E. HUXLEY, A.R. FARUGI, M. KRESS, J. BORDAS, and M.H.J. KOCH, *J. Mol. Biol.,* **158,** 637 (1982).

[45] N.V. SAMOSUDOVA, R.G. LYUDKOVSKAYA, G.M. FRANK, *Biophys.,* **20,** 449 (1975).

[46] M.M. DEWEY, R.J.C. LEVINE, D. COLFLESH, B. WALCOTT, L. BRANN, A. BALDWIN and P. DRINK, *Structural changes in thick filaments during sarcomere shortening in Limulus striated muscle* in: *Cross–bridge Mechanism in Muscle Contraction,* H. SUGI and G.H. POLLACK (Editors) University Park Press, Baltimore (1979) pp. 3-22.

[47] H.L.M. GRANZIER, J.A. MYERS, and G.H. POLLACK, *J. Musc. Res. Cell. Motil.*, **8,** 242 (1987).

[48] R. SIMMONS, *Nature (London)*, **351,** 452 (1991).

[49] J.A. RALL, E. HOMSHER, A. WALLNER, W.F.H.M. MOMMAERTS, *J. Gen. Physiol.*, **68,** 13 (1976).

[50] E. ERNST and J. TIGYI, *Acta Physiol. Acad. Sci. Hung.*, **2,** 253 (1951).

[51] E. ERNST and J. TIGYI, *Acta Physiol. Acad. Sci. Hung.*, **2,** 261 (1951).

[52] D. LORINCZI and Z. FUTO, *Acta Biochim. Biophys. Acad. Sci. Hung.*, **9,** 371 (1974).

[53] D. LORINCZI, *Acta Biochim. Biophys. Acad. Sci. Hung.*, **9,** 383 (1974).

[54] D. LORINCZI and J. TIGYI, *Acta Biochim. Biophys. Acad. Sci. Hung.*, **11,** 311 (1975).

[55] J. TIGYI, *MTA Biol. Oszt. Közl.*, **21,** 15-35 (1978) (Hungarian).

[56] K.A.P. EDMAN, G. ELZINGA, M.I.M. NOBLE, *J. Physiol.*, **281,** 139 (1978).

[57] J. TIGYI, *Acta Physiol. Acad. Sci. Hung.*, **16,** 129-137 (1959).

[58] J. TIGYI, *1st Internat. Biophys. Congress*, Stockholm, Abstract p. 259 (1961).

Comments to Prof. Tigyi's lecture

I think that it was excellent of prof. Tigyi to point to a number of unsolved problems relating to the sliding hypothesis of muscle contraction. However, I cannot agree with his view that for those reasons the hypothesis should be rejected, particularly not since there is no alternative one to replace it. At the moment I feel that the most fruitful approach is to develop and to refine our experimental techniques with the purpose of modifying and clarifying uncertain aspects of the sliding model.

Stephen Thesleff

THERMODYNAMICS OF IRREVERSIBLE PROCESSES APPLIED TO BIOLOGICAL SYSTEMS: A GENERAL SURVEY

DIETER WALZ

Biozentrum, University of Basel,

Basel (Switzerland)

Contents

Bioelectrochemistry IV
Edited by B.A. Melandri *et al.*, Plenum Press, New York, 1994

1.   Introduction

A characteristic of living systems is a continuous running of processes, while an equilibrium state for *all* processes occurs only when the system is dead. Therefore, life has to be supported by a regular input of energy in form of nutrients or light, which is converted into other forms of energy needed by the system. Energy conversion between processes thus plays a vital role in biology.

The consequences of the above statements are twofold. Firstly, thermodynamics dealing with equilibrium states only (*i.e. classical* thermodynamics or *thermostatics*) is inappropriate for the assessment of living systems. By the same token *reversible processes, i.e.* processes which proceed through a sequence of states all at or very close to equilibrium, can be assumed at most for some processes in a living system only under certain conditions. As long as the system is alive there are processes which proceed irreversibly. Hence, thermodynamics of irreversible processes (or *non-equilibrium thermodynamics*) is the appropriate tool. Secondly, a biological system or any part of it cannot be treated as if it would *not* exchange energy and/or matter with its surroundings (*i.e.* as an isolated system in the thermodynamic sense). On the other hand, an isolated system is the most suitable for thermodynamics. The first step in the treatment of a system therefore consists of a translation of the experimental system into a thermodynamic system which should be isolated.

2.   Translation of an experimental system into a thermodynamic system

A thermodynamic system is composed of a number of conceptual elements each of which is well defined in terms of exchange of matter, heat and work with its surroundings. The exchange occurs *via* the boundaries or walls of the elements, which may be classified according to the scheme in Table 1. Note that no wall exists which permits the exchange of matter but prohibits the exchange of heat. The classification given in Table 1 is also applied to the thermodynamic system itself. In this case the outermost wall of the system, *i.e.* that which separates it from the rest of the universe, determines the classification. The most appropriate system for thermodynamic consideration is an *isolated* system, because it is not subject to uncontrolled disturbances from the outside world. Indeed this is exactly what one attempts to do in properly designing an experiment.

In biological systems the elements usually consist of phases, compartments, and reservoirs. Compartments usually contain a single phase, separated from other compartments by a boundary (*e.g.* another phase such as a membrane). A reservoir is an element whose capacitance for heat or matter or electric charge (or whatever else applies) is very much larger than that of any other element. Clearly such a reservoir is

TABLE 1. CLASSIFICATION OF ELEMENTS IN A THERMODYNAMIC SYSTEM

Element	Permitted[1] (+) or prohibited (-) exchange [through a wall] of		
	matter	heat	work[2]
open	+	+	+
closed	-	+	+
adiabatic	-	-	+
isolated[3]	-	-	-

[1]Permitted means allowed but *not necessarily* required.

[2] Work includes the action of electric and/or magnetic fields.

[3] An isolated element is also said to be surrounded by rigid adiabatic walls since the rigidity of the wall prevents the exchange of work.

surrounded by appropriate walls and acts as source or sink for heat or matter *etc.* In constructing a thermodynamic system, the elements of the experimental system are substituted by equivalent elements of the thermodynamic system. The actual topology of the experimental system need not be retained.

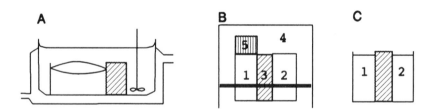

FIG. 1. Translation of an experimental system (A) into a thermodynamic system (B). In B two compartments with an aqueous phase (1, 2) are separated by a membrane (3) and surrounded by two reservoirs, one for heat (4) and one for mechanical work (5). The two-compartement system in C is the part of B in which the processes of interest take place. When constructing the thermodynamic system, the analytical devices needed to follow the processes in the experimental system have been neglected. This is legitimate since monitoring the parameters of the system should not disturb the processes. For further explanation see text.

As an example, consider an isolated muscle which is immersed in an appropriate aqueous solution with known composition (bathing solution, see Fig. 1A). The muscle is mounted in a special holder which allows a stimulation of the muscle and then either keeps the muscle's length constant (*isometric* contraction) or imposes a constant force on the muscle (*isotonic* contraction). In order to eliminate effects of a varying ambient temperature the vessel containing the solution is placed in a thermostat which should guarantee a constant temperature in the system. Moreover, an effective stirring of the bathing solution prevents the formation of gradients. Since the vessel is open the atmospheric pressure acts as a barostat, *i.e.* the system is under constant pressure.

The first step in translating such an experimental system into a thermodynamic system consists of defining a *heat reservoir* which acts as a heat source or heat sink, *i.e.* heat supplied to or absorbed from the rest of the system does not change the temperature (see Fig. 1B). Similarly, *a reservoir for mechanical work* is defined which provides the energy for the operation of the special holder and the stirring device. In a second step the phases present in the system are detected, a phase being defined by the chemical nature of its major constituent. The bathing solution but also the phase in the interior of the cells are aqueous phases, while a second phase is constituted by the membranes of the muscle cells. Assuming that the cells are identical in composition, all membranes can be combined into only one extended and, *e.g.*, plane membrane. Similarly, the aqueous phases in the internal spaces of the cells can also be collected and placed into one compartment. This compartment is separated by the membrane form another compartment which also contains an aqueous phase[1]. Note that the aqueous phases in both compartments in general have different composition in terms of solutes and hence have different physico-chemical properties. Therefore, in physical chemistry, these aqueous solutions are considered as two *different phases*. In biology, however, they are considered as representatives of the same phase but being present in *different compartments*. The reader should be aware of this conceptual difference between phases and compartments.

The thermodynamic system thus constructed is totally isolated from its surroundings and comprises all elements of the experimental system. Neither energy nor matter need to be supplied from the outside world during the experiment. The walls between the reservoirs and the compartments are selectively permeable to heat and mechanical energy. Note that the heat formed by the stirring of the sample is absorbed by the heat reservoir. Due to the barostatic effect of the atmosphere a constant pressure throughout the system can be assumed. The aqueous phases are homogeneous, *i.e.* temperature, composition, and electrical potential are the same throughout a given compartment. In the bathing solution this is achieved by stirring. The internal spaces of the cells are obviously not stirred. However, due to the small dimension of these spaces, diffusion of matter and heat are usually much faster than the processes which cause a change in composition and give rise to a production of heat. Hence the compartment representing the combined internal spaces can be treated as if it were stirred. It should be kept in mind, however, that in a thin layer adjacent to a charged surface

---

[1] This is not the only way how a muscle can be translated into elements of a thermodynamic system. Depending on how detailed the description should be, additional compartments separated by membranes from the compartment representing the inner space of the muscle could be added which would correspond to the internal space of the mitochondria and the sarcoplasmic reticulum. Alternatively, the whole muscle could be represented by one compartment and the membrane separating it from the bathing solution would then represent the connective tissue called the epimysium. However, elements and organelles of the cell which are not involved in the processes of interest (*e.g.* DNA in the nuclei or the GOLGI apparatus) need not to be included as additional phases or compartments.

homogeneity is not necessarily found. These interfacial domains play an important role if membrane phenomena are investigated, but shall be neglected here. Examples where interfacial domains are essential will be discussed in the subsequent chapter.

3.   Thermodynamic functions and the first law

The state of a system is uniquely defined by its internal energy, E_{int}, which is the sum of all sorts of energies present such as mechanical energy, heat, or chemical energy. In particular, E_{int} includes the electrical energy arising from charges and electrical potentials. The internal energy E_{int} (S, V, l, n_i) is a unique function of the variables volume V, elongation l (which is taken as a measure for perfoming mechanical work), chemical composition as indicated by the mole numbers n_i of the i th species, and a thermodynamic function called entropy S [1-4]. It is important to note that E_{int} and S are thermodynamic potentials, *i.e.* their value is uniquely defined for a given state of the system independent of the path on which this state was reached. In a thermodynamic system like that depicted in Fig. 1B, the internal energy $E_{int,k}$ can be defined separately for the kth element of the system (the elements being the reservoirs, the compartments and the membrane). The internal energy of the whole system is then given by

$$E_{int} = \sum_k E_{int,k}(S_k, V_k, l_k, n_{i,k})$$ (1)

where the sum has to be taken over all elements.

The processes occurring in a system cause a transition of the system from one state to another. In the following discussion this transition is dissected into small incremets which allows to use differentials for the variations of the parameters of the system. The change in internal energy $dE_{int,k}$ of a given element of the system due to such an incremental transition amounts to[2]

$$dE_{int,k} = TdS_k - pdV_k + X_k dl_k + \sum_i \tilde{\mu}_{i,k} dn_{i,k}$$ (2)

where p and T denote, respectively, the pressure and the absolute temperature in the system which are both constant as discussed in the preceding section. When writing equation (2) the electrochemical potential $\tilde{\mu}_{i,k}$ of the i th species in the k th element was introduced which is defined as the partial derivative of $E_{int,k}$ with respect to $n_{i,k}$ at constant $S, V, l, n_{i,k}$ (in all elements)

---

[2] The sign convention used in physics has been adopted here. According to this convention, work performed by an element of the system and heat withdrawn from it are counted negative, while work and heat absorbed by an element of the system are counted positive.

$$\tilde{\mu}_{i,k} = \left[\partial E_{int,k} / \partial n_{i,k} \right]_{S,V,l,n_{j,k}} \tag{3a}$$

and the mechanical force[3] X_k (arising from a mechanical potential) similarly defined as

$$X_k = \left[\partial E_{int,k} / \partial l_k \right]_{S,V,l,n_{j,k}} \tag{3b}$$

Again, $\tilde{\mu}_{i,k}$ is a thermodynamic potential like E_{int} and S.

The first law of thermodynamics states that energy cannot be created or annihilated but only converted from one form into another. Since the total system is isolated (see Section 2) it does not exchange energy or matter with its surrounding, therefore, E_{int} = constant or dE_{int} = 0 for an incremental transition. It then follows from equations (1) and (2) that

$$\sum_k \left[TdS_k - pdV_k + X_k dl_k + \sum_i \tilde{\mu}_{i,k} dn_{i,k} \right] = 0 \tag{4}$$

Since the total volume of the system is constant, $\sum_k p\, dV_k = 0$. No change in chemical composition occurs in the reservoirs shown in Fig. 1B, i.e. $dn_{i,4} = dn_{i,5} = 0$. Moreover, $dl_k \neq 0$ only for $k = 2$ and $k = 5$. Hence, for the system in Fig. 1B, equation (4) reads

$$T\sum_k dS_k + \sum_{k=1}^{3} \sum_i \tilde{\mu}_{i,k} dn_{i,k} + X_2 dl_2 + X_5 dl_5 = 0 \tag{4a}$$

If temperature and pressure are constant, a new thermodynamic function, G, called free enthalpy (or GIBBS free energy) can be defined as

$$G = E_{int} - TS + pV \tag{5}$$

This quantity measures the system's capability to perform useful work. Its change in the kth element of the system becomes [cf. equation (2)]:

$$dG_k = dE_{int,k} - Td\,S_k + pdV_k = X_k\,dl_k + \sum_i \tilde{\mu}_{i,K} dn_{i,k} \tag{6}$$

It is important to realize that dG_k in equation (6) merely indicates how much the free enthalpy changes in a transition. The actual fate of dG_k is considered in the following section.

---

[3] The force X_k and the elongation l_k as well as the velocity v_k defined as the time derivative of l_k are vectors. In the present context these vectors are parallel, hence a scalar notation can be used for these quantities.

4.   The uniqueness of entropy and the second law

Among the thermodynamic functions entropy has a unique feature expressed by the second law of thermodynamics. This law states that the change in entropy of a system has to be positive (or zero in a special case, see Section *8.1*) when the system changes form one state to another. Rearranging equation (4a) and introducing equation (6) yields

$$T d S_{tot} = T \sum_k d S_k = -X_2 d l_2 - X_5 d l_5 - \sum_{k=1}^{3} \sum_i \bar{\mu}_{i,k} d n_{i,k} = -d G_{tot} \geq 0 \tag{7}$$

where $d S_{tot}$ and $d G_{tot}$ are the change in entropy and free enthalpy, respectively, for the total system. Thus, the mandatory increase in entropy is covered by a decrease in free enthalpy.

It is possible to further analyze the entropy changes $d S_k$ for each element of the system [1-4]. It is then recognized that $d S_k = \delta_{ex} S_k + \delta_{in} S_k$, where $\delta_{ex} S_k$ is called the *exchange* contribution and $\delta_{in} S_k$ the *internal* contribution[4]. The exchange contribution is mandatory and is present even if the transition of the system is conceptually made to occur from one equilibrium state (or very close to it) to another, *i.e.* if all changes in the system associated with the transition are brought about by so-called reversible processes. In contrast, the internal contribution arises only if irreversible processes are involved in the transition. It is found that $\sum_k \delta_{ex} S_k = 0$, which means that $d S_{tot} = 0$ for a reversible transition. Thus, no free enthalpy has to be spent for an entropy change [cf. equation (7)] in a reversible transition and all can in principle be used in energy conversion by a suitable device. However, reversible transitions are hypothetical limiting cases virtually never realized in practice because they imply that the processes associated with them occur at an infinitely small rate. Processes of interest proceed at finite rates and therefore always give rise to an expenditure of free enthalpy and an increase in entropy which amounts to $d S_{tot} = \sum_k \delta_{in} S_k$. As a consequence, the free enthalpy available for the energy conversion is always less in real processes than expected from the corresponding hypothetical reversible transitions.

5.   The dissipation function

The change in entropy $d S_{tot}$ in equation (7) includes the contribution $X_5 d l_5$ arising from the reservoir for mechanical work. This contribution is of no relevance to the processes we are interested in. Therefore, in what follows, we consider only the en-

---

[4] Note that exchange and internal entropy change $\delta_{ex} S_k$ and $\delta_{in} S_k$ are not total differentials, in contrast to their sum $d S_k$

tropy change dS due to the processes of interest, which is equivalent to singling out the two-compartment system shown in Fig. 1C from the total thermodynamic system. When taking the time derivative of dS, it follows from equation (7) that

$$\Phi = TdS/dt = - X_2 dl_2/dt - \sum_{k=1}^{3} \sum_i \tilde{\mu}_{i,k} dn_{i,k}/dt \geq 0 \qquad (8)$$

The quantity Φ is called *dissipation function* and indicates the rate of loss (or dissipation) of free enthalpy due to irreversible processes in the two-compartment system.

In order to determine the dissipation function in equation (8) at any state of the system in the course of an experiment, it suffices to estimate the electrochemical potentials $\tilde{\mu}_{i,k}$ and the pertinent time derivatives d$n_{i,k}$/dt for the constituents of the system. No knowledge of the actual processes which take place in the system and cause d$n_{i,k}$/d$t \neq 0$ is required. Hence, these processes may be defined in any convenient form with the only limitation that a given set of definitions has to comply with the dissipation function. In what follows, it is assumed that the processes in the membrane phase [index $k = 3$ in equation (8)] have reached a steady state (or, strictly speaking, a pseudo-steady state, see Section 8) so that d$n_{i,3}$/d$t \approx 0$. Then only the processes occurring in the two compartments of Fig. 1C have to be considered.

6.   Flows and forces for chemical reactions, transport and mechanical processes

The jth chemical reaction, which occurs in compartment k, is given the index j,k in order to distinguish it from other reactions. It converts the initial reactants[5] $S_{s(j,k)}$ ($s = 1,2, \ldots$) with stoichiometric coefficients $v_{Ss(j,k)}$ into final reactants $P_{p\ (j,k)}$ ($p = 1,2 \ldots$) with stoichiometric coefficients $v_{Pp(j,k)}$ and *vice versa:*

$$\sum_s v_{Ss(\ j,k)} S_{s(\ j,k)} \rightleftharpoons \sum_p v_{Pp(\ j,k)} P_{p(\ j,k)} \qquad (9)$$

The notation *initial* and *final* reactants introduces the (arbitrarily chosen) *positive direction for the reaction* when going from initial to final reactants. Mass balance imposes a strict relation on the mole numbers, $n_{Ss(j,k)}$ and $n_{Pp(j,k)}$, of initial and final reactants of a reaction, respectively. It is most conveniently expressed by means of a quantity called degree of advancement, $\xi_{j,k}$, defined as [1]

$$\xi_{j,k} = \left[n_{Ss(j,k)}(0) - n_{Ss(j,k)}(t) \right] / v_{Ss(j,k)} = \left[n_{Pp(j,k)}(t) \cdots n_{Pp(j,k)}(0) \right] / v_{Pp(j,k)} \qquad (10)$$

where the arguments t and 0 indicate the mole numbers, respectively, at time t and at

---

[5] Initial and final reactants are also called *substrates* and *products* of the reaction, respectively.

the beginning ($t = 0$) of the experiment. Note that the expression for initial reactants in equation (10) becomes identical with that for final reactants if the following sign convention is introduced: *the stoichiometric coefficients for initial reactants have a negative sign and those for final reactants have a positive sign.* Taking the time derivative of $\xi_{j,k}$ in equation (10) yields a measure for the rate of the chemical reaction. It is called the flow, $J_{j,k}$, of the jth reaction in compartment k and becomes

$$J_{j,k} = d\xi_{j,k} / dt = (dn_{R_{r(j,k)}} / dt) / \nu_{R_{r(j,k)}} \tag{11}$$

where the symbol $R_{r(j,k)}$ ($r = 1,2,\cdots$) now stands for both initial and final reactants which need no longer be distinguished if the above sign convention is used. When introducing the time derivatives for the mole numbers in equation (11) into equation (8) the corresponding $\tilde{\mu}_{R_{r(j,k)}}$ can be collected and used to define a quantity called the affinity [1], $\mathcal{A}_{j,k}$, of the jth reaction in compartment k :

$$\mathcal{A}_{j,k} = -\sum_r \nu_{R_{r(j,k)}} \tilde{\mu}_{R_{r(j,k)}} \tag{12}$$

The sum in equation (12) has to be taken over all reactants with stoichiometric coefficients according to the sign convention mentioned above. The quantity $\mathcal{A}_{j,k}$ is an example of a thermodynamic force.

When dealing with transport of species across the membrane, it is necessary to arbitrarily choose a *positive direction of transport which has to be the same for all transport processes.* Here the direction from compartment 1 to compartment 2 (see Fig. 1C) is chosen as positive. By means of the mole numbers $n_{i,1}$ and $n_{i,2}$ of the ith species in compartment 1 and 2, respectively, the flow[6] J_i for the transport of this species can be defined as

$$J_i = - dn_{i,1} / dt = dn_{i,2} / dt \tag{13}$$

where the second part of equation (13) aries from mass balance. Introducing the time derivatives into equation (8) again shows that the $\tilde{\mu}_{i,k}$'s can be collected and used to define the thermodynamic force for transport[7], $\Delta\tilde{\mu}_i$, which is the difference in electrochemical potential between the compartements:

$$\Delta\tilde{\mu}_i = \tilde{\mu}_{i,1} - \tilde{\mu}_{i,2} \tag{14}$$

The flow of the kth mechanical process, J_k, is simply the velocity, v_k, determined

---

[6] In contrast to the flow of a chemical reaction, which is a scalar, the flow of a transport process is a vector. The direction of this vector is defined by the chosen positive direction for transport and is perpendicular to the membrane surface. Hence a scalar notation for this flow can be used.

[7] It is customary in non-equilibrium thermodynamics to use $\Delta\tilde{\mu}_i$ and $\mathcal{A}_{j,k}$ as the thermodynamic force for transport and chemical reaction, respectively. Some authors consider a transport process also as a chemical reaction which converts the ith species in compartment 1 with a stoichiometric coefficient of -1 into the species in compartment 2 with a stoichiometric coefficient of +1. Accordingly, the affinity $\mathcal{A}_i = \Delta\tilde{\mu}_i$ is assigned to the transport process. Although formally correct this notion is prone to confusion and should be avoided.

by the time derivative of the pertinent length,

$$J_k = v_k = dl_k/dt \tag{15}$$

Since work performed by an element is counted negative [see footnote 2 and equation (3b)], the force for the kth mechanical process is

$$X_k = -F_k \tag{16}$$

where F_k denotes the external mechanical force *exerted* on the element.

7.   Flows and forces in the dissipation function; flow-force relations

Chemical reactions and transport processes are connected by the effect of their respective flows on the mole number of the species involved. Thus, for the ith species which is transported and takes part as reactant $R_{r(j,1)}$ in the jth chemical reaction in compartment 1 and as reactant $R_{r(j',2)}$ in the j'th chemical reaction in compartment 2,

$$dn_{i,1}/dt = -J_i + \sum_j v_{Rr(j,1)} J_{j,1} \tag{17a}$$

and

$$dn_{i,2}/dt = J_i + \sum_{j'} v_{Rr(j',2)} J_{j',2} \tag{17b}$$

The changes in mole number given by equations (17) have to be identical with those in the dissipation function, and the chosen set of definitions of transport processes and chemical reactions is then appropriate. Introducing equations (15)–(17) into the dissipation function [see equation (8)] written for compartments 1 and 2 (since $dn_{i,3}/dt \approx 0$, see above) yields upon rearranging and in view of equations (12) and (14)

$$\Phi = \sum_i J_i \Delta\bar{\mu}_i + \sum_{k=1}^{2} \sum_j J_{j,k} A_{j,k} + v F \geq 0 \tag{18}$$

where the sums have to be taken over all transported species and all chemical reactions occurring in both compartments. The index 2 for the mechanical process was omitted in equation (18) because there is only one such process (*i.e.* the muscle in the special holder). The force F is the the tension (usually denoted by P) applied to the muscle which tends to stretch the muscle, while the flow v is the rate of expansion of the muscle (which is equal to the negative rate of contraction). Note, however, that it has become customary in muscle research to define contraction as the positive direction of the mechanical process which causes the force to be the negative tension.

When using the general notation J_p and X_p for the flow and the conjugate force, respectively, of the pth process (which is a transport, a chemical reaction, or a mechanical process) equation (18) simply reads

$$\Phi = \sum_p J_p X_p \geq 0 \qquad\qquad (18a)$$

It is then evident that the dissipation function is a sum of products of flows and conjugate forces[8]. The flow of a process has the same sign as its force when it runs *downhill*, *i.e.* when it occurs spontaneously, and the product of flow and conjugate force is positive. If flow and conjugate force have opposite signs the flow-force product is negative and the process is driven *uphill* by another process (or processes). This is the thermodynamic expression of the energy conversion due to some form of coupling between processes. It is possible because only the sum of all flow-force products has to be positive according to equation (18a) and not each product by itself. Thus, a muscle can contract against an external force ($v\,F < 0$) if this mechanical process is coupled to a chemical reaction, *e.g.* the hydrolysis of ATP. The flow-force product of this reaction has then to be larger than $-v\,F$.

The above conclusion was arrived at in most general terms without considering any mechanisms by which transport or chemical reactions are brought about. This is the great advantage of thermodynamics but, at the same time, its severe limitation. Thermodynamics cannot tell how flows are related to their conjugate forces and, in the case of energy conversion due to coupling, to the forces of other processes. Answers to this question can in general only be obtained on the basis of molecular or kinetic schemes, as extensively discussed in Ref. 5. It is found that the flow-force relation of a process is in general ambiguous. However, if certain constraints apply, it can be converted to an unambiguous relation which has the form of a hyperbolic tangent (see Fig. 2.). It is evident from Fig. 2 that, in a limited range of force located around the inflection point marked by X_o, the dependence of the flow of the ith process on the conjugate force can be approximated by a linear relation

---

[8] The quantities flow and force should be understood in a generalized sense. Flows comprise material transport and conversion defined as mole/time, as well as velocities in the physical sense defined as length/time. Similarly, forces comprise thermodynamic forces defined as energy/mole = mass x lenght^2/[time^2 mole] as well as mechanical forces defined as mass x lenght/time^2. The flow-force product of any process yields its power defined as energy/time = mass x length^2/time^3.

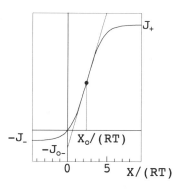

FIG. 2. Flow-force relation of a process under constraint. The dependence of the flow J (in arbitrary units) on the conjugate force X of a process follows a hyperbolic tangent function (solid line) whose parameters are determined by the extreme flows J_+ and J_- as explained in Ref. 5. The broken line shows a linear approximation of the flow-force relation valid in the force range of about ± 1.75 kJ/mol around the inflection point whose location is marked by X_0. J_0 indicates the intercept of the linear approximation on the J-axis.

$$J_i = L_i X_i + J_{oi} \qquad (19)$$

Here J_{oi} corresponds to the intercept of the linear approximation on the J-axis shown in Fig. 2. In the case of the jth pair of coupled processes,

$$J_{j,1} = L_{j,11} X_{j,1} + L_{j,12} X_{j,2} + J_{oj,1} \qquad (20a)$$

$$J_{j,2} = L_{j,21} X_{j,1} + L_{j,22} X_{j,2} + J_{oj,2} \qquad (20b)$$

The coefficents L are called generalized permeabilities of the processes and comprise the rate constants of the processes [5]. The terms J_o can vanish under certain conditions [5]. If they are finite, they cause an offset of the flows from the origin, which immediately shows that such relations can be valid only in the range of forces not encompassing the points $X_i = X_{j,1} = X_{j,2} = 0$ since all flows have to vanish at equilibrium (see Section 8.1). Nevertheless, many experimental systems were found to comply with such relations even with $J_o = 0$ and the symmetry condition $L_{j,12} = L_{j,21}$ [4]. In particular, muscle contraction often obeys linear flow-force relations over a wide range of forces, and a self-regulatory process was therefore proposed which should cause the extended linearity [3,4]. It thus appears that the phenomenological relations given in equations (19) and (20) can serve as useful approximations, as postulated by the formalism of linear non-equilibrium thermodynamics. However, they cannot be taken for granted, and their validity has to be checked carefully for every new system or whenever the conditions for a known system are altered.

8.   Steady states

Steady states are states in which the system has become stationary in the sense that its parameters are no longer time-dependent. Under certain circumstances the time-dependence does not vanish but it becomes negligibly small, in which case the state is called a pseudo-steady state. In general the parameters of a system can be constrained by clamping, which causes at least some of the forces to be clamped, too. If for technical reasons the forces cannot be totally clamped, it may nevertheless be possible to keep them approximately constant, *i.e.* to allow the system to reach a pseudo-steady state, by choosing appropriate chemical capacitances. The chemical capacitance[9] of a particular compartment for a given species relates the change in mole number to the change in concentration of that species in the particular compartment. From this definition it can be shown that the chemical capacitance is just the volume of the compartment which is formally increased when a buffering system for the species is added [5]. A clamping of flows is in general not feasible, except for the flow of the mechanical process. The special holder may be constructed such that it imposes the velocity v on the muscle, the easiest case being $v = 0$ due to a rigid holder.

If all forces are fixed, *i.e.* the maximum number of constraints is applied, the steady state is fully defined, since no more degrees of freedom are left. If some of the forces are fixed, the remainder will reach values in the steady state such that their conjugate flows become zero. For a pair of coupled processes this steady state is known as *static head* (abbreviated by *s. h.*) and is characterized by the maximal absolute value of the non-clamped force attainable under the given conditions. If no constraints at all are applied, the forces will all tend to decrease until the system eventually reaches the equilibrium state (see Section 8.1).

A vanishingly small change in the parameters of a compartment can also come about if the sum of the flows given on the right-hand side of equation (17) is close to zero. This happens if the given compartment has considerably smaller capacitances for the species present than the other compartment(s) in the system. A particular element which always statisfies this condition is a membrane which reaches a steady state (actually a pseudo steady state) long before the other elements of the system. Then the flows of charged species are restricted by

$$\sum_i z_i J_i = 0 \qquad\qquad (21)$$

where z_i stands for the electrical charge (or the electrochemical valence) of the ith species. Equation (21) arises from the condition of electroneutrality in the compart-

---

[9] Note that the term chemical capacity is frequently used instead of the correct notation chemical capacitance.

ments and can serve to estimate the difference, $\Delta\varphi$, between the electrical potentials, φ_k, in the two compartments separated by the membrane[10]. Thus, if only univalent ions can permeate through the membrane whose transport is described by the integrated NERNST-PLANCK equation [5], one arrives at the relation known as the GOLDMAN equation:

$$\Delta\varphi = \varphi_1 - \varphi_2 = (RT/\mathcal{F}) \ln \left[\sum_i P_i a_{i,2} + \sum_j P_j a_{j,1} \right] / \left[\sum_i P_i a_{i,1} + \sum_j P_j a_{j,2} \right]$$

for $z_i = 1, z_j = -1$ \hfill (22)

Here R and $\mathcal{F}$ are the gas constant and the FARADAY constant, respectively. P_i and $a_{i,k}$ denote the permeability and the activity [see equation (24)] of the ith species in the kth compartment, respectively, while φ_k is the electrical potential in that compartment.

8.1   The equilibrium state

This state is a special steady state characterized by $\Phi = 0$ with the corollary that all processes have ceased. It also implies that all flows vanish because all forces of unconstrained processes are zero. The latter condition allows to introduce relations pertinent to the equilibrium state. Before doing so, the electrochemical potential of the ith species in the kth compartment or phase is expanded into three terms

$$\tilde{\mu}_{i,k} = \mu^o_{i,k} + RT\ln\{a_{i,k}\} + z_i \mathcal{F}\varphi_k$$ (23)

The standard chemical potential, $\mu^o_{i,k}$ is a substance-specific constant which also depends on the type of phase (hence index k is retained) and, in general, on pressure and temperature. The next term in equation (23) comprises the activity of the species which is related to its concentration by

$$a_{i,k} = y_{i,k} c_{i,k}$$ (24)

where the factor $y_{i,k}$, called the activity coefficient, expresses the interaction of the ith

---

[10] The quantity $\Delta\varphi$ is commonly known as the *membrane potential*. However, this notation is incorrect because $\Delta\varphi$ is in general a potential *difference*. Since it is customary to choose one of the compartments in an experimental system as the reference point for potential measurements the electrical potential in that compartment is assumed to be zero, e.g. $\varphi_1 = 0$. the potential difference $\Delta\varphi$ is then equal to $-\varphi_2$ [see equation (22)] which led to the incorrect notation membrane potential.

species with all other constituents of the phase. The last term in equation (23) expresses the contribution of the electrical potential of the phase to $\tilde{\mu}$.

8.2   Equilibrium constants

The vanishing force for transport at equilibrium means $\Delta\tilde{\mu} = 0$ or $\tilde{\mu}_{i,1} = \tilde{\mu}_{i,2}$ [see equation (14)]. This condition can be generalized to the statement that, at equilibrium, $\tilde{\mu}_i$ has the same value in all phases and compartments into which the ith species can move. The qualification of movement in this statement is important, although at equilibrium nothing moves. An element in the system which sets a restriction to the movement of the species (*e.g.* an impermeable membrane) may cause different electrochemical potentials at equilibrium in the phases or compartments separated by the restrictive element. It is therefore important to specify which possible constraints exist in a given equilibrium state [6].

The condition of equal electrochemical potentials applied to the transport of the ith species between two phases (or compartments) with index k and k' yields, by virtue of equation (23),

$$K_{pi,kk'} = \exp\{(\mu^\circ_{i,k} - \mu^\circ_{i,k'})/RT\} = \left[a_{i,k'}/a_{i,k}\right]_{eq} \exp\{z_i \mathcal{F}(\varphi_{k'} - \varphi_k)/RT\} \quad (25)$$

The quantity $K_{pi,kk'}$ is the equilibrium constant for partitioning of a species called the *partition coefficient*. It relates the activities of the species in the two phases at equilibrium. Note that these activities are affected by a possible difference in electrical potential between the phases. Obviously, $K_{pi,kk'} = 1$ for partitioning of a species between two equal phases (*e.g.* between the aqueous phases in two compartments).

The condition $\mathcal{A}_{j,k} = 0$ for the jth chemical reaction in compartment k at equilibrium is transformed by means of equation (23) into

$$K_{cj,k} = \exp\{-\Delta G^\circ_{j,k}/RT\} = \prod_r \left[a_{Rr(j,k)}^{\nu_{Rr(j,k)}}\right]_{eq} \quad (26a)$$

where the standard free enthalpy of the reaction (usually called the standard Gibbs free energy), $\Delta G^\circ_{j,k}$, is defined as

$$\Delta G^\circ_{j,k} = \sum_r \nu_{Rr(j,k)} \mu^\circ_{Rr(j,k)} = \Delta H^\circ_{j,k} - T\Delta S^\circ_{j,k} \quad (26b)$$

It can be related to the standard enthalpy and the standard entropy of the reaction, $\Delta H^\circ_{j,k}$ and $\Delta S^\circ_{j,k}$, respectively. $K_{cj,k}$ is the equilibrium constant[11] for the reaction which determines the ratio of the reatants' activities at equilibrium, as indicated by the mass

---

[11] Note that equilibrium constants (if not dimensionless) have to be given in concentration units.

action ratio on the right hand side of equation (26a).

It has to be stressed that equilibrium constants which emerge from standard chemical potentials are always valid irrespective of the state of the system. The term equilibrium in their name merely refers to the fact that they relate the activities of reactants at equilibrium. Therefore, they can be used to cast the relations for the thermodynamic force of chemical reactions and transport [cf. equations (12) and (14)] into a form which contains the activities of the reactants explicitly. In view of equations (23, (25) and (26a) it then follows that for a chemical reaction:

$$\mathcal{A}_{j,k} = RT \ln\{K_{c,j,k} / [\prod_r a_{R_r(j,k)}{}^{\nu_{R_r(j,k)}}]\} \tag{27}$$

For the transport of the ith species from the k'th phase (or compartment) to the k'th phase (or compartment)

$$\Delta\bar{\mu}_{i,kk'} = RT \ln\{K_{pi,kk'} \cdot a_{i,k} / a_{i,k'}\} + z_i \mathcal{F}(\varphi_k - \varphi_{k'}) \tag{28}$$

which reduces to

$$\Delta\bar{\mu}_i = RT \ln\{a_{i,1} / a_{i,2}\} + z_i \mathcal{F} \Delta \cdot \varphi \tag{28b}$$

for the transport of the i th species between the acqueous phases of compartment 1 and 2 ($K_{pi,kk} = 1$, $\varphi_1 - \varphi_2 = \Delta\varphi$). As it should be, equation (28b) is also obtained upon insertion of equation (23) into equation (14).

9.   Efficiency and efficacy of energy conversion

The *efficiency*, η, of energy conversion between two coupled processes can be expressed by means of the flow-force product of the driven process (index *out* for output) and the driving process (index *in* for input) [3,4]. It is defined as

$$\eta = - J_{out} X_{out} / (J_{in} X_{in}) \tag{29}$$

The negative sign in equation (29) is introduced in order to obtain a positive efficiency since for the driven process $J_{out} X_{out} < 0$ (see Section 7). The efficiency η relates the output power to the input power, and represents the fraction of the entropy production of the driving process which is consumed by the driven process due to energy conversion. It should be understood that the feature of driven and driving process is not inherent in the system but depends on the conditions imposed on it (clamping of forces, chemical capacitances of species involved *etc.*).

It the force of the driven process is not clamped static head is attained in which case $J_{out} = 0$ (see Section 8) and, according to equation (29), $\eta = 0$. A more appropriate quantity for this case is the *force efficacy* [3,4] ε_X defined as

$$\varepsilon_X = - X_{out} / (J_{in} X_{in})$$
(30a)

It measures how much input power is needed to support the maximal force of the driven process and is also appropriate to assess the performance of a muscle under the condition of isometric contraction. In the not very frequent case of a tight coupling between the energy-converting processes the flow of the driving process also vanishes at static head, and ε_X may then be replaced by the force ratio $[X_{out} / X_{in}]_{s.h.}$ which is equal to the molecular stoichiometry of the coupling.

If the force of the driven process is clamped to zero the steady state attained is called *level flow* and characterized by the maximal absolute value for the flow of the driven process. The appropriate quantity for this case is the *flow efficacy* [3,4] ε_J defined as

$$\varepsilon_J = J_{out} / (J_{in} X_{in})$$
(30b)

It indicates how much input power is needed to sustain the maximal flow of the driven process and can be used to assess the performance of a muscle contracting against no external force.

Since in general the positive direction for the coupled process may be chosen independently (except for two coupled transport processes), and depending on the molecular mechanism, the coupling between the processes can be positive or negative. For positive coupling J_{in} and J_{out} have the same sign which causes opposite signs for X_{out} and X_{in}, while for negative coupling the forces have the same sign and the flows have opposite signs. Hence, ε_X and ε_J can be positive or negative according to the sign of the coupling.

To end this survey on efficiency of energy conversion the idealized case of reversible transitions shall be considered. As mentioned in Section 4, no free enthalpy is then to be spent on entropy production and all can be used in energy conversion. However this has to occur at infinitely small rates which means that $J_{in} \approx 0$ and $J_{out} \approx 0$. Therefore, despite of the maximal conversion in terms of forces, the efficiency of a reversible energy conversion is zero or, in other words, no power can be gained from such a conversion. Needless to say that no biological system could survive under such conditions.

References

[1] I. PRIGOGINE, *Introduction to Thermodynamics of Irreversible Processes,* Interscience Publishers, New York (1967).

[2] A. KATCHALSKY and P.F. CURRAN, *Non–equilibrium Thermodynamics in Biophysics,* Harvard University Press, Cambridge, MA (1965).

[3] S.R. CAPLAN in *Current Topics in Bioenergetics,* D.R. SANADI (Editor), Academic Press, New York (1971) Vol. **4**, pp. 2-77.

[4] S.R. CAPLAN and A. ESSIG, *Bioenergetics and Linear Non–equilibrium Thermodynamics: The Steady State,* Harvard University Press, Cambridge, MA (1983).

[5] D. WALZ, *Biochim. Biophys. Acta,* **1019**, 171 (1990).

[6] D. WALZ and S.R. CAPLAN, *Biochim. Biophys. Acta,* **859**, 151 (1986).

THERMODYNAMICS OF IRREVERSIBLE PROCESSES
APPLIED TO BIOLOGICAL SYSTEMS: SELECTED EXAMPLES

DIETER WALZ

Biozentrum, University of Basel,
Basel (Switzerland)

Contents

Bioelectrochemistry IV
Edited by B.A. Melandri *et al.*, Plenum Press, New York, 1994

This chapter is intended to illustrate the general principles presented in the preceding chapter by means of examples. Most examples pertain to the topics discussed in the previous chapters, and the reader is referred to these chapters for further details. Some examples are related to such topics but are taken from a different field.

1.   Non-equilibrium thermodynamic description of a muscle cell

Several authors have investigated the processes occurring in an isolated muscle which is electrically stimulated. An important parameter of the system is the concentration of Ca^{2+} in the myoplasm which changes dramatically with time during and after electrical stimulation (see Chapter 12). Figure 1A shows a sketch of a part of a muscle cell which includes all relevant elements. This experimental system is translated into the thermodynamic system depicted in Fig. 1B. The compartments numbered 1 to 3 represent, respectively, the myoplasm, the lumen of the sarcoplasmic reticulum, and the suspending medium. The membrane separating compartments 1 and 2 is the sarcoplasmic reticulum, while that between compartments 1 and 3 is the membrane of the muscle cell.

In order to assess the transport processes between compartments shown in Fig. 1B a positive direction has to be chosen (see Section 6 in the preceding chapter). Here it is chosen to be the direction out of compartment 1, *i.e.* transport from compartment 1 to 2, and from comparment 1 to 3 is positive. Accordingly, there are the Ca^{2+}-flows $J_{Ca,1}$ and $J'_{Ca,1}$ (see Fig. 1B) associated with the Ca^{2+}-ATPase in the sarcoplasmic reticulum and the cell membrane, respectively (see Fig. 1A). Similarly, the flows $J_{Ca,2}$ and $J'_{Ca,2}$ arise from the gated Ca^{2+}-pores in the two types of membranes, while the flow $J'_{Ca,3}$ is mediated by the Ca^{2+}, Na^+-exchanger in the cell membrane. Moreover, there are the flows $J_{Ca,4}$ and $J'_{Ca,4}$ due to the passive permeability of both membranes. Ca^{2+} is also involved in several binding reactions. When choosing the association as positive direction (see Section 6 in the preceding chapter) these chemical reactions are represented by the general scheme

$$Ca^{2+} + B \underset{\leftarrow}{\overset{\rightarrow}{}} Ca^{2+}.B \qquad\qquad (1)$$

The pertinent flows are denoted by J_p, J_T, and J_D for binding of Ca^{2+} to parvalbumin, to troponin C, and to the fluorescent dye, respectively (cf. Fig. 1). Besides these flows of Ca^{2+} there exist flows for transport of other species (Na^+, Cl^- *etc.*) and for chemical reactions (*e.g.* ATP-synthesis and ATP-hydrolysis, binding of Mg^{2+} ions to parvalbumin, *etc.*) which are omitted here since interest is focussed on Ca^{2+} movements.

According to equations (17) in the preceding chapter these flows cause the mole number $n_{Ca,k}$ of Ca^{2+} ions in compartment k to change with time. Since the pertinent concentration $c_{Ca,k}$ is defined as

$$c_{Ca,k} = n_{Ca,k} / V_k \qquad (2)$$

where V_k is the (constant) volume of the kth compartment $c_{Ca,k}$ also changes with time as shown in the following relations:

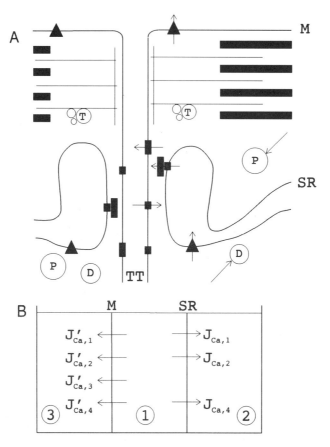

FIG. 1. Sketch of part of a muscle cell (A), and the corresponding thermodynamic system (B). The cell membrane (marked with M in A) has tubular invaginations called transverse tubules (TT). The sarcoplasmic reticulum (SR) forms a tubular structure called terminal cisternae which are adjacent to the transverse tubules. The regular array of thick and thin lines represents the contractile elements (thick and thin filaments with Z line, all not drawn to scale). The Ca^{2+}-ATPase is symbolized by closed triangles, while the closed square and T-shaped symbol represent the gated Ca^{2+}-pores known as the DHP receptor and the ryanodine receptor, respectively. The closed rectangle indicates the Ca^{2+}, Na^{+}-exchanger. Open circles with letters represent parvalbumin (P), troponin C (T) and a fluorescent dye (D) which all bind Ca^{2+}. Arrows indicate the direction of net Ca^{2+}-flows occurring upon electrical stimulation of the muscle. In the thermodynamic system shown in B, the two membranes separate three compartments which represent the myoplasm (1), the lumen of the sarcoplasmic reticulum (2) and the suspending medium (3). The arrows indicate the *positive direction* chosen for the flows of the transport processes. For further explanation see text.

$$dc_{Ca,1}/dt = -[J_{Ca,1} + J'_{Ca,1} + J_{Ca,2} + J'_{Ca,2} + J'_{Ca,3} + J_{Ca,4} + J'_{Ca,4} + J_P + J_T + J_D] / V_1 \quad (3a)$$

$$dc_{Ca,2}/dt = [J_{Ca,1} + J_{Ca,2} + J_{Ca,4}]/V_2 \quad (3b)$$

$$dc_{Ca,3}/dt = [J'_{Ca,1} + J'_{Ca,2} + J'_{Ca,3} + J'_{Ca,4}]/V_3 \quad (3c)$$

The flows in equations (3) are not constant but change with the state of the muscle. Thus, in a resting muscle $J_{Ca,2}$ (and also $J'_{Ca,2}$) is rather small due to a very low permeability of the gated Ca^{2+}-pore. It becomes very large during the stimulation of the muscle and returns to a small value immediately after excitation. Similar changes are observed for all other flows which respond to the tremendous variation of $c_{Ca,1}$ brought about by $J_{Ca,2}$. It should be added that equations (3) are written for the Ca^{2+}-ATPases in a steady state.

It is important to note that a given flow for transport may have rather different effects on the concentrations in the two compartments between which transport occurs. This is true if the volume of the phases in the two compartments are very different. Thus $c_{Ca,3}$ can be assumed as almost constant because the volume of compartment 3 is very much larger than that of compartment 1. Moreover, due to the chosen sign conventions for processes, it is unnecessary to split the net flows included in equations (3) into "forward flows" and "backward flows" which are used in a balance of "input components" and "output (or removal) components" in order to determine the concentration change in a given compartment (cf. chapter 12).

2.   Some aspects of binding reactions

Binding of a ligand L to a component B is an important chemical reaction and therefore worth to be considered in some detail. Examples are the reactions according to equation (1) where the ligand is Ca^{2+} and B is parvalbumin, troponin C, or the fluorescent indicators, which provide the means for estimating Ca^{2+} concentrations (see Chapter 2). Another example is the binding of Ca^{2+} to a buffer such as EGTA (see Section 3.2 in chapter 12) which is used to hold its concentration approximately constant.

2.1.   Thermodynamic parameters of binding

When characterizing binding, it is important to state which direction of the reaction is chosen as positive. Depending on this choice the parameters pertain either to association (indicated by the subscript *b* in Table 1) or to dissociation (subscript *d*). Table 1 also shows the relationship between the two sets of parameters. The last row of Table 1 concerns the notion of "binding affinity" which, although frequently used in biochemistry, is rather unfortunate since it is prone to confusion. It has become cu-

TABLE 1. THERMODYNAMIC PARAMETERS OF BINDING

Parameter	Association is positive direction	Dissociation is positive direction
Reaction	$L + B \underset{\longleftarrow}{\rightarrow} LB$	$LB \underset{\longleftarrow}{\rightarrow} L + B$
Standard free enthalpy [see eqn. (26b) in prec. chapter]	$\Delta G^o_b = \mu^o_{LB} - \mu^o_L - \mu^o_B$ $= -\Delta G^o_d$	$\Delta G^o_d = \mu^o_L + \mu^o_B - \mu^o_{LB}$ $= -\Delta G^o_b$
Equilibrium constant [see eqn. (26a) in prec. chapter]	$K_b = [a_{LB}/(a_L\,a_B)]_{eq}$ $= \exp\{-\Delta G^o_b/(RT)\}$ $= 1/K_d$	$K_d = [a_L\,a_B/a_{LB}]_{eq}$ $= \exp\{-\Delta G^o_d/(RT)\}$ $= 1/K_b$
Thermodynamic affinity [see eqn. (27) in prec. chapter]	$\mathcal{A}_b = RT\ln\{K_b\,a_L\,a_B/a_{LB}\}$ $= -\mathcal{A}_d$	$\mathcal{A}_d = RT\ln\{K_d\,a_{LB}/(a_L a_B)\}$ $= -\mathcal{A}_b$
"Binding affinity" high	$\Delta G^o_b \ll 0 \quad K_b \gg 1M^{-1}$	$\Delta G^o_d \gg 0 \quad K_d \ll 1M$
low	$\Delta G^o_b \gg 0 \quad K_b \ll 1M^{-1}$	$\Delta G^o_d \ll 0 \quad K_d \gg 1M$

stomary to qualify binding with a relatively large *negative value* for ΔG^o_b and a large value for K_b as having a "high affinity". Accordingly binding with a relatively large positive value for ΔG^o_b and a small value for K_b has a "low affinity". This affinity should not be mistaken for the affinity $\mathcal{A}$ which is the thermodynamic force of the reaction (see Section 6 in the preceding chapter). This force is determined by the actual values of the reactants' activities with respect to the equilibrium constant (see Table 1). Hence, irrespective of the binding affinity in the biochemical sense, $\mathcal{A}_b$ or $\mathcal{A}_d$ can attain any value in the range from very large negative to very large positive values.

2.2.   Kinetic parameters of binding

The flow of the binding reaction can be expressed by means of the rate constants k_b and k_d which determine the rate of the unidirectional association and dissociation reaction, respectively [1]. With association chosen as the positive direction for binding the flow J_b becomes (cf. equation 11 in the preceding chapter)

$$J_b = -\,dn_L/dt = -\,dn_B/dt = dn_{LB}/dt = V\,[k_b\,a_L\,a_B - k_d\,a_{LB}] \qquad (4)$$

where V is the volume of the compartment (or phase) in which the reaction takes place. Detailed balancing [1] requires that

$$K_b = k_b/k_d \tag{5}$$

Accordingly, the larger K_b (*i.e.* the higher the "binding affinity" in the biochemical sense) the larger k_b has to be with respect to k_d. However, since the association requires the collision of the two species L and B, there is an upper limit to k_b arising from the rate of diffusion of the species in the given medium. For low molecular species such as Ca^{2+} this limit is of the order of 10^9 M^{-1} s^{-1}. Hence, if K_b attains a value of some 10^8 M^{-1} (in the case of parvalbumin $K_b = 2.5$ x 10^8 M^{-1} or $K_d = 4nM$) k_d has to be of the order of 10 s^{-1}. As a consequence effective binding in such a case sets in at very low concentrations of the ligand and association occurs relatively fast, however, the dissociation of the complex is rather slow. Therefore, a sensitive fluorescent indicator which can detect low concentrations of Ca^{2+} reports an increasing concentration more or less in real time but may substantially lag behind the time course of a decreasing Ca^{2+} concentration. On the other hand an indicator which reports both increasing and decreasing concentrations in real time cannot be very sensitive.

2.3.   *Binding at pseudo-equilibrium*

If the rate constants k_b and k_d are large enough the binding reaction relaxes to a steady state (see Section 8 in the precedent chapter) much faster than all other processes in the system. If neither the component B nor the complex LB are involved in additional processes except binding of L this steady state is an equilibrium state. A changing concentration of L then disturbs this equilibrium state but, owing to the short relaxation time, binding almost instantaneously follows this change in L. As a consequence the affinity $\mathcal{A}_b$ (or $\mathcal{A}_d$) never deviates substantially from zero, and the reaction is said to be at a pseudo-equilibrium which can be satisfactorily approximated by the equilibrium condition listed in Table 1. The fast responding indicator referred to above is an example for such a situation, and a buffer should also fulfil this condition in order to work effectively [1].

An application of pseudo-equilibrium pertains to diffusion of a ligand in the presence of a binding component B. For the present purpose it suffices to assume a unidirectional gradient of L which occurs in the direction of a space coordinate x. Moreover, activity coefficients [see equation (24) in the preceding chapter] are set to unity. In this case the diffusional flow of L, J_L, is given by

$$J_L(x) = - A D_L \, dc_L/dx \tag{6}$$

where D_L is the diffusion coefficient for L, and A denotes the area of a plane perpendicular to x. In the slab between two such planes located at x and $x + dx$ the three species L, B, and LB are present at the concentrations $c_L(x)$, $c_B(x)$, and $c_{LB}(x)$, respectively, which are subject to the mass balances

$$c_{L,\,tot}(x) = c_L(x) + c_{LB}(x) \text{ and } c_{B,\,tot}(x) = c_B(x) + c_{LB}(x) \tag{7}$$

If $c_{B,\,tot}(x) \gg c_{L,tot}(x)$ holds true, $c_{LB}(x) \ll c_{B,tot}(x)$ and $c_B(x) \approx c_{B,tot}(x)$. It then follows from equation (7) and the pseudo-equilibrium condition that

$$c_L(x) = c_{L,tot}(x)/[1 + K_b\,c_{B,tot}(x)] \tag{8}$$

Introducing the space derivative of c_L calculated from equation (8) into equation (6) yields

$$J_L(x) = -A\,D\,'_L\,dc_{L,tot}/dx \tag{9a}$$

where D'_L is an apparent diffusion coefficient which is related to D_L by

$$D'_L = D_L/[1 + K_b\,c_{b,tot}(x)] \tag{9b}$$

Thus, if binding of L to B occurs, diffusion of total L appears to be slowed down because the concentration of free L is reduced (see also Chapter 12); the diffusion of free L, however, is not affected. Note that $c_B(x) \approx$ const and $c_{B,tot}(x) \approx$ const due to the condition $c_{B,tot}(x) \gg c_{L,tot}(x)$, but $c_{LB}(x)$ forms a gradient in the same direction as that of $c_L(x)$. The diffusion of LB down this gradient can be neglected if D_{LB} is assumed to be much smaller than D_L. In the general case all these simplifications do not apply and the apparent diffusion coefficient for L also depends on D_B and D_{LB}.

3.   Interfacial domains and the electric field in membranes

The phases in a thermodynamic system are considered to be homogeneous, *i.e.* composition and electrical potential are the same within a given phase (see Section 2 in the preceding chapter). Strictly speaking, this is only true for the bulk of a phase but not for a very thin layer on its boundary. In this layer, which is called interfacial domain, the electrical potential and the concentration (or activity) of charged species are in general not constant. Interfacial domains are essential if electrical phenomena in biological membranes are investigated because they have an effect on both the concentration of charged species at the membrane surface and the electric field within the membrane.

3.1   *Diffuse double layers and surface potentials*

Consider a plane membrane which carries fixed charges on its surfaces and is inter-spersed between two aqueous phases. The concentrations of ions in the interfacial do-mains adjacent to the membrane deviate from those in the bulk phase such that a net charge is formed in the interfacial domains which neutralizes the fixed charges on the membrane surfaces. Fixed charges on the membrane surface and the ions in the interfacial domain are then said to form a diffuse double layer. The GOUY-CHAPMAN theory provides a quantitative description of this phenomenon (see *e.g.* Ref. 1 or 2) but is too complex to be presented here. The major assumptions used in this theory are:

(1) All fixed charges on a membrane surface are summed up and the resulting net charge

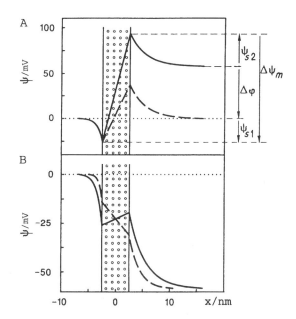

FIG. 2. Electrical potential profiles for a membrane with charged surfaces. The electrical potential ψ was cal-culated according to the GOUY-CHAPMAN theory. It is plotted with respect to an *x*-coordinate system whose direction is perpendicular to the plane membrane surface and whose origin is at the center of the membrane. The quantities ψ_{s1} and ψ_{s2} are the surface potentials at the membrane surfaces facing compartment 1 (on the left) and compartment 2 (on the right), respectively. The potential differences $\Delta\varphi$ and $\Delta\psi_m$ are the membrane potential and the transmembrane potential difference, respectively. Density of surface charges are -0.02 C/m^2 and 0.01 C/m^2 on the surfaces facing compartment 1 and 2, respectively; the dielectric constant of the membrane was assumed to be 5 and that of the aqueous phase 80; temperature 20 °C. Salt concentrations are 100 mM and 10 mM in compartments 1 and 2, respectively. Panel A pertains to a uni-univalent salt and $\Delta\varphi = -58.2$ mV (solid line) or 0 mV (broken line). Panel B is for $\Delta\varphi = 58.2$ mV and a uni -univalent salt (solid line) or a salt with a divalent cation and a univalent anion (broken line).

is equally distributed (or smeared) over the surface.

(2) The electrochemical potential of an ion is constant from the bulk of the phase up to the surface of the membrane, *i.e.* the equilibrium condition (or pseudo-equilibrium condition, see preceding section) is assumed for the diffuse double layers.

(3) There is no charge within the membrane, *i.e.* the electric field in the membrane is constant.

By means of the GOUY-CHAPMAN theory the space-dependence of the electrical potential and the concentration of ions can be calculated. Fig. 2 presents some examples of the electrical potential profile along the direction indicated by the x-coordinate which is perpendicular to the surfaces of the plane membrane. The features to be learned from these examples can be summarized as follows:

(1) The electrical potential at the membrane surface deviates from that in the bulk phase and thus forms what is called the *surface potential* [1] (marked by Ψ_{s1} and Ψ_{s2} in Fig. 2). Surface potentials decrease with increasing salt concentrations.

(2) Divalent ions, particularly with a sign opposite to that of the fixed surface charge, are more effective than univalent ions.

(3) The electrical potential difference $\Delta\varphi = \varphi_1 - \varphi_2$ between the bulk of the phases (*i.e.* the membrane potential) has only minor effects on the surface potentials. Under no circumstances can the surface potentials be inferred from the membrane potential.

(4) The surface potentials strongly affect the transmembrane potential $\Delta\psi_m$ and thus the electric field within the membrane. Under certain conditions this field can even have a polarity opposite to that expected from the polarity of the membrane potential.

As is evident from Fig. 2 the thickness of diffuse double layers is of the order of a few nanometers. In several biological systems the distances between membranes is of the same order of magnitude. Typical examples are the plane membranes of thylakoids in chloroplasts, the cristae formed by the inner membrane of mitochondria, the disk-shaped protrusions of the membrane in the outer segment of retinal rods, as well as the mem-brane pair formed by the transverse tubules and the terminal cisternae of the sarcoplasmic reticulum in muscle cells (see Fig. 1A). In all these cases the diffuse double layers of adjacent membranes overlap, as shown in Fig. 3 for thylakoids, and a bulk phase does not exist between the membranes. As a consequence the concentration of a charged species is not constant between the membranes, and a membrane potential $\Delta\varphi$ across such membranes can no longer be defined.

3.2   *Effect of the electric field on components in the membrane*

The electric field associated with the transmembrane potential difference $\Delta\psi_m$ (see Fig. 2) affects the susceptible parameters of a component in the membrane. Thus, a

---

[1] This surface potential is analogous to but not identical with the surface electric potential χ of a phase which arises from a dipole layer on the surface of the phase.

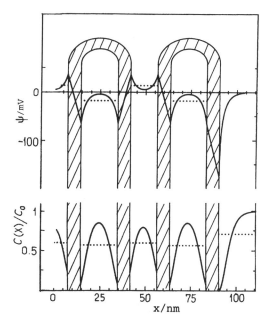

Fig. 3. Profiles for the electrical potential ψ and the concentration c of univalent ions in a stack of thylakoids isolated from chloroplasts. The upper panel shows the electrical potential ψ and the lower panel the concentration $c(x)$ of those univalent ions (with respect to the bulk phase concentration c_0 = 10mM) whose valence has the same sign as the charges on the adjacent membrane surfaces. Both parameters are plotted with respect to an x-coordinate whose direction is perpendicular to the plane membrane surfaces and whose origin is at the center of the thylakoid stack. Surface charge densities (in C/m²) are − 0.02 and 0.01 on the luminal and the outer surface of the thylakoid membranes, respectively, and − 0.2 on the surface of the stack. Membrane tickness is 7 nm, the distance of the membranes in the thylakoids is 20 nm, while that between thylakoids is 15 nm. Dielectric constants are 5 and 80 for membrane and aqueous phase, respectively, and temperature is 20 °C.

change in the absorption spectrum of electrochromic[2] pigment molecules such as chlorophylls or carotenoids which are embedded in the membrane indicates that the electric field in the membrane has changed due to an altered Δφ and/or changes in the surface potentials (see *e.g.* Ref. 3). Another example pertains to charged proteins incorporated in the membrane if the protein can adopt different conformations with a concomitant rearrangement of its charges.

Fig. 4 presents a sketch of a membrane-bound protein which can adopt m different conformations in the sense defined by HILL [4]. The protein carries k charges such that in

---

[2] Electrochromic means that the absorption spectrum of a pigment is affected by an electric field to which the molecule is exposed. Electrochromism arises from shifts in the electronic energy levels due to the electric field as well as from the effect of the field on permanent and induced dipoles of the molecule.

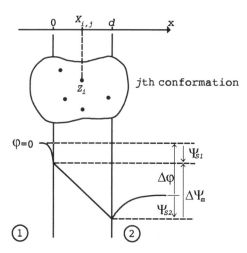

FIG. 4. Sketch of a charged membrane-bound protein, and the electrical potential profile. The membrane with thickness d is represented by the two vertical lines. It separates two compartments numbered 1 and 2. The membrane-bound protein has several charges, all within the membrane, as indicated by the dots. In the j th conformation of the protein the i th charge with valence z_i is located at $x_{i,j}$ with respect to the x-coordinate system shown on the top. For the meaning of the symbols denoting different parts of the potential profile see legend to Fig. 2.

the j th conformation the i th charge with valence z_i is at the position x_{ij} with respect to the x-coordinate shown in Fig. 4. The electric field in the membrane is assumed to be constant which causes the potential ψ in the membrane to depend linearly on x:

$$\psi(x) = \psi_{s1} - x\,\Delta\psi_m/d \tag{10}$$

Here, ψ_{s1} and $\Delta\psi_m = \psi(0) - \psi(d)$ denote the surface potential on the membrane side facing compartment 1 and the transmembrane potential difference, respectively, while d is the membrane thickness (cf. Fig. 4). Note that the reference point for the electrical potential has been chosen to be in the bulk phase of compartment 1 [5]. The quantity $\Delta\psi_m/d$ indicates the electric field strength, E_m, in the membrane. The electrochemical potential of the protein in the j th conformation then becomes [3]

$$\tilde{\mu}_j = \mu^\circ{}_j + RT\ln\{P_j\} + \mathcal{F}\sum_{i=1}^{k} z_i\psi(x_{ij}) = \mu^\circ{}_j + RT\ln\{P_j\} + \mathcal{F}[z_p\psi_{s1} - z_j\Delta\psi_m] \tag{11}$$

[3] When calculating $\tilde{\mu}_k$ the electrical potential in the membrane is assumed to be unaffected by the charges of the protein. If this approximation is not done the relations for the electric terms become intractably complex.

where μ°_j and P_j denote the standard chemical potential and the probability of the jth conformation, respectively. The right hand side of equation (11) is obtained when substituting $\psi(x)$ from equation (10). The quantity $\bar{z}_j$ is an effective valence of the jth conformation defined as

$$\bar{z}_j = \sum_{j=1}^{k} x_{i,j} / d \qquad (12)$$

while z_p is the net valence of the protein which amounts to

$$z_p = \sum_{j=1}^{k} z_j \qquad (13)$$

Each conformation has the same electrochemical potential if the partitioning of the protein over the different conformations is at equilibrium (see Section 8.2 in the preceding chapter). It then follows from equation (11) that

$$P_j = P_1 \, f_j \, (\Delta\psi_m) \qquad (14a)$$

with the abbreviation

$$f_j \, (\Delta\psi_m) = K_{1,j} \exp \{ \mathcal{F} (\bar{z}_j - \bar{z}_1) \, \Delta\psi_m / (RT) \} \qquad (14b)$$

$K_{1,j}$ is the equilibrium constant for the partitioning of the protein between conformations 1 and j, which is determined by the pertinent standard chemical potentials:

$$K_{1,j} = \exp \{ (\mu^{\circ}_1 - \mu^{\circ}_j) / (RT) \} \qquad (15)$$

The probabilities of all conformations have to sum up to unity:

$$\sum_{j=1}^{m} P_j = 1 \qquad (16)$$

By means of equations (14) and (16), one obtains for the probability of the jth conformation

$$P_j = f_j \, (\Delta\psi_m) / [1 + \sum_{j'=2}^{m} f_{j'} (\Delta\psi_m)] \qquad (17)$$

Suppose that the numbering of conformations is such that for a suitable value of the transmembrane potential difference denoted by $\Delta\psi_{m,o}$ all $f_j (\Delta\psi_{m,o})$ vanish. The protein is then entirely in the first conformation. Changing this potential difference from $\Delta\psi_{m,o}$ to any appropiate value $\Delta\psi_m$ leads to a repopulation of the conformations with index higher than 1. Concomitantly the charges in the protein are displaced thus giving rise to a displacement current which is positive when flowing from compart-

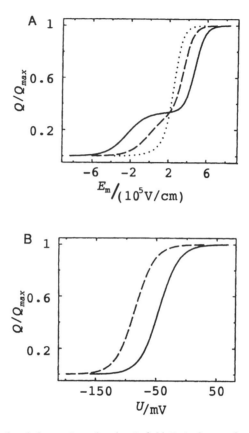

FIG. 5. Dependence of displaced charge Q on the electric field E_m in the membrane (A) and on the membrane potential U (B). The values for Q are normalized with the maximal value Q_{max} attained at a sufficiently large value for E_m or U. The curves in A were calculated with equation (18) and the following parameter values: d = 5nm, m = 3, $\bar{z}_2 - \bar{z}_1$ = 0.5, $\bar{z}_3 - \bar{z}_1$ = 1.5, $K_{1,3}$ = 0.001, and $K_{1,2}$ = 10 (solid line), 1 (broken line), 0.1 (dotted line). The curves in B were calculated with equation (20) and the following parameter values: $\bar{z}_1 - \bar{z}_2$ = 1.6, $K_{1,2}$ = 200, ψ_{s1} = 0, and ψ_{s2} = − 40 mV (solid line), 0 (broken line).

ment 1 to compartment 2. The integral of this current over time is equal to the total charge $Q(\Delta\Psi_m)$ displaced during the repopulation of the conformations. Since the jth charge is moved over the distance $x_{i,j} - x_{i,1}$ and not across the entire membrane the fraction $z_i \mathcal{F}(x_{i,j} - x_{i,1})/d$ only contributes to the total charge, hence

$$Q(\Delta\psi_m) = n_p \mathcal{F} \sum_{j'=2}^{m} P_{j'} \, (\bar{Z}_{j'} - \bar{Z}_1) = n_p \mathcal{F} \sum_{j'=2}^{m} (\bar{Z}_{j'} - \bar{Z}_1) f_{j'} \, (\Delta\psi_m) / [1 + \sum_{j'=2}^{m} f_{j'} (\Delta\psi_m)] \tag{18}$$

where n_p denotes the mole number of the protein. Fig. 5A presents examples of the dependence of Q on the electric field strength E_m ($=\Delta\psi_m/d$) for a protein which can adopt three conformations ($m = 3$). It is evident from these examples that, for given values of $\bar{z}_{j'}$ the transitions between the subsequent conformations appear as separate steps in the $Q(E_m)$-curves only if the equilibrium constants $K_{1,j}$ have sufficiently different values.

The quantity $\Delta\psi_m$ in equation (18) is the transmembrane potential difference which is related to the experimentally accessible membrane potential $\Delta\varphi$ according to (see Fig. 4)

$$\Delta\psi_m = \Delta\varphi + \psi_{s1} - \psi_{s2} \tag{19}$$

Substituting equation (19) into equation (18) written for $m=2$ yields

$$Q(\Delta\psi_m) = \frac{-n_p \, \mathcal{F}(\bar{z}_1 - \bar{z}_2)}{1 + \exp\{\mathcal{F}(\bar{z}_1 - \bar{z}_2) \, (\Delta\varphi + \psi_{s1} - \psi_{s2})/(RT)\}/K_{1,2}}$$

$$= \frac{Q_{max}}{1 + \exp\{-(U - \bar{U})/\mathcal{K}\}} \tag{20}$$

The second expression in equation (20) is the form used by RIOS[4] (see Chapter 12) for the analysis of the potential-sensitive DHP-receptor. It is obtained by means of the following substitutions

$$\mathcal{K} = RT/[\mathcal{F}(\bar{z}_1 - \bar{z}_2)], \quad Q_{max} = n_p \, RT/\mathcal{K},$$

$$U = -\Delta\varphi, \quad \bar{U} = \psi_{s1} - \psi_{s2} - \mathcal{K}\ln\{K_{1,2}\} \tag{21}$$

where U is the membrane potential defined with the opposite sign convention than $\Delta\varphi$. Accordingly, the displacement current is positive when flowing from compartment 2 to compartment 1, and $Q(\Delta\psi_m)$ has the opposite sign. The mean potential difference $\bar{U}$ is determined by the equilibrium constant for the partitioning of the protein between the two conformations, but also by the surface potentials ψ_{s1} and ψ_{s2}. While ψ_{s1} can be assumed to be constant, ψ_{s2} strongly depends on the concentration of Ca^{2+} in compartment 2 (cf. Fig. 2) which increases with decreasing U due to the opening of the DHP-receptor (see explanation for Q in section 3.4 of Chapter 12). The situation is even further complicated by the close apposition of the membranes of the transverse tubules

---

[4] This author uses the notion BOLTZMANN term for the exponential in equation (20). It refers to the fact that, in terms of statistical mechanics, equation 11 describes a BOLTZMANN distribution.

and the sarcoplasmic reticulum. Conditions comparable to those shown in Fig. 3 for thylakoid membranes have then to be expected which leads to a considerably more complex relation between ψ_m and $\Delta\varphi$ as that given in equation (19).

Typical values for the parameters in equation (21) at about 5 °C were found to be (see Ref. 6): $K = 15$ mV, $\bar{U} = -40$ mV, and $Q_{max}/C_m = 25$ nC/μF, where C_m is the electrical capacity of the membrane. With an average value of 1 μF/cm² for C_m [7] one obtains $Q_{max}/A = 25$ nC/cm², where A is the membrane area. From these data, and by means of equation (21), it can be derived that $\bar{z}_1 - \bar{z}_2 = 1.6$, $n_p/A \approx 1.6$ nmol/m² (i.e. 1000 molecules per μm²), and $K_{1,2} \approx 200$ if $\psi_{s1} = 0$ and $\psi_{s2} = -40$ mV. Fig. 5B presents the dependence of Q on U calculated with these parameter values, and for $\psi_{s2} = 0$ and -40 mV. It is worth mentioning that the shift of the $Q(U)$-curve to lower U-values in the case of depolarized cells (see Section 3.3.1. in Chapter 12), which is explained by an activation of the receptor, could at least in part be attributed to a lower ψ_{s2} due to a much higher Ca²⁺ concentration in the myoplasm after prolonged depolarization.

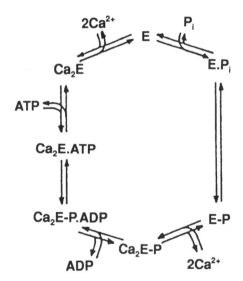

FIG. 6. Kinetic scheme for the Ca²⁺-ATPase. The enzyme adopts different states due to the following transitions (from top center in counterclockwise direction): the free enzyme E binds first 2 Ca²⁺ and then 1 ATP from the myoplasm, the bound ATP phosphorylates the enzyme (E-P) and the resulting ADP is released to the myoplasm, the enzyme releases 2 Ca²⁺ to the lumen of the sarcoplasmic reticulum, the enzyme is dephosphorylated yielding the form E.P$_i$ which finally releases inorganic phosphate to the myoplasm.

4.   The Ca²⁺–ATPase

An important enzyme for Ca^{2+} - homeostasis is the Ca^{2+} - ATPase residing in the membrane of organelles such as the sarcoplasmic reticulum. It pumps Ca^{2+} across these membranes on the expense of ATP-hydrolysis, *i.e.* the reaction

$$ATP \rightleftarrows ADP + P_i \tag{22}$$

where P_i denotes (inorganic) phosphate. If activity coefficients are set to unity the affinity (or thermodynamic force) of this reaction is [cf. equation (27) in the preceding chapter]

$$\mathcal{A}_{ph} = RT \ln\{K_{ph}\, c_{ATP}/(c_{ADP}\, c_{P_i})\} \tag{23}$$

K_{ph} (1.7 x 10^5 M) is the equilibrium constant of the reaction, while c_{ATP} (8 mM), c_{ADP} (40 μM), and c_{P_i} (8 mM) are the concentrations of ATP, ADP, and phosphate, respectively. Typical values for these concentrations in a cell are given in parentheses (see *e.g.* Ref. 8) and yield, when introduced into equation (23), $\mathcal{A}_{ph}/(RT)$ = 22.2. It is generally accepted that the Ca^{2+}-ATPase pumps 2 Ca^{2+} ions per ATP hydrolyzed. Hence, with the force just calculated, the enzyme can *in principle* build up a Ca^{2+}-concentration ratio of about 10^5 at static head in the absence of a membrane potential [2 $\Delta \tilde{\mu}_{ca}$ = $- \mathcal{A}_{ph}$ and equation (25)].

4.1.   Kinetic scheme

A minimal kinetic scheme which accounts for the stoichiometry of 2 Ca^{2+} per ATP hydrolyzed was proposed by PICKART and JENCKS [9] and is depicted in Fig. 6. In the present context it suffices to analyze this scheme for a constant force $\mathcal{A}_{ph}$. Moreover, the membrane potential $\Delta\varphi$ is assumed to be zero due to large enough permeabilities of other ionic species. For the enzyme in steady state the flow of pumped Ca^{2+} ions, $J_{Ca,E}$ (positive from compartment 1 to compartment 2) and the flow of ATP -hydrolysis, J_{ph}, can be calculated by means of the cycle diagram method according to HILL [4] and are found to be

$$J_{Ca,E} = 2\, J_{ph} = 2\, n_E \frac{(c_{Ca,1}/c_{Ca,2})^2 \exp\{\mathcal{A}_{ph}/(RT)\} - 1}{B_0 + B_1\,(c_{Ca,1}/c_{Ca,2})^2 + c^2_{Ca,1}/K^*_1 + K^*_2/c^2_{Ca,2}} \tag{24}$$

for $\Delta\varphi$ = 0. Here, n_E denotes the mole number of enzyme, while $c_{Ca,k}$ is the concentration of Ca^{2+} in the kth compartment. B_0, B_1, K^*_1, and K^*_2 in equation (24) are abbreviations for terms comprising $\mathcal{A}_{ph}$, the concentrations of ATP, ADP, and phosphate as well as the transition probabilities between different states of the enzyme cycle [1,4].

4.2.   *Investigation in isolated vesicles*

If the performance of the Ca^{2+}-ATPase is investigated in a system consisting of isolated membrane vesicles in suspension the Ca^{2+} concentration $c_{Ca,1}$ in the suspending medium (*i.e.* in compartment 1) can be controlled. When neglecting a possible membrane potential $\Delta\varphi$ the concentration $c_{Ca,2}$ in the vesicles is related to the thermodynamic force of Ca^{2+} transport, $\Delta\tilde{\mu}_{Ca}$, and to $c_{Ca,1}$ by [cf. equation (28b) in the preceding chapter with $\Delta\varphi = 0$]

$$c_{Ca,2} = c_{Ca,1} \exp\{- \Delta\tilde{\mu}_{Ca}/(RT)\} \tag{25}$$

Introducing equation (25) into equation (24) yields a relation which can be rewritten such that the dependence of the flows on $\Delta\tilde{\mu}_{Ca}$ adopts the form of a hyperbolic tangent (cf. Fig. 2 in the preceding chapter).

The total flow of Ca^{2+} observed in a vesicular system amounts to

$$J_{Ca,tot} = J_{Ca,E} + J_{Ca,l} \tag{26}$$

where $J_{Ca,l}$ is the leak flow. It arises from the permeability, $P_{Ca,l}$, of the membrane for this ion and, in the absence of a membrane potential, is driven by the difference in

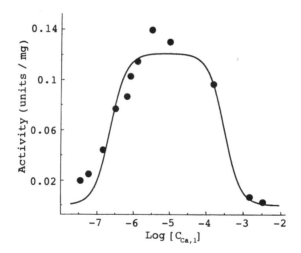

FIG. 7. Dependence of ATP-hydrolysis on $c_{Ca,1}$ for the Ca^{2+}-ATPase of chicken cerebellum microsomes at pseudo-level flow. Experimental data (●) taken from Fig. 3 of Ref. 10 were fitted to equation 24 (solid line) under the level-flow condition $c_{Ca,2} = c_{Ca,1}$ and $\Delta\varphi = 0$. $\mathcal{A}_{pb}/(RT)$ was assumed to be 22.2, while values of parameters fitted are $(B_0 + B_1)/(fn_E) = 3.6 \times 10^{10}$ mg/units, $K^*_1 fn_E = 2.5$ mM^2 units/mg, $K^*_1/(fn_E) = 1820$ mM^2 mg/units. The factor f converts the authors' "units/mg" to the flow-units mol/s. The differences between the experimental points and the calculated curve in the range of $c_{Ca,1} \leq 10$ μM are most likely due to the pseudo-level flow condition pertaining in the experiment, whose deviation from the true level flow condition used in the parameter fitting procedure becomes the more effective the lower $c_{Ca,1}$.

Ca²⁺ concentration between the two compartments [1],

$$J_{Ca,l} = P_{Ca,l} \left(c_{Ca,1} - c_{Ca,2} \right)$$ (27)

The instrinsic permeability of vesicular membranes is usually rather small. By the addition of an agent called *ionophore* this permeability can be substantially increased. In an ideal case the permeability could be made so large that any flow $J_{Ca,E}$ generated by the Ca²⁺-ATPase would be compensated by a leak flow $J_{Ca,l}$ [*i.e.* $J_{Ca,tot} \approx$ 0, see equation (26)] which is driven by a very small value for $c_{Ca,2} - c_{Ca,1}$. The enzyme would then operate very close to *level flow* (see Section 8 in the preceding chapter). In practice, however, this situation can rarely be accomplished since the achievable permeabilities do not suffice for an effective clamping of the Ca²⁺ concentration difference to nearly zero. Hence, a finite but more or less small difference as compared to the maximal value mentioned above remains even in the presence of a substantial amount of an ionophore, and the enzyme may then be said to operate under pseudo-level flow conditions.

The dependence of J_{pb} on $c_{Ca,1}$ under such conditions was measured by MICHE-LANGELI *et al.* [10] in chicken cerebellum microsomes, and their data are re-plotted in Fig. 7. Equation (24) with the simplifying assumption $c_{Ca,1} = c_{Ca,2}$ used in a non-linear parameter fitting program yielded the result indicated by the solid line in Fig. 7 with values for parameters as given in the legend. The bell-shaped curve arises essentially from the two parameters K^* and K^* which comprise the equilibrium constants for the dissociation of Ca²⁺ from the enzyme into compartment 1 and 2. The increase in activity of the Ca²⁺-ATPase observed at $c_{Ca,1}$ in the nanomolar range reflects the usual dependence of an enzyme on the concentration of its reactants (in this case Ca²⁺ in compartment 1) which eventually reaches saturation. The decrease in activity at $c_{Ca,1}$ in the millimolar range indicates that the enzyme is increasingly subject to *kinetic control*. Since the rate of association at such high concentrations becomes much faster than the corresponding rate of dissociation the enzyme is more and more pushed into the states with bound Ca²⁺ (cf. Fig. 6) which effectively slows down the turnover of the whole enzyme cycle.

An entirely different situation arises if the system is allowed to attain static head (see Section 8 in the preceding chapter) with respect to the movement of Ca²⁺ at a physiological (and approximately constant) value for $c_{Ca,1}$, and in the absence of an ionophore. Due to the low value of $P_{Ca,l}$ a substantial concentration difference $c_{Ca,1} - c_{Ca,2}$ is now built up which is much larger than in the presence of an ionophore. As a consequence $J_{Ca,E}$ is progressively reduced [cf. equation (24)] until it balances the still relatively small leak flow. At this point $J_{Ca,tot}$ vanishes and the system has reached static head. If the Ca²⁺-ATPase is now fully blocked by a rapid addition of an inhibitor J_{pb} = $J_{Ca,E}$ = 0, and the flow of Ca²⁺ *initially* observed just after the inhibition of the en-

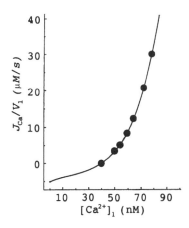

FIG. 8. Dependence of the total flow J_{Ca} on the Ca^{2+} concentration $c_{Ca,1}$ in the myoplasm. Experimental data (●) taken from Fig. 4 of Ref. 11 were fitted to equations 24, 26, 27, and 31 (solid line). Values of parameters: $n_E/V_1 = 200$ μM, $A_{pb}/(RT) = 22.2$, $B_O = 450$ s, $B_1 \approx 9 \times 10^8$ s, $K_1^* \approx 440$ nM^2 s^{-1}, $K_2^* = 6.5$ mM^2 s, $P_{Ca}/V_1 = 0.0011$ s^{-1}, $V_1/V_2 = 5$, $c_{Ca,tot} = 5.9$ mM. Values of parameters describing binding of Ca^{2+} in compartment 1 were taken from Ref. 11.

zyme is approximately equal to $J_{Ca,K(sb)} = -J_{Ca,E(sb)} = -2 J_{pb(sb)}$ [cf. equations (24) and (26) with $J_{Ca,tot} = 0$], where the index sb refers to static head. It is important to note that the flow $J_{pb(sb)}$ observed at static head has no relation to the flow shown in Fig. 7 at the given value for $c_{Ca,1}$ because there is a tremendous difference in $c_{Ca,2}$ at static head as compared to the pseudo-level flow condition.

4.3.   Investigation in the muscle

The fluorescent dyes which bind Ca^{2+} make it possible to estimate the Ca^{2+} concentration directly in the myoplasm (see chapter 12). KLEIN et al. [11] have used this technique to monitor the time course of this concentration during the recovery phase of the muscle after excitation. In the thermodynamic system shown in Fig. 1B this concentration is $c_{Ca,1}$, while $J_{Ca,1}$ denotes the Ca^{2+}-flow due to the Ca^{2+}-ATPase in the sarcoplasmic reticulum. If all primed flows in equation (3a) are neglected[5] it follows from this equation that

$$J_{Ca,tot} = J_{Ca,1} + J_{Ca,2} + J_{Ca,4} = -[V_1 \, dc_{Ca,1}/dt + J_P + J_T + J_D] \qquad (28)$$

The flows due to Ca^{2+}-binding (J_p, J_T, and J_D) were described by relations according to equation (4). This leads to a set of differential equations whose solutions provide

[5] This is legitimate because $c_{Ca,3}$ = constant (see section 1) in contrast to $c_{Ca,2}$, which is substantially decreased during excitation of the muscle.

these flows as a function of time [11]. Introducing these flows into equation (28) then yields $J_{Ca,tot}$ as a function of time which, when combined with the time course of $c_{Ca,1}$, enabled the authors to calculate the dependence of $J_{Ca,tot}$ on $c_{Ca,1}$ as shown in Fig. 8.

$J_{Ca,1}$ in equation (28) is the same as $J_{Ca,E}$ in equation (26), and the left hand side of equation (28) then becomes identical with equation (26) if $J_{Ca,l}$ is equated to $J_{Ca,2}$ + $J_{Ca,4}$. This composite leak flow can still be described by equation (27) if $P_{Ca,l}$ is interpreted as an average permeability arising from the (inactive) Ca^{2+}-pore and the intrinsic membrane permeability. KLEIN et al. adopted the simplest MICHAELIS-MENTEN type of rate-law for the assessment of $J_{Ca,E}$ which reads

$$J_{Ca,E}/V_1 = v_{max}\, c_{Ca,1}^n/[K + c_{Ca,1}^n] \tag{29}$$

Here, v_{max} is "the maximum pump rate at saturating Ca^{2+} concentrations and K is the nth order dissociation constant" [11]. Introducing equations (27) and (29) into equation (26) yields under the simplifying assumptions stated in equation (30):

$$J_{Ca,tot}/V_1 = B_E\, c_{Ca,1}^n + B_L \text{ for } c_{Ca,1}^n \ll K \text{ and } c_{Ca,1} \ll c_{Ca,2} \approx const \tag{30}$$

with the abbreviations $B_E = v_{max}/K$ and $B_l = (P_{Ca,l}/V_1)\, c_{Ca,2}$. KLEIN et al. [11] fitted equation (30) to the experimental data (such as those shown in Fig. 8) and obtained a value for n which was about 4. Since only 2 Ca^{2+} ions are transported by one enzyme molecule (see above) they concluded that the functional Ca^{2+}-ATPase in the sarcopalsmic reticulum operates as a dimer.

If $c_{Ca,2}$ in equation (24) is set to zero equation (29) is obtained. It thus becomes evident that the rate-law used by KLEIN et al. is a special case which applies only if $c_{Ca,2} = 0$. this condition, however, is certainly not fulfilled. In fact, $c_{Ca,2} \gg c_{Ca,1}$ which was used by the authors when approximating $J_{Ca,l}$ by a constant leak flow [cf. equation (30)]. Hence, the general relation given by equation (24), together with equation (27), has to be used in equation (26) for an appropriate analysis of the data. The concentration $c_{Ca,2}$ included in the resulting relation can be calculated from the mass balance of Ca^{2+} and amounts to

$$c_{Ca,2} = c_{Ca,tot} - (V_1/V_2)\, [c_{Ca,1}\, (1 + r_{b,1}) + n_{Ca,E}/V_1] \tag{31}$$

Here, $c_{Ca,tot} = n_{Ca,tot}/V_2$ denotes the total Ca^{2+} concentration which would arise if all Ca^{2+} ions in the system (represented by the total mole number $n_{Ca,tot}$) would be in compartment 2. The quantity $r_{b,1}$ is the ratio of bound over free Ca^{2+} in compartment 1, where bound is the sum of all species $Ca^{2+}.B$ [see equation (1) with B = parvalbumin, troponin C, fluorescent dye]. The mole number of Ca^{2+} bound to the enzyme, $n_{Ca,E}$, can be calculated by means of the cycle diagram method according to HILL [4].

An analysis based on equations (24), (26), (27), and (31) when applied to the experimental data shown in Fig. 8 yielded the result represented by the solid line. Values of parameters are listed in the legend to this figure and emerged from a non-linear fitting procedure. Since this result was obtained with the stoichiometry of 2 Ca^{2+} per enzyme [cf.equation (24) and Fig. 6] it clearly demonstrates that there is no need for invoking a dimer as functional unit of the Ca^{2+}-ATPase. This example also demonstrates the necessity for a consistent thermodynamic and kinetic description of a given system. Otherwise fallacies and unwarranted conclusions are liable to occur.

References

[1] D. WALZ, *Biochim. Biophys. Acta*, **1019**, 171 (1990).

[2] J.Th.G. OVERBEEK in *Colloid Science*, H.R. KRUYT (Editor), Elsevier, Amsterdam (1952), Vol **1**, pp 126-146.

[3] H.T. WITT, *Quart. Rev. Biophys.*,**4**, 365 (1971).

[4] T.L. HILL, *Free Energy Transduction in Biology*, Academic Press, New York (1977).

[5] D. WALZ and S.R. CAPLAN, *Bioelectrochem. Bioenergetics*, **28**, 5 (1992).

[6] E. RIOS and G. PIZARRO, *Physiol. Reviews*, **71**, 849 (1991).

[7] P. MÜLLER and D.O. RUDIN in *Current Topics in Bioenergetics*, D.R. SANADI (Editor), Academic Press, New York (1971), Vol.**3**, pp 157-242.

[8] D. WALZ and S.R. CAPLAN, *Cell Biophys.*, **12**, 13 (1988).

[9] C.M. PICKART and W.P. JENCKS, *J. Biol. Chem.*, **259**, 1629 (1984).

[10] F. MICHELANGELI, F. DI VIRGILIO, A. VILLA, P. PONDINI, J. MELDOLESI and T. POZZAN, *Biochem. J.*, **275**, 555 (1991).

[11] M.G. KLEIN, L. KOVACS, B.J. SIMON and M.F. SCHNEIDER, *J. Physiol.*, **441**, 639 (1991).

INDEX